BEAR AND FOXES

THE INTERNATIONAL RELATIONS OF THE EAST EUROPEAN STATES, 1965 – 1969

RONALD HALY LINDEN

EAST EUROPEAN QUARTERLY, BOULDER
DISTRIBUTED BY COLUMBIA UNIVERSITY PRESS
NEW YORK

1979

EAST EUROPEAN MONOGRAPHS, NO. L

Ronald Haley Linden is Assistant Professor of Political Science
at the University of Pittsburgh

Library of Congress Catalog Card Number 78063276
ISBN 0-914710-45-1

Printed in the United States of America

To N

PREFACE

This work is a hybrid. It is a study of an area, East Europe, but within the rubric of a more generalized discipline, international relations. As a comparative study of foreign policy it seeks to make a contribution to the development and improvement of that subdiscipline. As a substantive investigation, it seeks answers to some specific historical questions about the foreign policies of the East European states during the second half of the sixties. In its entirety, the aim of this work is to learn something both about the integration of an international region and at the same time about the historical genesis of the anomalous foreign policy of one of its members.

If these purposes seem too disparate, too distant from each other to offer a reasonable chance of successful achievement, I may be guilty of aiming too high, but certainly innocent of perpetuating artificial (and dysfunctional) distinctions within the discipline. As I will argue early in this work, and hopefully prove at its completion, studies of East Europe need not leave political science behind, and political scientists need not have a mental map of the world which labels East Europe "unexplored."

If the precise genesis of this work cannot be pinpointed exactly, the catalytic milieu in which it was spawned and supported can certainly be properly acknowledged. For me this meant, originally, contact with political scientists committed to pursuing rigorous analytic work on the Soviet Union and East Europe. In short, other hybrids. Thus I am indebted intellectually to Walter C. Clemens, William Zimmerman and Zvi Gitelman, though they are strangers to the final manuscript, and to Carl Beck, whose enthusiasm for this work has been as encouraging to me as the generous financial support provided by the University of Pittsburgh Center for International Studies of which he is Director.

In the research and writing of this work I have been fortunate in securing a wide measure of scholarly and financial assistance. My thanks must go to Edward Morse, now of the Department of State, who not only encouraged me to pursue this project as a dissertation at Princeton University when others were skeptical, but backed up his encouragement with substantive suggestions and worthwhile criticism. Jeffrey Hart and Michael Kagay of Princeton were careful readers and generous discussants, particularly on the methodological aspects of the work. For helping me to bring a recalcitrant computer to heel, I would also like to thank the kind and patient staff of the Princeton Computer Center.

In addition, Jeanne Laux of the University of Ottawa and Jan F. Triska of Stanford University both read parts of the manuscript in an earlier form and offered substantive counsel, for which I am grateful. This work has benefitted as well from discussions with Andrew Axline, John Sigler, David Seidman, Patrick McGowan, Paul Cocks and Richard Brody, and from an opportunity to present chapters I and II as a paper at a meeting of the Midwest Political Science Association on a panel chaired by Maurice East. This study is better for having heeded the suggestions of these and other advisors and has ignored their advice to its detriment, I am sure. Nevertheless, I am quite unwilling to absolve my discussants from their responsibility for any errors that remain, as it certainly must have been their advice and not my execution that was at fault!

The Department of Political Science at the University of Pittsburgh has not only proven a comfortable and supportive place in which to work, but has provided teaching opportunities which have allowed me to further explore—I hope fruitfully for myself and my students—the various subdisciplines of political science represented in this work.

Early financial support for this work was generously forthcoming from the International Research and Exchanges Board, which made it possible for me to do research in the United States, Canada, and in Romania, and in addition, supported much of my Romanian language training. The dissertation from which this work is derived was written at Princeton University under support from an NDFL language fellowship and the Russian Studies Program. Finally, the Russian and East European Studies Center of the University of Pittsburgh granted me funds for work on the book's graphics. Much of this work was done by Frank Garrity of Pittsburgh, whose patience and careful attention I sincerely appreciate.

Research accomplished in Romania would certainly not have been possible without the assistance of Dr. Nicolae Fotino of the Association of International Law and Relations in Bucharest, and of the Romanian Academy of Political and Social Sciences, under the auspices of the National Council for Science and Technology.

I owe a special thanks to my three coders, Mary C. Lyons, Barry Blufer and Laura D. Tyson, whose careful efforts allowed the content analysis of Chapter III to become a reality, and to Stephen Fischer-Galati, the editor of the present series, whose support and patience insured that the volume itself would achieve the same status. For rendering my cryptic notes and marginalia into finished manuscript I express my gratitude to Joan Tronto and Richard Hayden, to Martha Speer, Mary Hamler, Frieda Safyan, Lori Stuart and Henrietta Moss. I am grateful also to Thomas Magar and Kendall Stanley for their work on the book's Index.

Excusing myself to Ethan for being absent so much from his new young life serves no purpose for him—at least until he can read these words; but he will do so eventually and in the meantime it serves to assuage my guilt.

And I want to thank Nancy; for pretending that sharing a hasty yogurt at 9:45 is dinner; that being with someone who mumbles footnotes in his sleep is romantic; that living with someone who is writing a book is living.

TABLE OF CONTENTS

LIST OF ABBREVIATIONS

CC	Central Committee
CCP	Chinese Communist Party
CMEA	Council For Mutual Economic Assistance (Comecon)
CPR	Chinese People's Republic
CPSU	Communist Party of the Soviet Union
DRV	Democratic Republic of Vietnam (North Vietnam)
EEC	European Economic Community
EFTA	European Free Trade Association
FRG	Federal Republic of Germany (West Germany)
GATT	General Agreement on Tarriffs and Trade
GDR	German Democratic Republic (East Germany)
ID	Index of Deviance
IMF	International Monetary Fund
IP	Interaction Percentage
IS	Interaction Score
LCY	League of Communists of Yugoslavia
NATO	North Atlantic Treaty Organization
NFLSV	National Front for the Liberation of South Vietnam
NPT	Nonproliferation Treaty
PCC	Political Consultative Committee (Warsaw Pact)
PRG	Provisional Revolutionary Government (South Vietnam)
RCP	Romanian Communist Party (Rom: PCR)
RPR	Rumanian People's Republic
RSR	Romanian Socialist Republic
RWP	Rumanian Workers' Party (Rom: PMR)
SED	Socialist Unity Party (GDR)
WTO	Warsaw Treaty Organization (Warsaw Pact)

LIST OF TABLES

LIST OF TABLES
Continued

LIST OF FIGURES

LIST OF FIGURES
Continued

CHAPTER ONE

EAST EUROPE AND POLITICAL SCIENCE: BREACHING THE LAST FRONTIER

Perhaps no geographic area of the world remains as alien to the American political scientist as East Europe. It has only been in the last decade that students of communist politics have recognized the need to build a body of communicable knowledge based on rigorous comparative analysis, theoretically, or at least conceptually, guided. Obstacles to this progress, to mention a few, have included: 1) lack of a constructive dialogue between political science theorists and "area" specialists; 2) the inappropriateness of western-oriented internal and international political models for the study of the area; 3) the meager substantive or conceptual rewards expected from studying "Soviet-type" systems, compared with studying the Soviet system itself; and 4) formidable language and data problems hindering access to the grist of comparative study.

Still, progress along this path has been steady, marked by milestone, if largely hortatory, volumes such as Roger Kanet's *The Behavioral Revolution and Communist Studies*.[1] Frederic Fleron's, *Communist Studies and the Social Sciences*[2] and special sections, issues or indeed the entire existence of journals devoted to studying "comparative communism."[3] More recently the conceptual and methodological calls to arms have been fewer and the number of empirical battles joined- leaving aside the question of whether they have been victories or defeats- greater. Such edited works as Jan Triska, *Communist Party States*.[4] the same editor's eight volume series, *Integration and Community Building in Eastern Europe*,[5] Triska and Paul Cocks, *Political Development in Eastern Europe*,[6] Chalmers Johnson, *Change in Communist Systems*,[7] Carl Beck, *et al.*, *Comparative Communist Political Leaders*,[8] and Beck and Carmelo Mesa-Lago, *Comparative Socialist Systems*,[9] have offered, if not strictly enforced, a comparative framework for the study of those states classified as socialist or communist. These works have perforce included East Europe within their purview. And individual authors have not shrunk from attempting cross national and multi-thematic comparisons in East Europe, as the ambitious works of Ghita Ionescu, Richard Gripp and Triska and Johnson indicate.[10]

As for Foreign Policy . . .

If the comparative study of communist politics, i.e. internal politics, thus appears to be enjoying a vigorous childhood, the comparative study of the foreign relations of these states- in particular the East European states- seems still to be in gestation. In addition to those impediments mentioned above, the study of the foreign policies of East Europe has been slowed or put off by: 1) the persistence of "bloc" or "satellite" images, which further reduce the gains expected from studying the area; 2) the attractiveness of studying internal political developments due to their apparently greater significance for Soviet action and policy; 3) the even lower level of useful dialogue between students of East Europe and those specializing in international relations; and, not in the least by 4) the embryonic state of the enterprise of comparative study of foreign policy.[11]

Still, there are harbingers. Most of the contributions to Triska's *Communist Party States*, for example, focus on international relations. We have our hortatory volume, a special issue of *Studies in Comparative Communism*, which was itself the product of a conference on communist foreign policies.[12] A few edited editions have appeared, more thematically than comparatively structured, including: Charles Gati, (ed.), *The International Politics of Eastern Europe*.[13] Jane Shapiro and Peter Potichnyj (eds.), *From the Cold War to Detente*;[14] and Robert King and Robert Dean, (eds.), *East European Perspectives on European Security and Cooperation*.[15] And a handful of individual research efforts have touched on East European foreign policies in systematic conceptually guided studies.[16]

The present work is aimed at offering a contribution toward filling some gaps in the study of comparative communist politics, and, more particularly the comparative study of the foreign policies of some of these states. These needs, it seems to me, can be summed as follows:

1) *comparative-descriptive:* There is a basic paucity of studies whose purpose is a systematic, precise, carefully controlled and

conceptually guided description of the foreign policies of the states of East Europe. If the recognition is now widespread that it is, at the very least, inappropriate to approach the area as a unified group of satellites, blindly imitating all Soviet initiatives, investigations and documentations of this situation are much less in evidence. Description has been the "poor relation" of political science, yielding pride of place to causally oriented, theory building efforts. This has had unfortunate consequences for the study of East Europe. Offering a complex, detailed comparative study of significant aspects of these states' foreign policies would be more than "mere" description and would by itself advance the field substantively.[17]

2) comparative-analytical: The above remarks notwithstanding, we should not, in Charles Gati's phrase, lower the "high jumper's bar:" which marks the standards of acceptable social science.[18] We do need to try to understand why states differ from each other as well as how and in what dimension. As Roger Kanet has put it, "Few studies of the foreign policies of the Communist countries, for example, really analyze—rather than merely mention—the relationships between domestic and foreign policy. What impact do differing domestic conditions–whether cultural, socio-economic, geographic or political–have on the foreign policies of Communist states?"[19] This work addresses itself to several such questions in both its horizontal (comparative) and longitudinal (case study) investigations.

3) rigor and precision: Implicit in both of the above and explicit in the call for comparative studies of the internal and external politics of communist states, is the need to replace impressionistic and idiosyncratic approaches with a precision of definition and conception, and a rigor of investigation and explanation. Again, Kanet makes the case: "Most important, I believe, is the need to introduce into studies of Communist foreign policy some of the concern for precise, rigorous methods of analysis which is evident among students of general international relations, and the development and testing of generalizations concerning foreign policy in Communist politics."[20]

4) methodological pluralism: The above does not argue, nor do others suggest, that only comparative work is of real worth, or should replace traditional, narrative or even impressionistic work. This point hardly needs reiteration, having been made by peacemakers involved in methodological struggles throughout this and other disciplines.[21] Indeed part of the present work is a single country investigation which depends in part on others like it and is offered as a historical test of the hypotheses examined in the broader, comparative sections.[22]

It is not only to those hardy souls who work on the "other Europe" that this work is aimed, however. An equally strong purpose of this work is to make a contribution, both substantively and conceptually, to the broader comparative study of foreign policy.

First, as noted above, this field has largely ignored East Europe, hurting the development of substantive knowledge of the area[23] and the growth of the foreign policy subdiscipline as well. Several students of comparative foreign policy, for example, have noted the necessity for further refinements of James Rosenau's nation-state genotypes- a "pre-theoretical" grouping of states by size, development, accountability and degree of policy penetration, designed as building blocks for hypotheses on foreign policy.[24] Harf, Hoovler and James, for example, conclude their assessment of the effect of external and systemic variables on foreign policy with the following:

> Our results suggest that it is insufficient to examine the entire universe of nations in the same test. Rather, the key to understanding the effect of internal and external source variables may lie in studying their impact upon *selected subsets* of nations.[25]

Thus, the advantage for the comparative study of foreign policy of focusing on communist states—and on East Europe in particular—is that it allows one to hold several important variables, say size, type of one party rule, constant within rough parameters, while testing the explanatory power of other variables;[26] or on the other hand to delineate more sensitive variations in previously crude or dichotomous variables.

Finally, viewed as a study in international relations this work strives to make a contribution toward a fuller understanding of the nature of regional integration. Claims on this score—as with any based on area- and time-focused work—will have to be limited. Nevertheless, the particular conception and measurement of integration offered here is not so bound as to prevent its being usable or even desireable for examining regional integration in other contexts.

The study of regional integration has for the most part passed East Europe by.[27] This is a clear case in which models developed to study western, free-market, non-cocercive situations were simply not appropriate to the East European planned-economy hegemonic subsystems.[28] Focuses upon institutions and formal or neo-functionalist models reveal little about the substance of overall integration in the area,[29] while studies relying upon elite or public attitudes *a la* Deutsch[30] meet with either inaccesible or worthless data. This is not to say that integration is a poor way to approach a study of the region. It is rather to suggest that an alternative view of integration is required for studying a region like East Europe.

Following sociologists, students of international regions have come to recognize the necessity of disaggregating their conceptual and empirical examinations of integration.[31] There have been investigations into the levels and causes of three types of integration: social, or in sociologists terms, communicative integration;[32] functional economic and political integration;[33] and attitudinal or cultural integration.[34] Political scientists have virtually ignored a fourth type of integration, which Landecker labels "normative" integration. This refers to the relation between community or group standards and the individuals for whom they establish norms.[35]

What does normative integration mean as applied to world politics? It is the degree to which member states of an international group adhere in their behavior to some group norm or set of norms. International alliances invariably have as one of their goals coordination of some aspect of the actions of the member states. This is the *raison d'etre* of alliances. It may be an attempt to coordinate dispersal of a commonly held economic valuable, such as the OPEC nations' oil embargo, or an attempt to generate a broader degree of conformity in members' international actions toward key targets, e.g. NATO vis a vis the Soviet Union and Eastern Europe or the OAS *vs* Cuba. For an international political alliance then, one can compare the members' behavior to assess the degree of adherence to the appropriate group norm. Holsti, Hopmann, and Sullivan have utilized this approach to highlight alliance "mavericks."

> Nations which are members of an alliance establish characteristic patterns of relations with other nations, alliances, regional groupings, and international organizations. A bloc member in good standing

> thus adheres to certain predictable patterns of behavior with other members of the alliance, the leading nations in the bloc, nations within the opposing alliance and nonaligned states. In these terms deviation from bloc norms is the establishment of qualitatively and quantitatively different patterns of interaction with various international environments.[36]

This approach to integration presents us with some significant conceptual and definitional problems. What are the "norms" that will be used for comparison of the nations' behavior? What will constitute deviation from those norms, and more importantly, what is the significance of such deviation for the "normative integration" of the group? Lastly, is normative integration integration at all?

Beginning with the first question, norms, as defined in *A Dictionary of the Social Sciences* refer to:

> 1) a statistical standard of comparison constituted by what is in some sense the average or modal value of the variable on which the items in a population are being compared; 2) the average or modal, i.e. most typical, behavior, attitude, opinion, or perception found in a social group; 3) a standard shared by the members of a social group to which the members are expected to conform, and conformity to which is enforced by positive and negative sanctions.[37]

Which definition of norm is appropriate and useful for studying an international region? Statistical or modal standards of comparison, descriptive norms, are the regularities in a social system; norms of the "ought" type are prescriptive or prohibitive: they are the rules of action in the system. This last conception of norms constitutes the most common understanding of the term as utilizied in the study of world politics, though it is at the same time the most difficult to operationalize for empirical study.[38] All three are related however, and while type no.3 norms are perhaps the most significant for world politics, types no.1 and no.2 are important in themselves and moreover can be utilized in an investigation of type no.3 norms.

Prescriptive or prohibitive norms come into being through customary, verbalized or formalized patterns of action, or interaction, in an existing social system.[39] Investigation of the existence of norms supports the notion that such existence is, in today's world, a matter of degree. Evidence regarding norms has been frequently sought in statements that a rule does or should exist, and in formal legal documents which tend to create or are themselves international norms.[40] But evidence regarding the existence or creation of norms comes also from customary patterns of behavior.[41] Such behavior in world politics tends to create expectations that it will continue, which expectations may or may not be supported by desires that it continue. The form of behavior may be of greater or lesser significance and the causes for its regularity may differ widely. But it seems reasonable to assume, as does Goldmann,

> . . . that behavior patterns, developed because of a desire on the part of actors and/or sanctioners to obey or create a norm, *or for any other reason*, in time come to be regarded as mandatory.[42]

Therefore a systematic comparison of the international behavior of the members of a regional group, focusing upon typical or modal interaction patterns, can tell us: a) the degree of uniformity and deviation in the behavior of the group members, i.e. an assessment of type no.1 and no.2 norms, and hence, b) the degree to which type no.3 norms are being created and adhered to. Thus a comparative study of the foreign policy behavior of a group of states yields evidence, for descriptive and explanatory purposes, regarding the level of normative integration of that group.

Before proceeding to the formulation of hypotheses and the precise operationalization of the terms "norm" and "deviation," we must first assess the significance of that deviation. That is, what will we make of typical, "maverick" behavior on the part of a group member to be deviance? Sociology has traditionally distinguished deviation, statistically more frequent or less frequent behavior, from deviance, undesirable rule-breaking behavior, through the potential the latter carries with it for negative societal reaction, i.e., sanctions. Thus, Box defines deviant behavior as "behavior which places its perpetrator at risk of being punished by those who have the institutionalized power and occasionally the consensual authority, to do something to those who do not keep to the rules."[43] Students of norms and their infraction in world politics have generally followed suit and focused upon norm-violating behavior as it relates to potential sanctions.[44] Further, as Goldmann points out, it is not necessary to reconstruct the exact utility function of such potential sanctions in the minds of decision-makers. To consider their deviating behavior significant, we need only to be able to reasonably assume that in the group we are studying, "authorities do not regard the utility of sanctions to be trivial."[45] There seems to be ample evidence from the recent history of East Europe to support this assumption.[46]

Thus individual member deviation from group norms becomes significant as deviance because: a) these members' actions hinder the norm creation within the group through their failure to contribute to the creation of customary, expected patterns of action and b) these members are incurring upon themselves the possibility of sanctions from other group members.[47]

But is normative integration among a group of nations integration? According to Nye, it is not. He asserts that, in studying international integration, "similarity of national behavior does not necessarily mean interdependence."[48] Nye is supported in this position by Kegley and Howell, whose test of Nye's classification system for Southeast Asia found identity of policy positions–as measured by similarity of voting in the UN–to have "little or no relation to how these states are integrated societally, attitudinally, or with respect to intergovernmental cooperation."[49]

Responding first to Kegley and Howell's argument, one can use these authors own findings–i.e. that integration is multi-dimensional, and that its different dimensions are relatively independent of each other–to argue that policy identity or similarity may be still another type of integration, separate from the others, requiring separate investigative approaches. Kegley and Howell themselves state that the different dimensions of integration require this separate testing.[50]

As for Nye's quite unarguable statement on the relationship between behavioral similarity and interdependence, there are two possible responses. First, as Hansen states after reviewing the study of west European integration,

> There was too little recognition of the fact that policy coordination among member-states of the European community could *substitute* for greater supranational control. While economic integration does require that many economic policies be coordinated, it does not require formal political or quasi-political institutions to undertake that coordination.[51]

It seems clear that this statement can be applied to overall foreign policy as well, especially in a hegemonic context, i.e. East Europe.[52] Second, even if similarity of behavior has not substituted for supranational control, it may, as Leon Lindberg stresses, indicate its presence.[53] This is especially important in East Europe where informal, non-institutionalized forms of "control" have generally been more important for the region's integration than have formal organizations. For the study of regional integration then, the significance of norm-creating and norm-adhering behavior should remain at least as an option of empirical investigation, rather than an object of *a priori* exclusion.

In sum, the three interrelated purposes of this work are: 1) to increase the substantive bases of knowledge of the international relations of the East European states; 2) to offer suggestions relating to the concepts and methods used in the comparative study of foreign policy; and 3) do the same regarding the study of international regional integration. In the hopes that the execution is equal to these goals, the task is addressed.

CHAPTER TWO

INTERNATIONAL INTERACTIONS

This work is aimed at describing the similarity and dissimilarity in the foreign policies of the East European states, and thus the level of normative integration of the region, and also at assessing the relative merits of various causal explanations for those policies. In the latter connection several propositions will be tested once an index of foreign policy deviance has been constructed.

First, is deviance related to level of economic development? Will the more industrialized or less industrialized members of the alliance be found to be the most deviant? One might expect that the increasing rationalization, decentralization and technological improvement of a state's economy would lead it to search out its best business and trade ties and thus lead the more industrially advanced nations out of the relative stagnation of Comecon and toward greater ties with the West. According to this reasoning "(t)his issue of economic rationality inevitably becomes one of national independence."[1]

This might be viewed as an international analog of the domestic situation produced by increasing economic development, i.e. the heightened need for economic rationality, decentralization and reform, and the greater differentiation of economic and, therefore, political demands.[2]

On the other hand, one might not expect those nations which were benefitting most under the aegis of the alliance, say East Germany, to be seeking other partners. Rather, it would be the economically less developed members who would be expected to deviate the most due to: 1) their fear and suspicion of the domination and supranational planning efforts of the "core of industrially more advanced party-states;" [3] and 2) their desire to achieve the most rapid rate of industrialization, a drive necessitating use of the best available technology, much or most of which is available only outside the East.[4] Thus expectations as to the association between level of development and deviance are somewhat equivocal. As Phillip Uren points out, "[i] in certain respects the advantage for both groups lies elsewhere."[5]

We may be able to tip the scales of expectation, however, by consideration of the significance of interdependence. If it is true, as Morse suggests, that increased levels of modernization are accompanied by increased interdependence of national economies, then the more advanced members of the alliance will be more vulnerable to the use of economic pressure.[6] Such pressure constitutes a theoretically expected strategy utilized to improve alliance performance.[7] And both the vulnerability and the pressure are an empirically observable phenomenon in the case of the Soviet-East European alliance.[8] Hence, one might then expect the more economically advanced members of the group to be the least deviant. Miles and Gillooly have demonstrated that in economic terms, "within the Eastern European subsystem, there appears to be an inner clique of the industrially more advanced members: the Soviet Union, East Germany, Czechoslovakia, and Poland."[9] Will the junior members of this "inner clique" also be the least deviant in terms of overall international behavior?

Investigation of a second question may provide evidence bearing on the development-deviance proposition. Focusing specifically on the trade dependence of the East European states, we can ask if there is a relationship between this particular form of dependence, i.e. on the Soviet Union as main supplier of imports and main purchaser of exports, and foreign policy deviance. Students of international politics have consistently pointed to the significance of such dependence on states' foreign policies;[10] while several studies have verified the predominant position of the Soviet Union in the trade flows of the East European states.[11] If we can assume then, that the Soviet Union has both the capability and the inclination to use economic levers to insure conforming alliance behavior, then we can expect non-deviance to be positively associated with trade dependence upon the USSR.[12]

Finally, the literature on alliances in general has frequently stressed the importance of security in determining alliance behavior. This has usually taken the form of propositions relating the presence or absence of an external threat to alliance cohesiveness.[13] For the Soviet alliance system, however, it is important to consider the presence of an "internal" threat, i.e. the use of military force by the USSR against a nonconforming member state. If it is unarguable that the East European alliance system is hegemonic,[14] and if past Soviet behavior (in Hungary and Czechoslovakia) is at all relevant,[15] then military/geographic position *vis-a-vis the USSR* would very likely be an important factor in the member's behavior.[16] Thus, we expect that deviance will be positively related to favorable military position *vis-a-vis* the USSR.[17]

DEFINITIONS AND METHODOLOGY

Both the form of the nations' behavior to be compared and the norm to which they are compared must be operationalized. The unit of analysis employed in this section will be the *international behavior* of the individual states of East Europe. This consists of all legitimate, intentional, foreign interactions engaged in by the relevant political actors. This operationalization attempts to reduce the ambiguity and resulting disutility of the term "foreign policy" by using as an indicator of that concept "readily identifiable empirical referents"[18]. It is important in this regard to utilize a definition that allows us to separate for analytical and descriptive purposes the dependent from the independent variable within the phenomenon commonly referred to as foreign policy.[19] International behavior is offered as a definition which will produce "discrete units of action which can be observed with sufficient frequency to allow variations in their structure and in the sources and consequences of their behavior to become evident."[20]

The conception of foreign policy is narrowed for purposes of this study in two important ways. First, it is more focused than Rosenau's concept of an "undertaking" in that it 1) more clearly separates the act from decision-makers' purposes and goals;[21] and 2) it concerns interactions, that is, completed engagements between two or more nations. This latter is a function of utilizing the systemic level of analysis, whereas a national approach such as Rosenau's would be less likely to emphasize the target of the "undertakings."[22] Secondly, for purposes of this part of the study verbal international behavior, i.e. expressions of attitude, position pronouncements, etc. have not been included. International attitudes are examined in Chapter 3.

The *norm* against which we will compare the international behavior of the group members will be a statistical representation of the typical patterns of action for the states. It will be a "constructed type," an analytical "simplification of the concrete" which enables the observer to render quantitative comparisons and measurements of relative deviation.[23] This norm will be labelled the "average East European state." Its computation and use are more fully described below. Finally, degree and direction of *deviance* will be operationalized through quantitative comparison of the individual states behavior with the norm.

The spatial domain of this study encompasses the East European states of the Warsaw Treaty Organization as well as two non-members, Albania and Yugoslavia.[24] The Soviet Union is not included in this part of the study owing to the recognition evident in international relations literature and common sense notions that the actions pursued by the superpowers differ qualitatively and quantitatively from those of the smaller states.[25] International interactions of the type studied here represent an investment of national resources for their pursuance. This, therefore, makes the amount and direction of such interactions subject to constrictions which would be greater in smaller states than in superpowers. It seems unlikely, for example, that comparing Bulgaria's international interactions with those of the Soviet Union would yield valid or useful comparisons, while comparing those of Bulgaria and Romania would. On the other hand, it is the operant assumption of this study, as in others that there is an analyzable difference between the nature of an interaction and that of an attitude.[26] While a small nation may lack the resources or capability of engaging in multi-directional and/or costly interactions, it may, without serious resource investment take positions on a variety of international events and issues. Therefore, to compare the attitudes of the smaller countries with those of the Soviet Union seems entirely appropriate. This is done in Chapter 3.

The temporal domain of the study covers the period April, 1965 through April, 1969. Though all periodization is somewhat arbitrary, this time-frame is particularly well demarcated by and includes within it significant events in East European and world politics. Beginning with the ascension of Nicolae Ceausescu to party leadership in Romania at the end of March, 1965, and ending with a similar ascension by Gustav Husak in Czechoslovakia, in April 1969, it is the latest period of substantial length during which all of the party leaders of the states included remained the same, with one significant exception (Czechoslovakia). The period embraces the vigorous pursuit by Charles De Gaulle of a French "middle way" policy and the beginning of the West German *ostpolitik,* the differential effects of which can be observed.

The leadership change that did occur during this period, that in Czechoslovakia, provides this period with one of its two crises; the other begin the Arab-Israeli War of June, 1967. Both of these events can thus be kept in mind for consideration of their effects on the patterns of interaction. The period also embraces the renewal dates for all East European and Soviet Twenty-Year Treaties of Friendship, Cooperation and Mutual Aid, and the increasing intensity of Sino-Soviet hostility.

All bilateral, national-level interactions that occurred between April, 1965 and April, 1969 have been coded. Diplomatic visits of important persons, on virtually all levels of party and government have been encoded.[27] Any agreements or protocols initialed or signed in all areas of potential agreement—cultural, scientific, educational, economic—as well as political are included.[28]

In addition to identifying the type of interaction, the coding indicates the relative significance of the behavior by assigning explicit weights to each genus of interaction. This reflects the widespread recognition that not all events that occur between nations are equal and that therefore a measurement scheme must enable the researcher to tap both the frequency and the relative intensity of that interaction.[29] The weights are assigned according to the nature of the interaction, not the dyad membership. An economic agreement between Romania and Israel, for example, is given the same weight as one between Romania and Bulgaria. This allows the event patterns themselves and their analysis to reveal the interesting phenomena.[30] The coding method is listed in Table II.1.

The specific values assigned to certain kinds of events reflect the present observer's judgments, based on the relative significance attributed these events in the literature on East Europe and Soviet foreign policy. This source of scale validity is somewhat "softer" that the panel-of-judges approach used by Azar,[31] and Moses et. al.,[32] but it does rely similarly on expert opinion, and moreover has greater specific relevance to East Europe.[33] The scale employed is ordinal, i.e. an event coded as a 5 is considered to have more significance than one coded as a 4 or 3, but no determination is made as to how much more important one is than the other (an interval scale).[34] For our purposes the question is not, "Is a scientific and technical agreement equal to one-third of a long-term trade agreement?" It is rather: "If these values are posited and applied throughout for all the nations' interactions, what patterns occur?" Thus, it is important to reiterate that the coding is uniform for all possible dyadic combinations. Any distortion of foreign policy actions, therefore, will be uniform across the eight nations' actions.

Table II. 1: EVENT CODING SYSTEM

Coded Value	Type of Event
5	Establish or resume diplomatic relations; high-level visit, i.e. party leader, state government head or Foreign Minister
4	Sign or renew long-term multipurpose treaty; visit of Defense Minister, Deputy or Vice Premier or Deputy Prime Minister
3	Multipurpose trade agreement (3-5 year); extend credit; troop stationing treaty; visit of trade or aid delegation; visit of Foreign Trade Minister or other Ministers, e.g. Light or Heavy Industry, Labor, Engineering, Sports, etc.; visit of Planning Commission Chairman; visit of Deputy Ministers
2	Multipurpose trade protocol (1-2 year); license agreement; co-production agreement; tariff agreement; raise diplomatic legation to embassy level; visit of Planning Commission delegation, Trade Union Chairman (National), or other Central Committee, Politburo, or Presidium member; visit of military delegation
1	Scientific and technical exchange; cultural, consular, or visa-travel agreement; agreements on medicine, health, transportation or repayment of debts; one-item trade contract; establish trade-promoting company, commission, or bureau; agreements on "cooperation" in fields, e.g. education, economics

Events not coded are the following: religious visits (Cardinals, Rabbis); out-of-power politicians, party members or dignitaries from the west, (e.g. Richard Nixon, C.P. Snow, French socialists); hosting of cultural or educational fairs or conferences; youth, trade union or writers' conferences or delegations without the national Chairman of same; industrial exhibitions (though any contracts reported as a result are included); meeting of standing mixed commissions; visits to the United Nations or between Ministers therein; ambassadorial audiences by the host country; visits from non-ruling communist parties, with the exception of the Provisional Revolutionary Government of South Vietnam (or the N.L.F.) which is included under "Other Communist" states (below).

Events are discrete entities, coded for their value as such. Thus, a visit by a Foreign Trade Minister is coded as a 3, but one which culminates in a trade protocol of one year is coded as 3 plus 2 or 5.

If an event is an asymmetrical visit, e.g. Deputy Minister to a Foreign Minister, the higher level leader is considered to determine the value of the visit.

A clear bias of the scale is its focus on cooperative interaction only, as opposed to both cooperative and conflictual interaction. This reflects the contextual specificity of the approach as well as a desire to correct for biases in existing classifications. For East Europe most of the events categories suggested by the globally-oriented systems, e.g. WEIS,[35] DON,[36] COPDAP,[37] would not contain sufficient numbers of events to make their measurement and comparison meaningful. Such categories as "threat with force specified"[38] "severance of diplomatic relations"[39] or "Nations A and B engage in limited war activities"[40] simply do not occur with any significant frequency in East Europe. Moreover, insufficient nuance and specificity, plus the largely *sub rosa* nature of intra-communist conflict make events-based indicators relatively insensitive to conflict within East Europe.[41] Hopmann, and Hughes did find, however, that verbal based indicators, i.e. those based on content analysis of documents and speeches, were useful indicators of conflict,[42] and this approach is utilized in Chapter 3.

A second point is that existing scales and categorizations tend to be biased in favor of conflict;[43] a bias which would force a severe closure on data available for this study. Moreover, Hermann has demonstrated that in fact most of the events engaged in by all nations are neutral or co-operative, not conflictual.[44]

These considerations tend to support the cooperative focus of the present scale.

Finally, the present approach can be criticized on the grounds that it contains only five different weighting points for events, making it a rather crude instrument. This point is a valid one and reflects the introductory nature of this investigation. Still, the evidence is not yet persuasive that increasing the number of categories is either justified or productive. First, simply having more categories would not necessarily solve the sensitivity problem, as the multi-pointed scales referred to above indicate.[45] It is rather by improving the substance of the categories—by relating them more closely to their contextual base—that the sensitivity of the indicators will be increased.[46] Second, the productivity and reliability of the larger scales are themselves subject to some doubt.[47] The present scale, though probably not exhaustive of all possible types of events, does utilize mutually exclusive categories. It is employed here in the cautious hopes that its use does not exceed its limitations.[48]

Table II.2: INTERACTION PARTNERS

Albania (Alb)

Chinese People's Republic (CPR)

Other Communist (Other Comm): Principally North Vietnam, but including also Mongolia, North Korea, Cuba and the PRG/NLF of South Vietnam

USSR

East Europe (EE): Bulgaria, Czechoslovakia, G.D.R., Hungary, Poland, Romania

Yugoslavia (Yugo)

Nonaligned (Non):

European—Austria, Finland, Switzerland, Sweden
All Black African Nations

Mediterranean African Nations—Tunisia, Sudan, Morocco, Algeria

Mideast—Iraq, Saudi Arabia, Kuwait, Yemen

Asian—Principally India, Burma

United Nations—U. Thant visit only

Arab States (AR): Lebanon, Syria, U.A.R., Jordan

Israel (IS)

West (West):

European—France, U.K., Benelux countries, Norway, Denmark, Iceland

Mideast—Iran

Asian—Pakistan, Singapore, Indonesia, Australia, New Zealand, Japan

Western Hemisphere—Canada, Central and South America

Non-national actors—E.E.C., E.F.T.A., The Vatican, G.A.T.T., I.M.F.

United States (US)

Federal Republic of Germany (FRG)

The determination of target groups, i.e. whether a particular nation is associated with the "West" or "Nonaligned," is determined by formal alliance membership for the intervening years. The partners of East European interaction are grouped in Table II.2.

The sources for the data are two. The "Current Developments" section of the journal *East Europe* provided a comprehensive month-by-month survey of action and interactions of all the countries considered. In addition, *Keesings Contemporary Archives*, a standard data source for international relations, was employed. Neither source was considered primary. Each had events that the other had missed. The two sources thus provided both a balance against possible systematic bias of events reporting and a greater insurance of near totality of events data.[49]

In all, 1552 interactions were coded by the present writer with intra-coder reliability assessed at .86.[50] Data were checked for possible duplicate recording of events and to insure balanced recording when both members of the dyad were East European. Unsubstantiated reports of clandestine meetings (two only) were not included. Finally, the extended time period is expected to minimize the degree of distortion created by beginning-ending demarcations, though it is recognized that some distortion from this effect may still be present.

FINDINGS: SALIENT FEATURES

Perhaps the most efficient way to begin comparing the international behavior of the East European states is to illustrate their interaction patterns. When all of their interactions have been coded and summed by country and year, a four-year interaction chart for each can be drawn. (See Figures II.1-8). Some comparisons are immediately evident. Bulgaria and Romania are the two most active members of the WTO, surpassing even the reasonably active nonmember, Yugoslavia. The relative political isolation of the GDR and Albania is readily apparent. Note that while all the nations except the GDR have at least some interaction with China during this period, none except Albania have any substantial activity. Regarding the FRG, all except Albania have some interaction, but only Romania has a significantly greater amount. This is true to a lesser degree regarding the United States, where Romania is joined by Yugoslavia, both having much greater interaction levels than any of the others. Immediately noticeable also are the high interaction levels of Poland, Bulgaria and especially Romania with the West. All three of these have substantially greater interaction levels with the West than does nonaligned Yugoslavia. Certainly a key difference between Romania and all the other countries is its interaction balance with the Arab States/Israel, a balance which favors Israel. On interactions with Nonaligned states, Romania and Bulgaria are the most active of the allies, the latter exceeding even Yugoslavia in this regard. With regard to relations between Yugoslavia and the other states, Romania has the highest absolute level of interaction, one-fourth more than the next highest, Czechoslovakia, and more than one-third more than the next highest, Hungary. Even active Bulgaria was apparently not cultivating substantial relations with its Balkan neighbor.

With regard to their own East European partners, all have relatively high levels of interaction with each other. For Czechoslovakia, the GDR and Hungary—the middle belt of the WTO—these relations represent their modal interaction target. This is true also of Yugoslavia—yet another instance where Yugoslavia's behavior has been at least as conforming as any of the acutal WTO members. For Poland and Romania their interactions with their fellow allies take second place to their relations with the West; and for Bulgaria they are third, behind interactions with the nonaligned nations as well as with the West. More will be said below on individual country patterns. A further interesting item for comparison here is the EE/USSR interaction ratio for the states. Yugoslavia has overwhelmingly more interaction with the WTO nations than with their patron, the USSR. If one begins to suspect therefore that this ratio is characteristic of deviance, a perusal of the interaction charts reveals that Poland and also Romania have substantially more EE than USSR interactions. However the presence of similar unbalanced ratios for Hungary and the GDR confound this intuitive notion.

Lastly, in relationships with other communist nations, Romania has the highest level of interactions, but it does not differ substantially from that of the other allies. Relations with Albania are almost completely absent for all WTO members.

The Interaction Score chart is useful for pointing up salient differences. The phenomena of a highly active state such as Bulgaria or Romania is an important aspect of foreign policy deviance. But we can render the nations' interactions more comparable by controlling for absolute level of activity and examining instead the attention patterns that characterize those interactions that a nation does engage in. Figures II.9-16 display the interaction patterns of these nations using the familiar pie chart. We are asking now: "Given the level of activity sustained by each of these countries during these years, how did each divide its time? With whom did it interact and in what proportion?" This particular method of display,

EAST EUROPEAN INTERACTION SCORES 1965-69

Fig. 1 BULGARIA

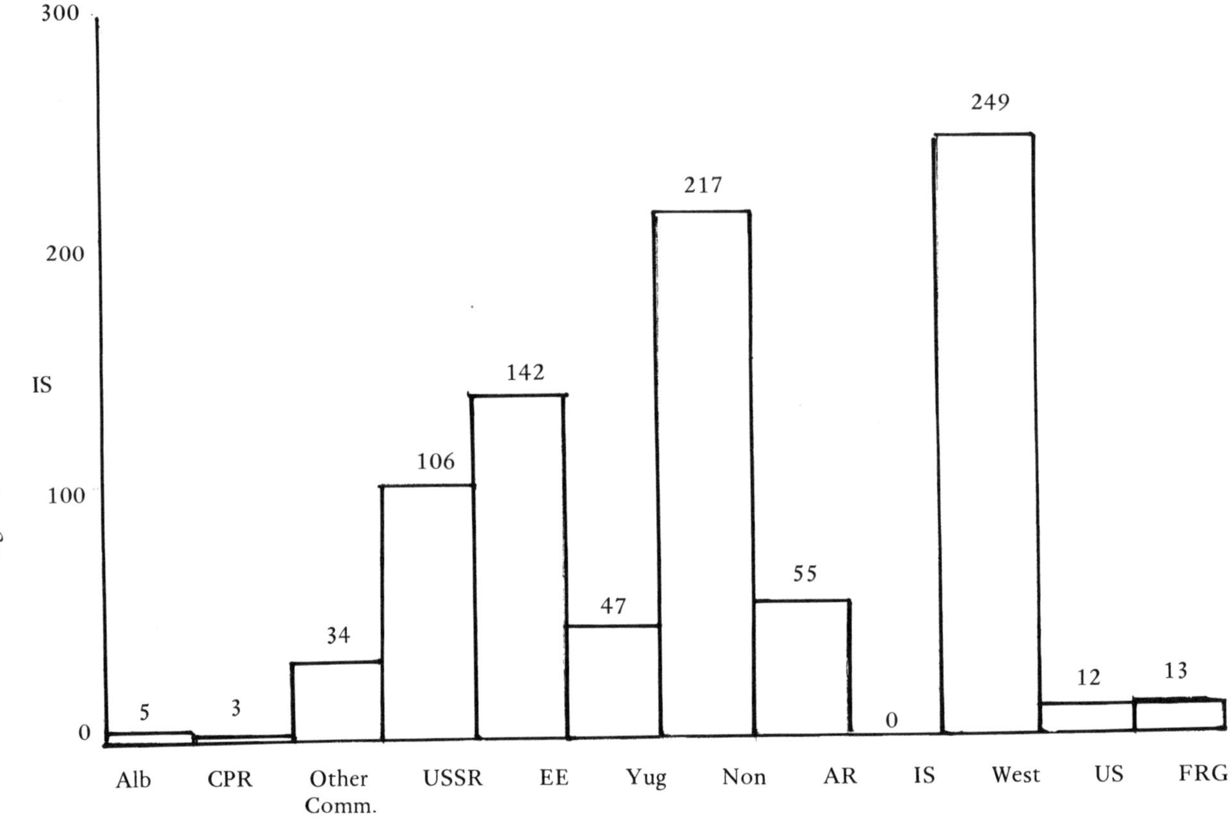

EAST EUROPEAN INTERACTION SCORES
1965-69

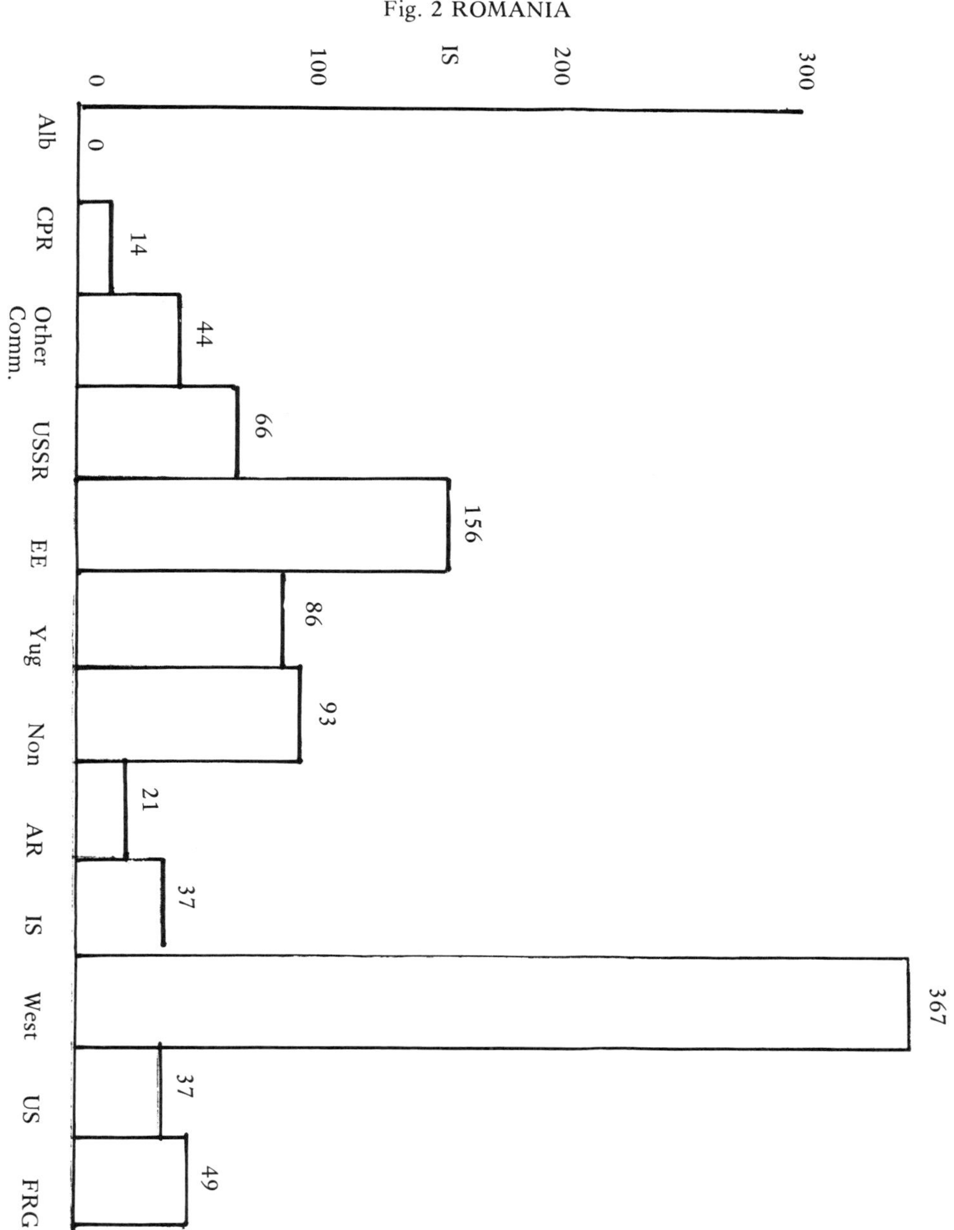

Fig. 2 ROMANIA

EAST EUROPEAN INTERACTION SCORES
1965-69

Fig. 3 HUNGARY

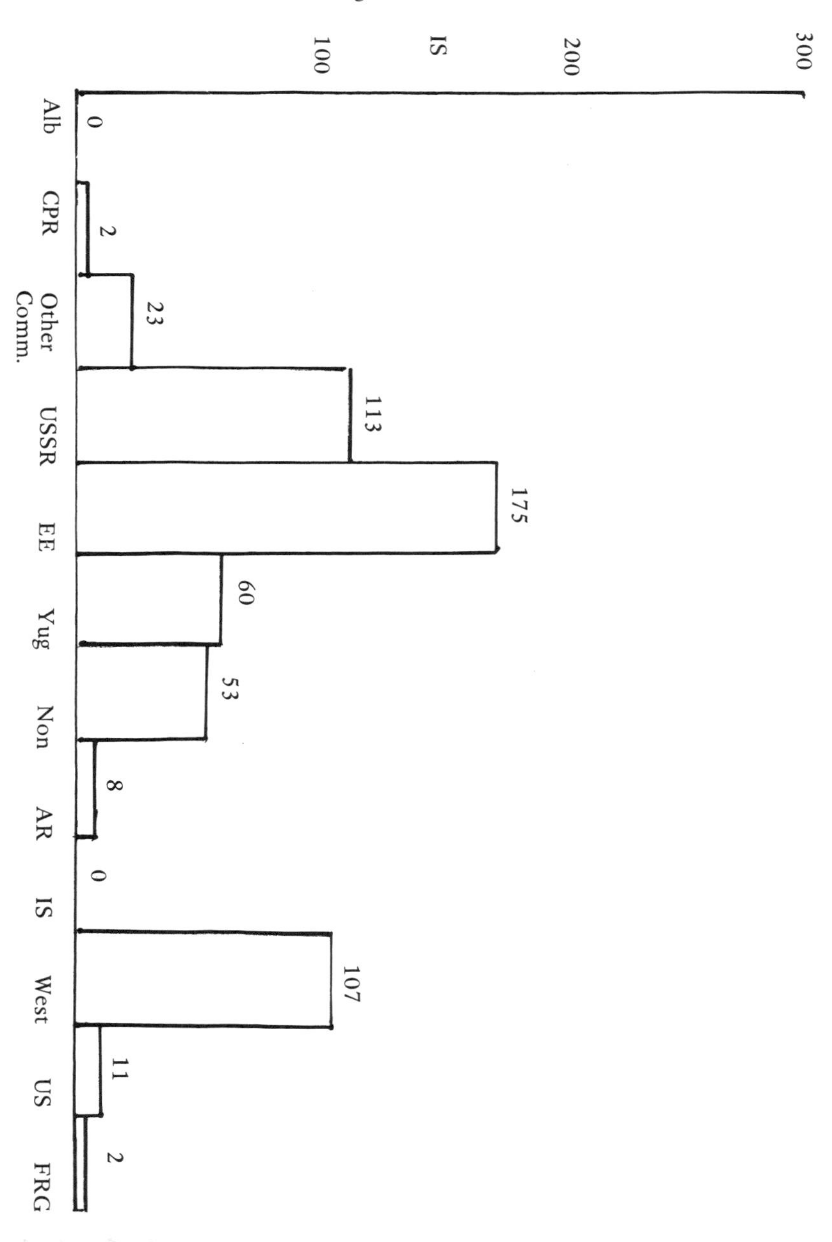

EAST EUROPEAN INTERACTION SCORES
1965-69

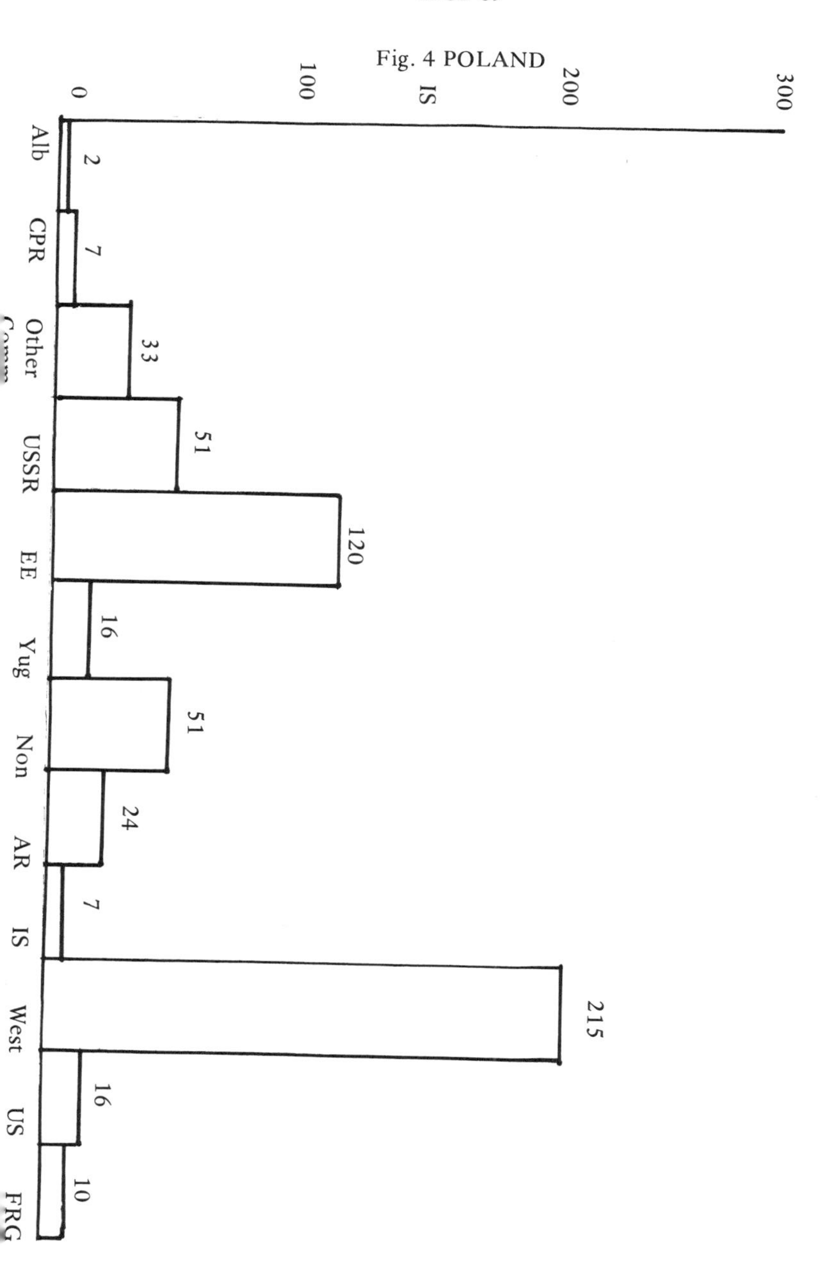

Fig. 4 POLAND

EAST EUROPEAN INTERACTION SCORES 1965-69

Fig. 5 CZECHOSLOVAKIA

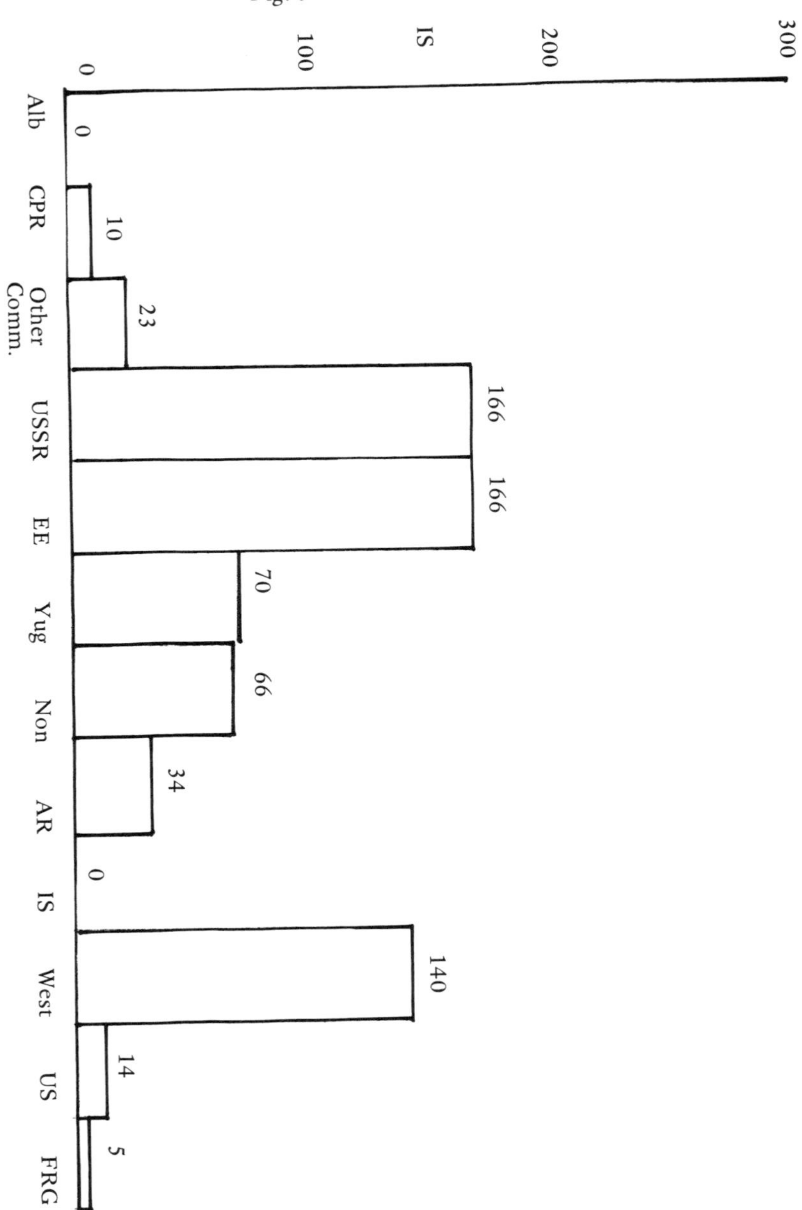

EAST EUROPEAN INTERACTION SCORES
1965-69

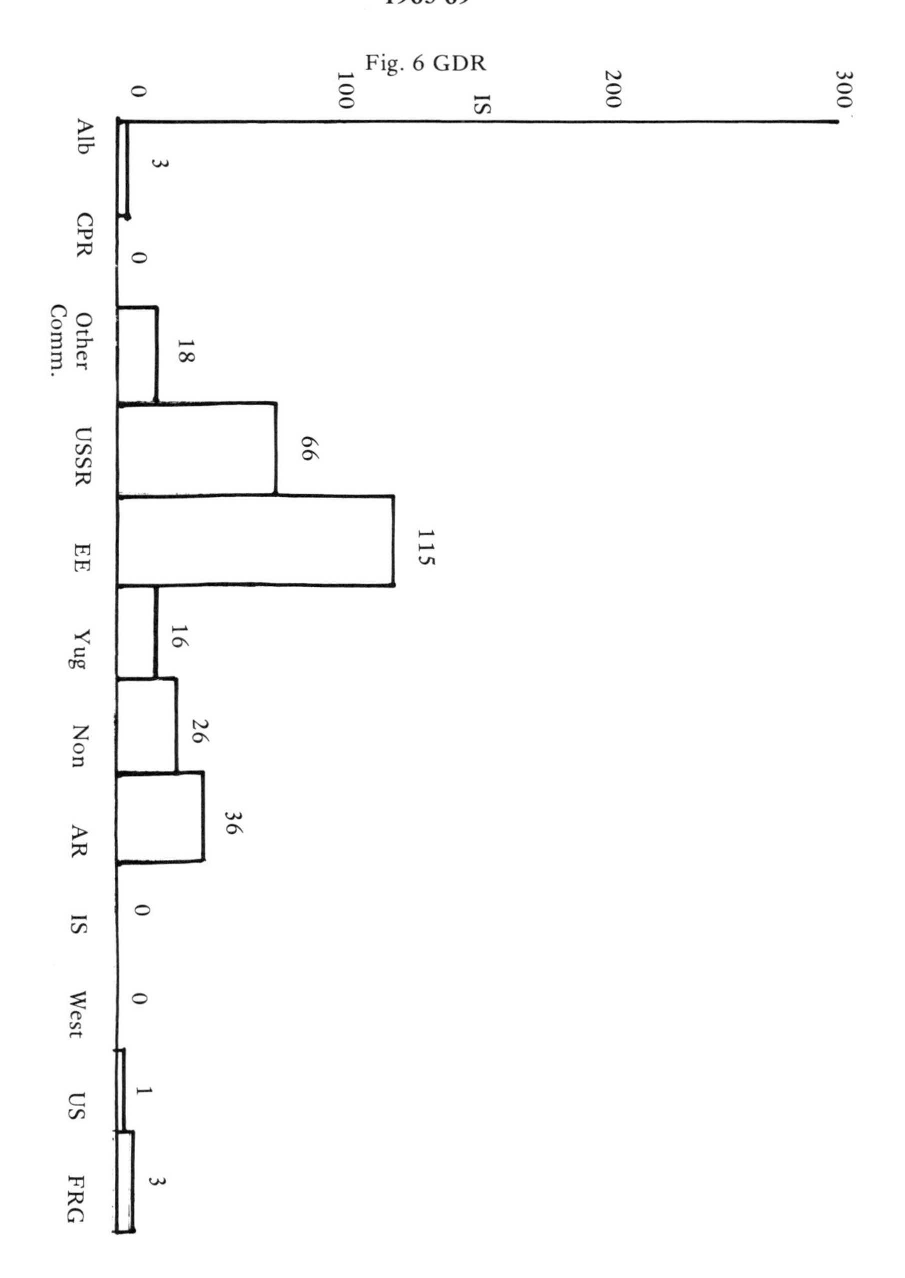

EAST EUROPEAN INTERACTION SCORES 1965-69

Fig. 7 YUGOSLAVIA

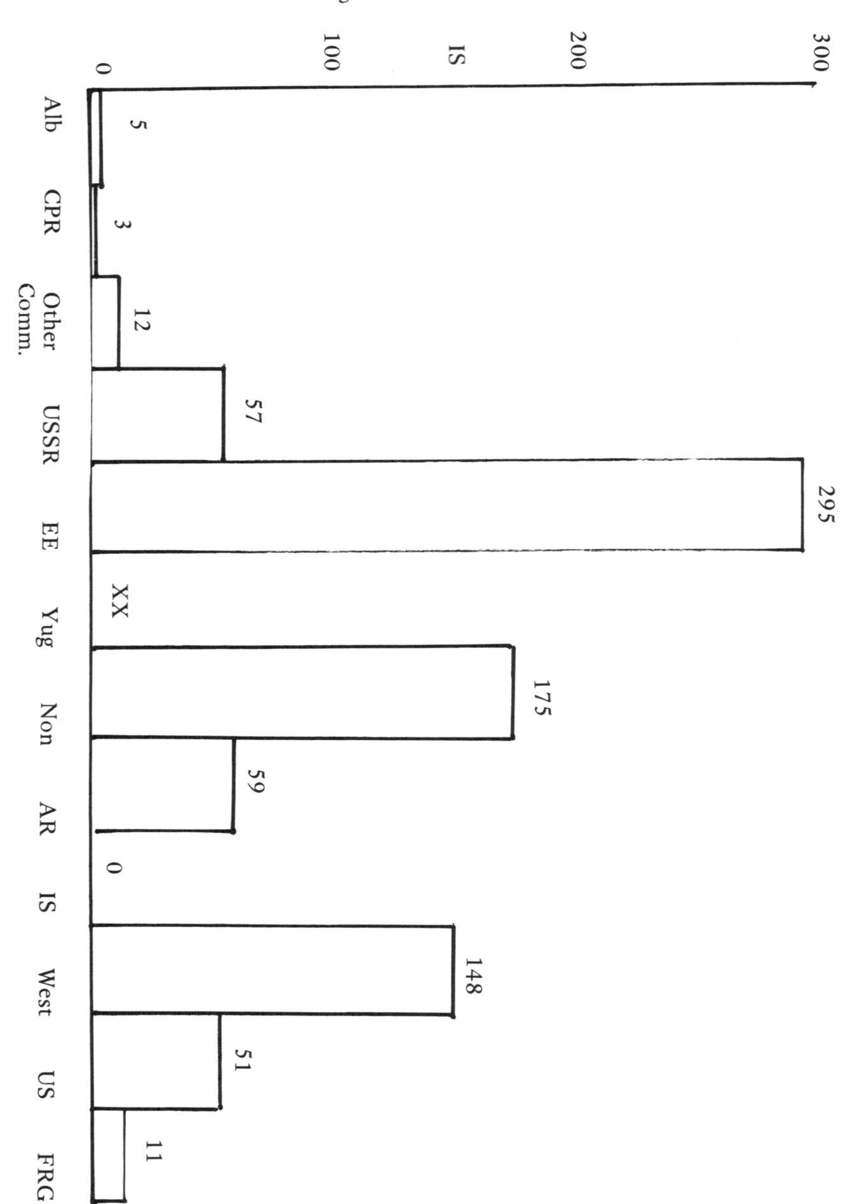

EAST EUROPEAN INTERACTION SCORES
1965-69

Fig. 8 ALBANIA

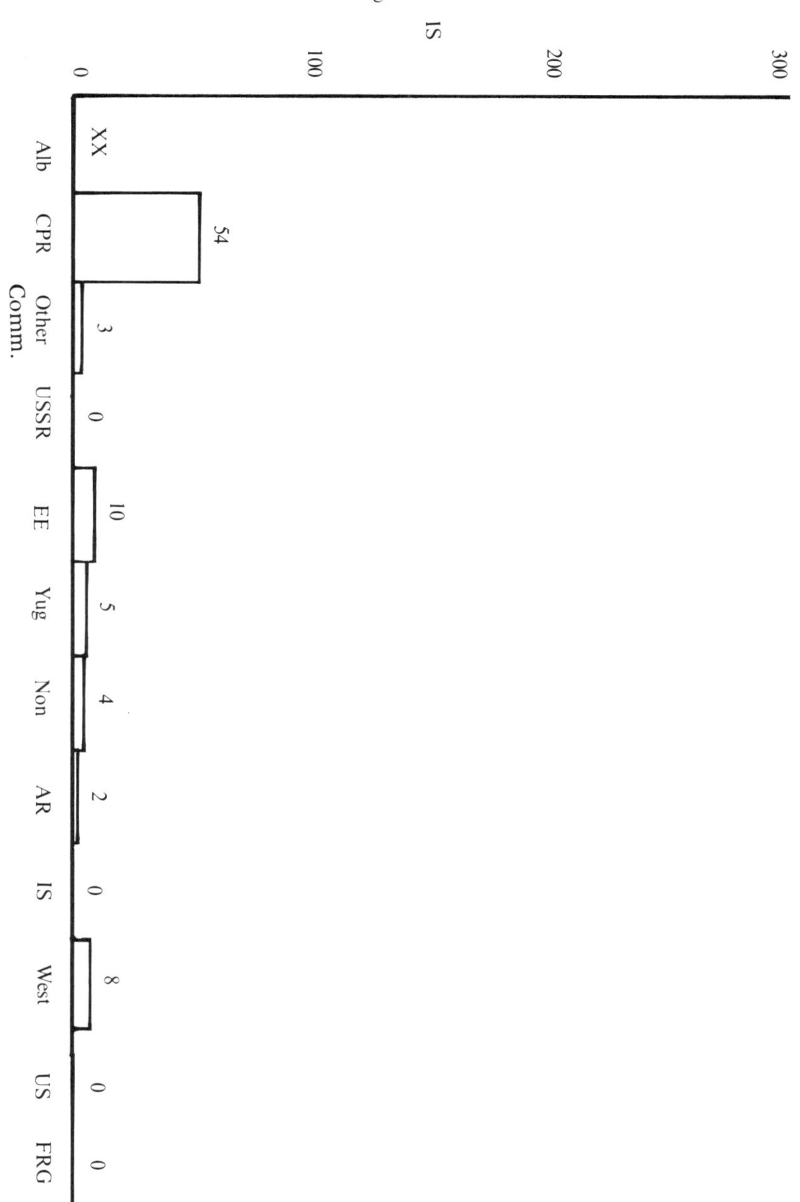

EAST EUROPEAN INTERACTION PERCENTAGES 1965-69

Fig. 9 BULGARIA

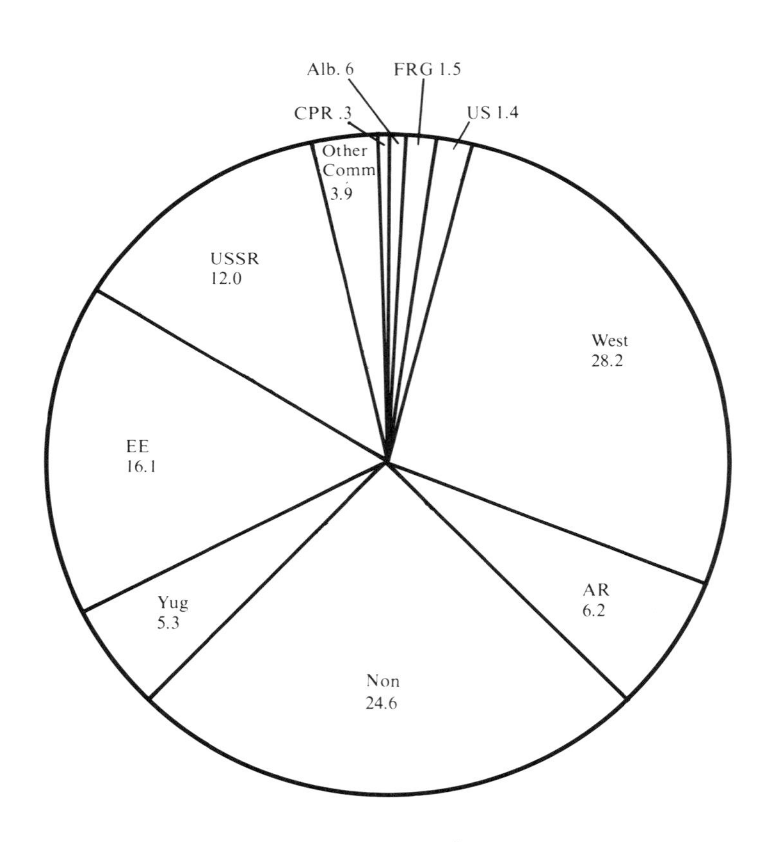

EAST EUROPEAN INTERACTION PERCENTAGES
1965-69

Fig. 10 CZECHOSLOVAKIA

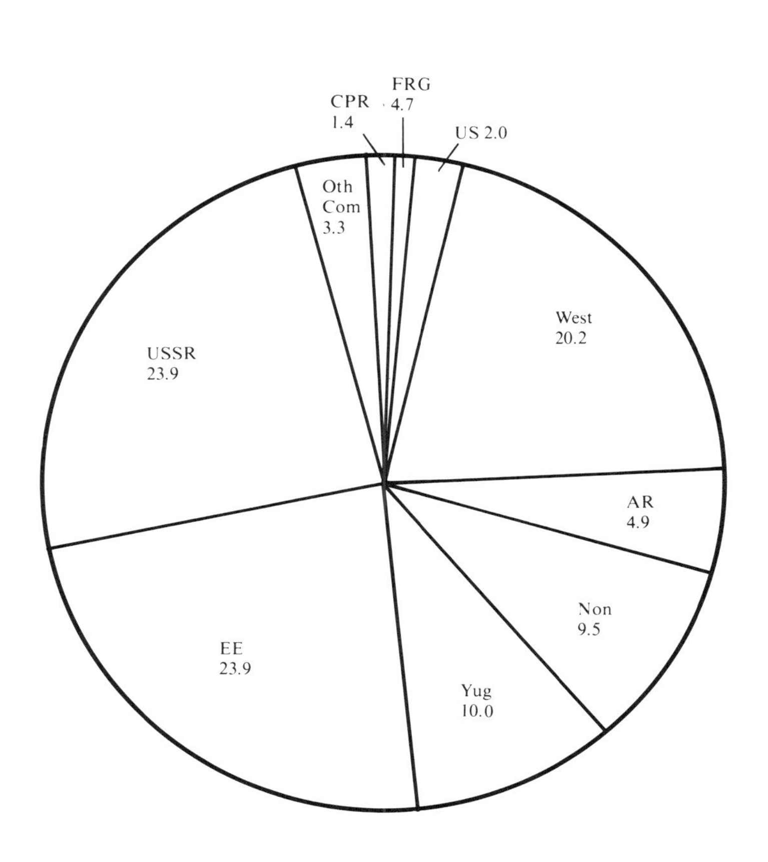

EAST EUROPEAN INTERACTION PERCENTAGES 1965-69

Fig. 11 GDR

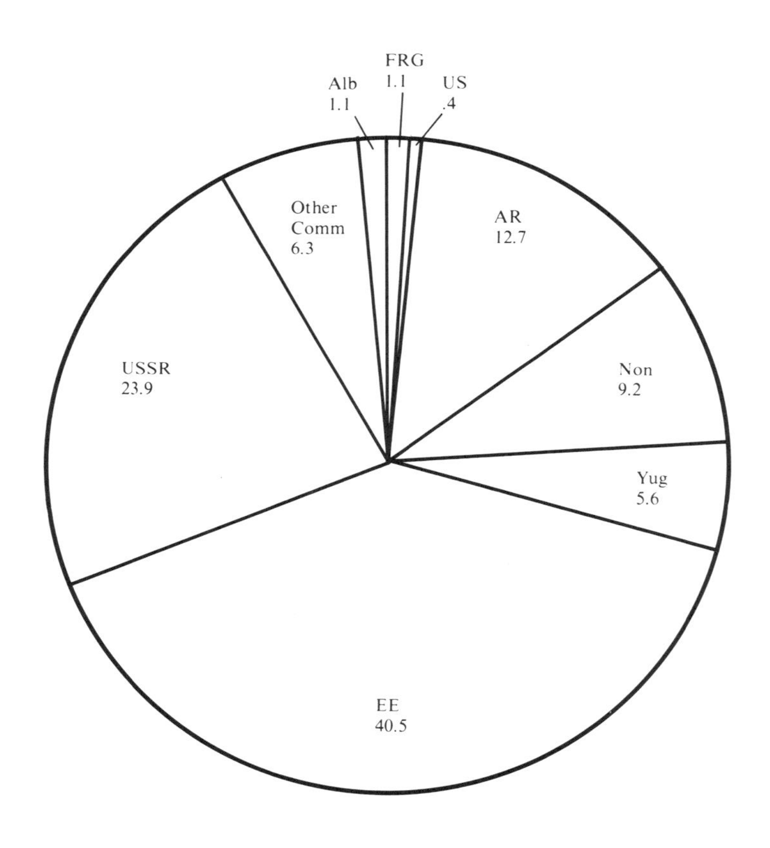

EAST EUROPEAN INTERACTION PERCENTAGES
1965-69

Fig. 12 HUNGARY

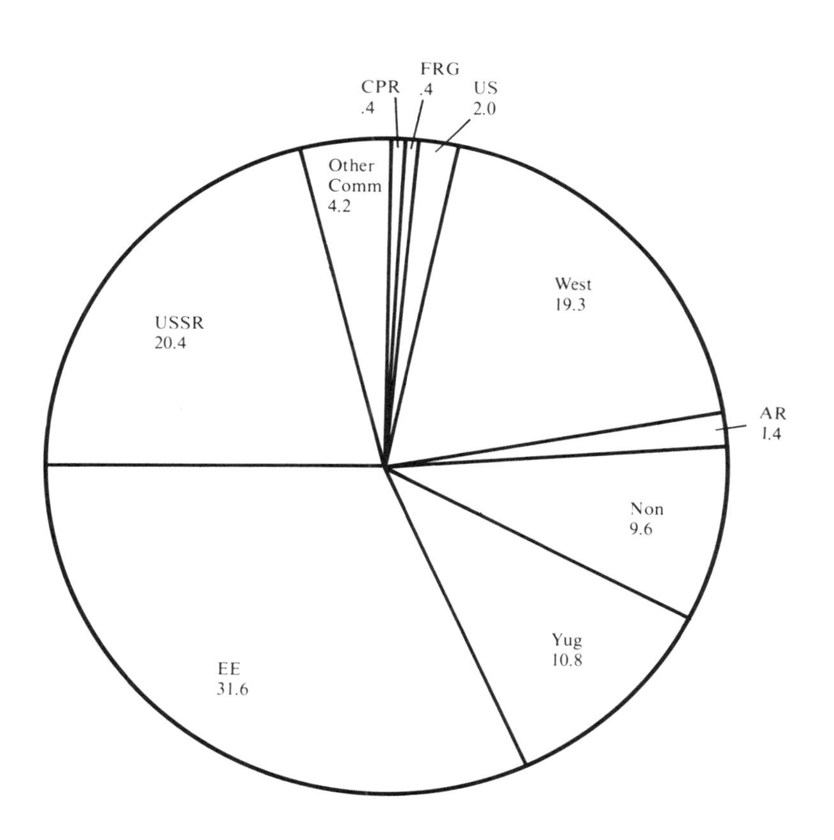

EAST EUROPEAN INTERACTION PERCENTAGES 1965-69

Fig. 13 POLAND

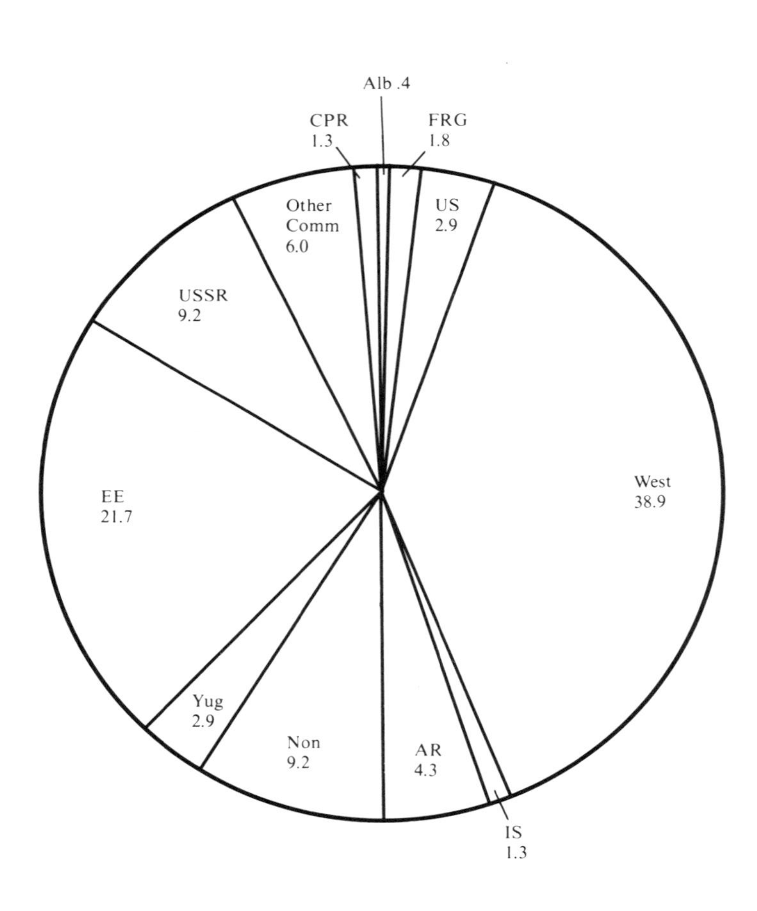

EAST EUROPEAN INTERACTION PERCENTAGES 1965-69

Fig. 14 ROMANIA

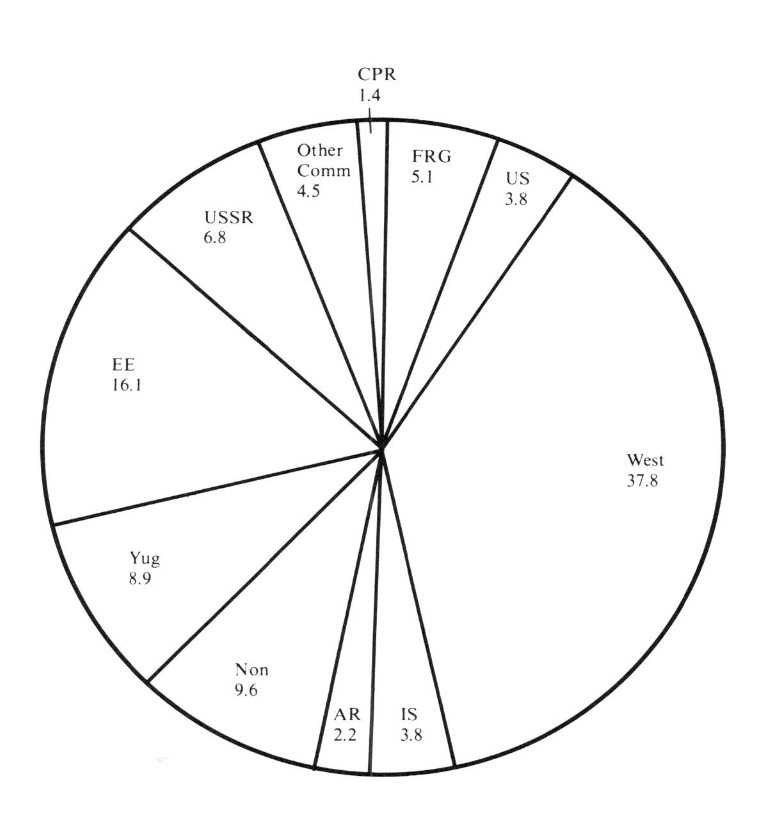

EAST EUROPEAN INTERACTION PERCENTAGES 1965-69

Fig. 15 YUGOSLAVIA

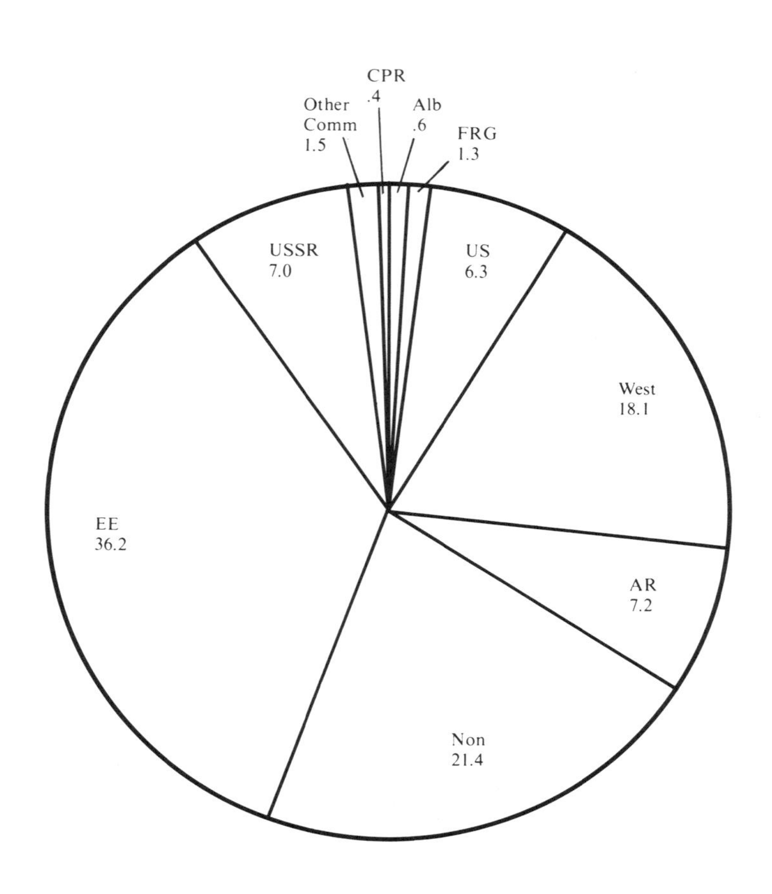

EAST EUROPEAN INTERACTION PERCENTAGES 1965-69

Fig. 16 ALBANIA

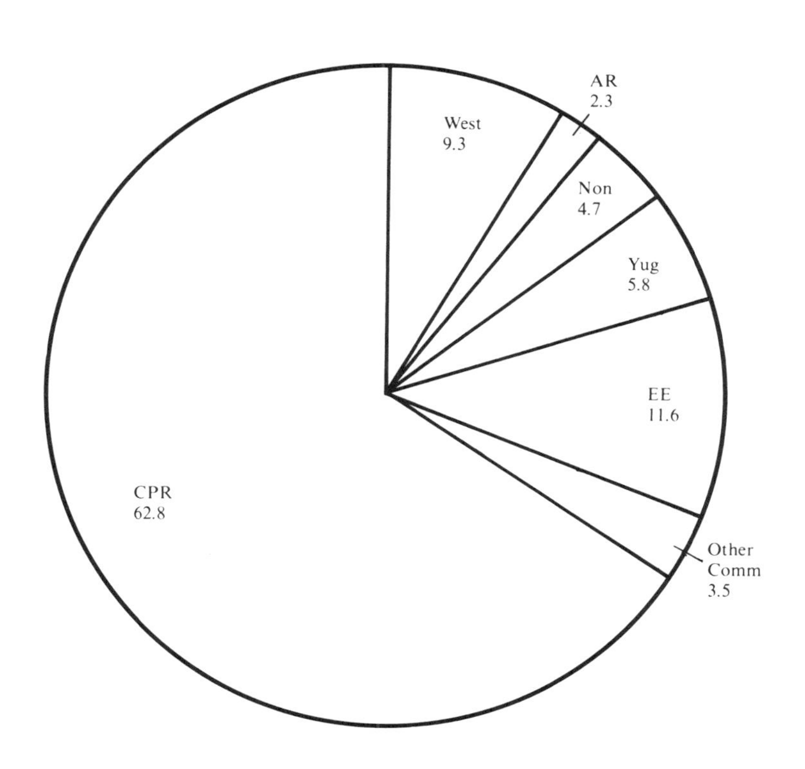

by rendering interaction patterns into percentage terms, facilitates mathematical comparison and manipulation. It is to this that we now turn.

THE "AVERAGE" EAST EUROPEAN STATE

Table II.3 lists the four-year percentage of interactions for each East European nation with all of the target nations or groups. For example, we see in the first column that between 1965 and 1969, 12.0% of Bulgaria's interaction score was achieved with the Soviet Union, 16.1% with other East European nations, etc. The column marked "WTO Average" indicates the horizontal average of those percentages for WTO members only. This provides an idea of the degree to which the "average" WTO member interacted with the particular target. While no such "average" state exists in reality, this standard will provide our type No. 1 and No. 2 behavioral norm (see page 36) against which to measure the existence and degree of deviation by group members.[51] A z-score measurement of relative deviation will be calculated for each Interaction Percentage (IP).[52] Thus we see in Table II.4 that Bulgaria's IP with the USSR (12.0%) is about one-half of a standard deviation unit below average, its IP with other EE states, more than nine-tenths of an S.D. unit below average.

Figure II.17 illustrates the dispersion of the East European states' z-scores for ten potential interaction partners.[53] The z-scores arranged by partner are in Table II.5. The place where Yugoslavia would fall in each spread is also indicated, though of course its values are not considered in calculating the WTO average. It is interesting nevertheless to compare the international behavior of this nation with that of the Warsaw Pact members, while also examining how well the alliance itself did in achieving relatively uniform behavior toward key targets.

With regard to the major targets we can see from these dispersions that Hungary, East Germany and Czechoslovakia deviate from the norm to the positive side in their interactions with the Soviet Union, while Romania, Poland, Yugoslavia, and to a lesser extent Bulgaria deviate to the negative side of the mean. The dispersion toward East Europe elaborates on the interesting anomaly mentioned previously.[54] East Germany and Hungary "overfulfill" the norm, and once again we see Romania and Bulgaria, and to a lesser extent Poland, as negative deviants; Czechoslovakia's rate almost exactly equals the norm. What is noteworthy is Yugoslavia's high positive score, indicating its active involvement with

TABLE II.3
EAST EUROPEAN INTERACTION PERCENTAGES [1]

East European States

Partners	BUL (883)[2]	CZE (694)	GDR (284)	HUN (554)	POL (552)	ROM (970)	WTO Ave	YUG (816)	ALB (86)
Alb.	.6	0	1.1	0	.4	0	.35	.6	XX
CPR	.3	1.4	0	.4	1.3	1.4	.80	.4	62.8
Other Comm.	3.9	3.3	6.3	4.2	6.0	4.5	4.70	1.5	3.5
USSR	12.0	23.9	23.2	20.4	9.2	6.8	15.88	7.0	0
EE	16.1	23.9	40.5	31.6	21.7	16.1	24.98	36.2	11.6
Yugo.	5.3	10.1	5.6	10.8	2.9	8.9	7.27	XX	5.8
Non	24.6	9.5	9.2	9.6	9.2	9.6	11.95	21.4	4.7
AR/IS	6.2/0	4.9/0	12.7/0	1.4/0	4.3/1.3	2.2/3.8	5.3/.85	7.2/0	2.3/0
West	28.2	20.2	0	19.3	38.9	37.8	24.07	18.1	9.3
US	1.4	2.0	.4	2.0	2.9	3.8	2.08	6.3	0
FRG	1.5	.7	1.1	.4	1.8	5.1	1.77	1.3	0

1. Columns indicate the percentage of the Interaction Score achieved with each partner. (Percentages may not add to 100 due to rounding.)
2. Total Interaction Score, from Table A.1, Appendix I.

TABLE II.4
Z-SCORES [1] of
EAST EUROPEAN INTERACTION PERCENTAGES
East European States

Partners	Bulg.	Czech.	GDR	Hungary	Poland	Romania	Yugo.[2]
Alb.	+ .559	– .783	+1.678	– .783	+ .112	– .783	+ .559
CPR	– .845	+1.014	–1.351	– .676	+ .845	+1.014	– .676
Other Comm.	– .669	–1.172	+1.339	– .418	+1.088	– .167	–2.678
USSR	– .512	+1.059	+ .966	+ .597	– .882	–1.199	–1.172
EE	– .932	– .113	+1.628	+ .695	– .344	– .932	+1.177
Yugo.	– .630	+ .905	– .534	+1.129	–1.398	+ .521	XX
Non	+2.040	– .395	– .444	– .379	– .444	– .379	+1.524
AR	+ .228	– .094	+1.837	– .961	– .243	– .763	+ .475
IS	– .553	– .553	– .553	– .553	+ .293	+1.921	– .553
West	+ .286	– .268	–1.667	– .330	+1.027	+ .951	– .353
US	– .576	– .068	–1.425	– .068	+ .696	+1.459	+3.579
FRG	– .158	– .625	– .392	– .801	+ .018	+1.946	– .275

1. $Z = \frac{x - \bar{x}}{\sigma}$

2. These values are not as statistically precise as those for WTO members since Yugoslavia's values were not included in the calculation of the mean and standard deviation. They are offered nevertheless as being useful for comparative purposes. No z-scores were computed

FIGURE II.17

EAST EUROPEAN Z-SCORE DISPERSIONS BY INTERACTION PARTNER

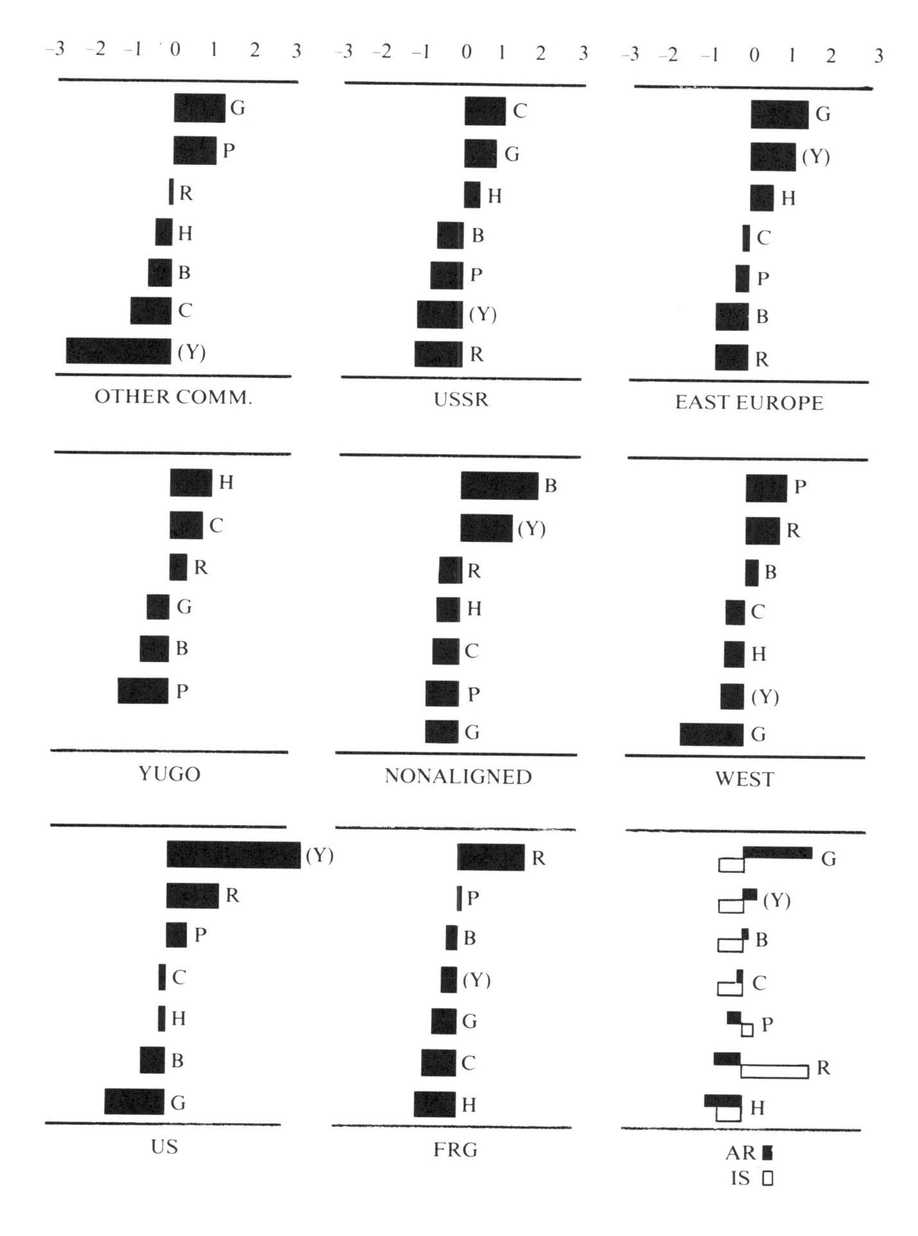

KEY: B=Bulg., C=Czech., G=GDR, H=Hung., P=Poland, R=Romania, (Y)=Yugo.

TABLE II.5
EAST EUROPEAN Z-SCORES BY INTERACTION PARTNER

Interaction Partner	East European Country	Z-Score	Interaction Partner	East European Country	Z-Score
Other Comm.	GDR	1.339	USSR	Czech.	1.059
	Poland	1.088		GDR	.966
	------------			Hungary	.597
	Romania	− .167		------------	
	Hungary	− .418		Bulgaria	− .512
	Bulgaria	− .669		Poland	− .882
	Czech.	−1.172		(Yugo)	−1.172
	(Yugo)	−2.678		Romania	−1.199
East Europe	GDR	1.628	Yugoslavia	Hungary	1.129
	(Yugo)	1.177		Czech.	.905
	Hungary	.695		Romania	.521
	------------			------------	
	Czech.	− .113		GDR	− .534
	Poland	− .344		Bulgaria	− .630
	Bulgaria	− .932		Poland	−1.398
	Romania	− .932			
Nonaligned	Bulgaria	2.040	West	Poland	1.027
	(Yugo)	1.524		Romania	.951
	------------			Bulgaria	.268
	Romania	− .379		------------	
	Hungary	− .379		Czech.	− .268
	Czech.	− .395		Hungary	− .330
	Poland	− .444		(Yugo)	− .353
	GDR	− .444		GDR	−1.667
United States	(Yugo)	3.579	FRG	Romania	1.946
	Romania	1.459		Poland	.018
	Poland	.696		------------	
	------------			Bulgaria	− .158
	Czech.	− .068		(Yugo)	− .275
	Hungary	− .068		GDR	− .392
	Bulgaria	− .596		Czech.	− .625
	GDR	−1.425		Hungary	− .801
Arab States	GDR	1.837	Israel	Romania	1.921
	(Yugo)	.475		Poland	.293
	Bulgaria	.288		------------	
	------------			Hungary	− .553
	Czech.	− .094		Czech.	− .553
	Poland	− .243		Bulgaria	− .553
	Romania	− .763		GDR	− .553
	Hungary	− .961		(Yugo)	− .553

NOTE: Line equals a z-score of zero, indicating identity of the interaction percentage with the mean.

the other East European states. (The "Yugo" chart indicates which of the EE states were most active in this movement.) This is a further indication that nonalignment may not always mean negative deviant behavior; that it may also indicate elements of accepted norm adhering–indeed overfulfilling–behavior.

Other notable points in the dispersion charts are: Bulgaria's very high positive score on the Nonaligned scale, exceeding even Yugoslavia; the high score for Poland and Romania in interaction with the West, as well as a minor deviance here by Bulgaria; similarly high scores by Poland and Romania on the U.S. scale, though far from highest, Yugoslavia.[55] Clearly evident also are Romania's deviations with regard to the FRG and Israel, in both cases having scores greatly separated from all other East European states. On the other hand, the GDR greatly exceeds the norm in its interactions with the Arab States, while Bulgaria and Yugoslavia are also above average.

Let us now view these interactions not from the perspective of the targets but from that of the East European states. Figure II.18 displays for each East European state its z-scores for ten targets.[56] These charts enable one to see at a glance the degree and direction of deviations from the norm that are included in each state's overall interaction pattern.

To understand the positive or negative dimension of each type of deviation, evaluations as to their effect on the overall cohesion of the alliance can be added as follows. Overfulfillment of the norm in interactions with the USSR, EE, and Other Communists, fits its mathematical designation (positive) in that it is seen as indicating adherence to the alliance, greater interaction with other alliance (or suitable) states, and the alliance core itself. Therefore less than normal interaction in these categories can fairly be judged as negative. Also interaction with the Arab States, as opposed to Israel is seen as desirable, hence overfulfillment is positive, underfulfillment, negative.

Concerning the relations of the individual WTO members with Yugoslavia, and the interaction of all seven nations with the West, the Nonaligned, the US, FRG and Israel, evaluation is reversed. In these cases underfulfillment of the norm is the behavior of a nation more tied to the alliance (positive deviance). Conversely, a high z-score, revealing as it does more than average interaction with non-alliance, non-Soviet, or pro-West states is indicative of a shift of attention away from the alliance (negative deviance).

FIGURE II.18

Z-SCORES FOR EAST EUROPEAN STATES WITH EVALUATED QUADRANTS

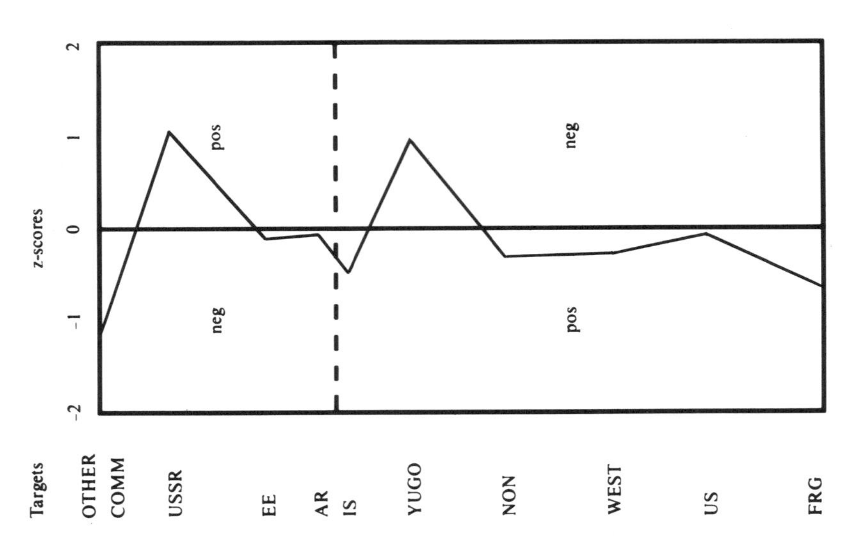

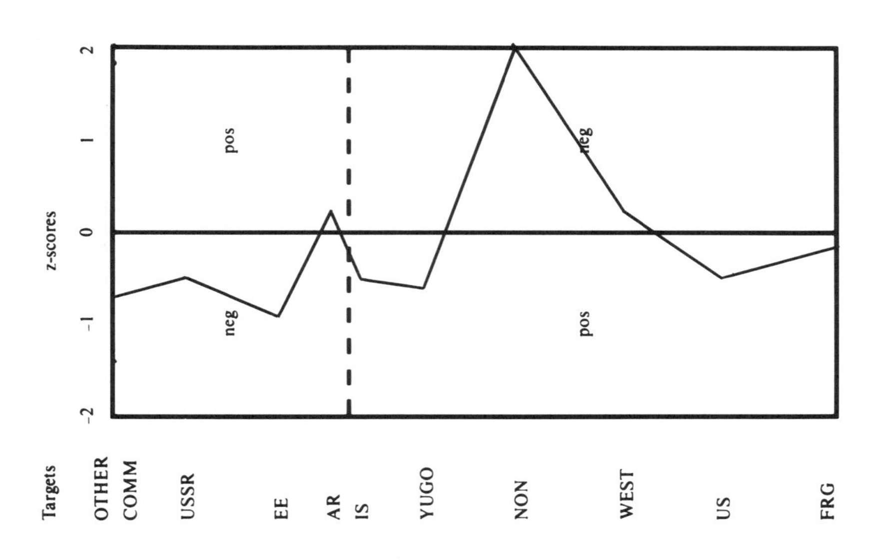

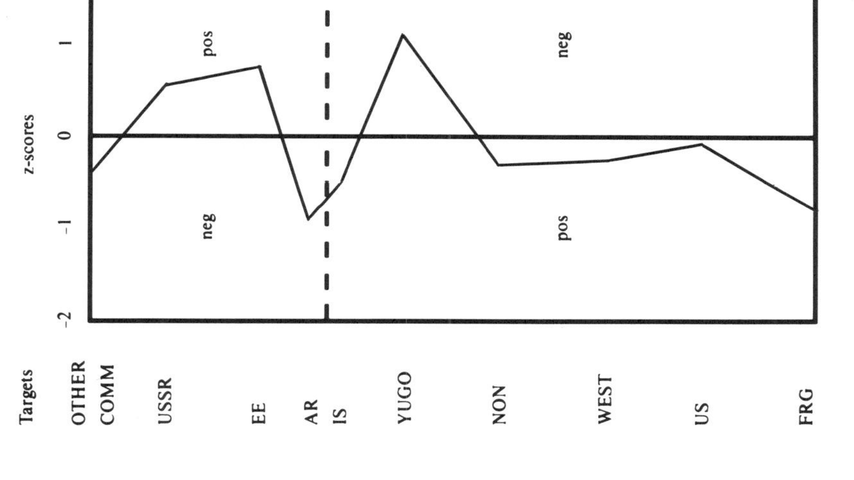
z-scores
-2
-1
0
1
2
pos
neg
neg
pos
Targets
OTHER COMM
USSR
EE
AR IS
YUGO
NON
WEST
US
FRG

HUNGARY

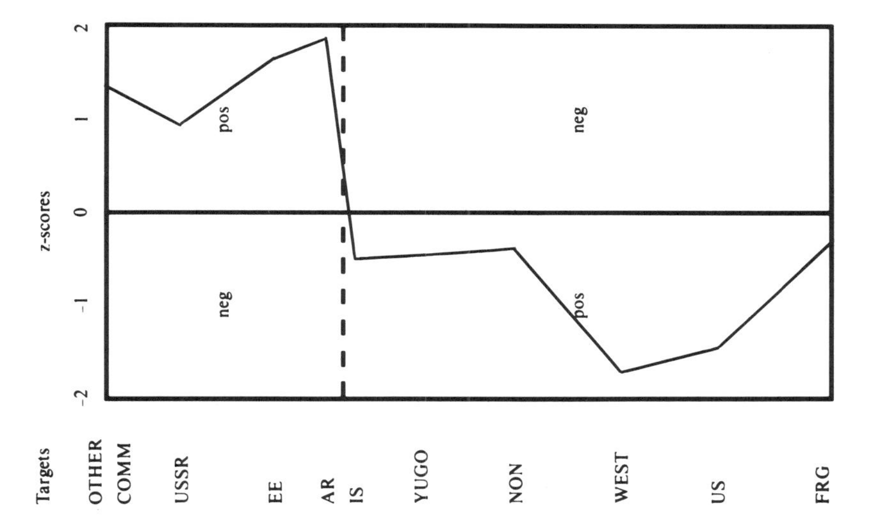
z-scores
-2
-1
0
1
2
pos
neg
neg
pos
Targets
OTHER COMM
USSR
EE
AR IS
YUGO
NON
WEST
US
FRG

GDR

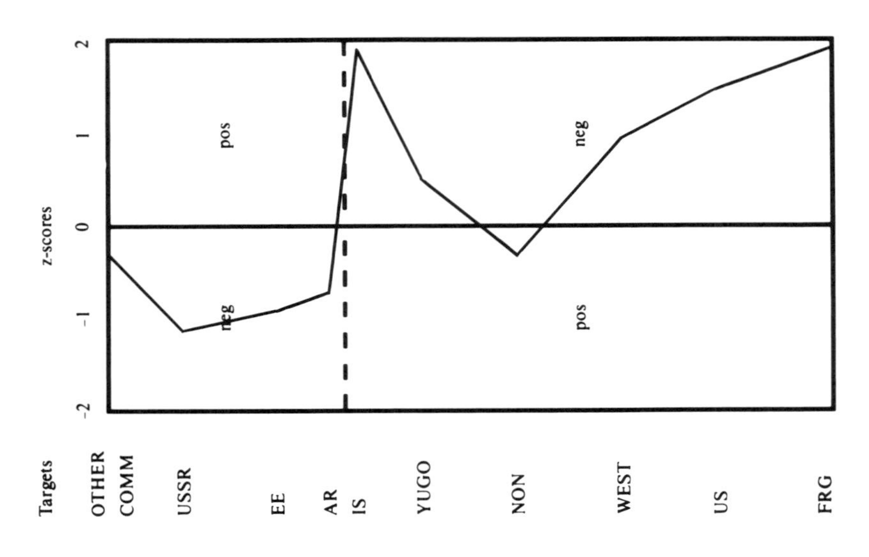
z-scores
-2
-1
0
1
2
pos
neg
neg
pos
Targets
OTHER COMM
USSR
EE
AR IS
YUGO
NON
WEST
US
FRG
ROMANIA

POLAND

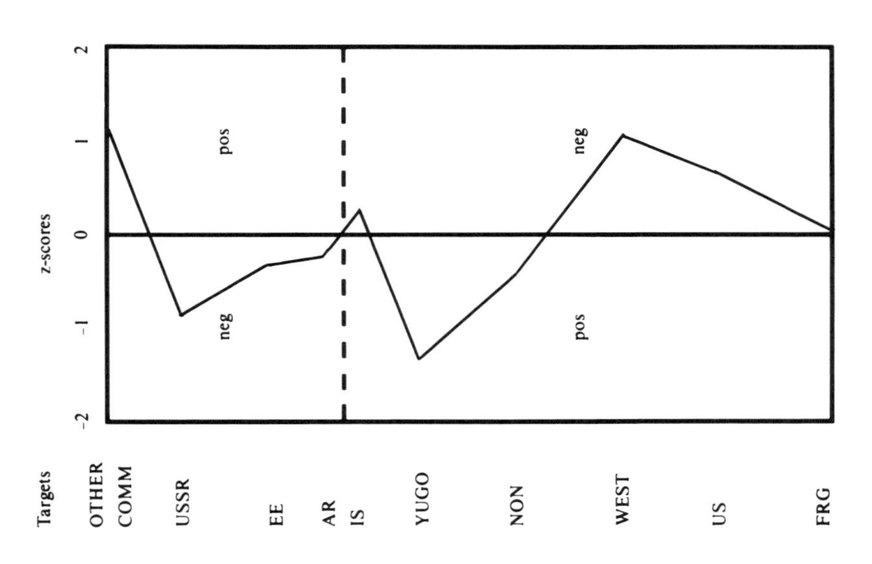
z-scores
-2
-1
0
1
2
pos
neg
neg
pos
Targets
OTHER COMM
USSR
EE
AR IS
YUGO
NON
WEST
US
FRG

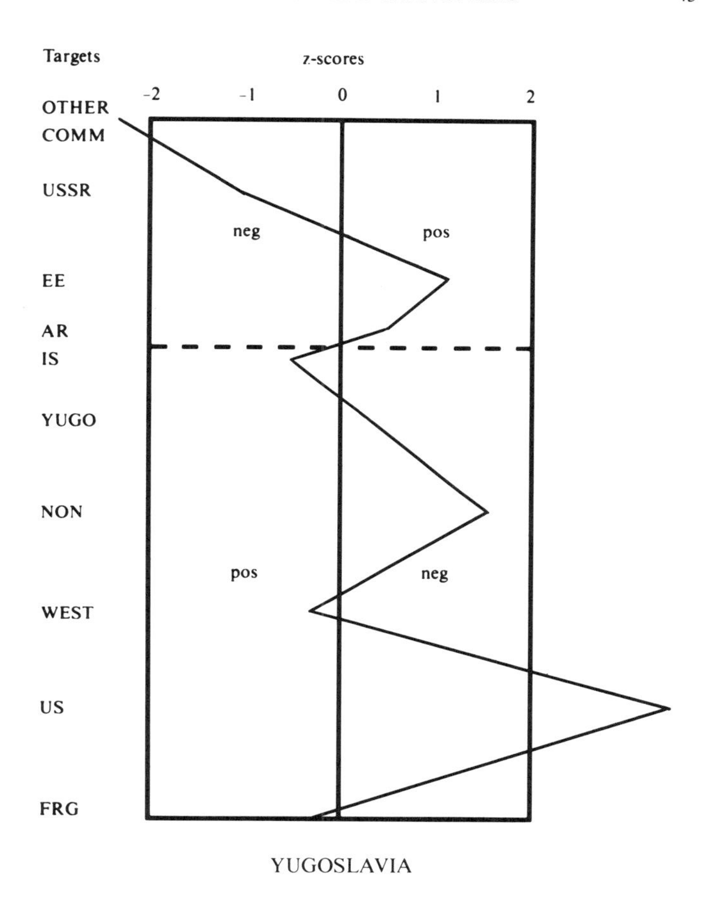

YUGOSLAVIA

These evaluations of positive or negative deviance are added to the chart by labelling the quadrants positive or negative according to the above judgements.[57]

Thus is pictured graphically the strong norm-adhering tendencies of Czechoslovakia and Hungary in almost all categories. "Enthusiastic" or positive deviant alliance behavior is illustrated by the GDR's greatly overfulfilling the norm on all the "proper" categories, while demonstrating below-norm activity patterns with less desirable partners. The opposite case is illustrated by Romania, whose line is tilted strongly in in northwest-southeast direction reflecting its mostly "negative" differences from the norm. We see in Poland and Bulgaria somewhat more complicated cases. The latter, considered by most to be highly adherent to the Soviet allegiance structure, actually demonstrates a substantial amount of negative deviance. Its interaction percentage lies below the norm on relations with Other Communist states, the Soviet Union, the rest of Eastern Europe, and Yugoslavia. Below normal deviation toward this last is positively evaluated in alliance terms as are its underfulfillment vis-a-vis the U.S. and Israel. But when one adds to the negative underfulfillment noted above the above-norm IP with the West and much larger above-norm IP with the Nonaligned states, the result is an ally not nearly as "aligned" as expected. In fact its behavior more nearly approaches that of Yugoslavia, a mixture of selected and nearly balancing deviations.

To obtain a measure of the overall performance of each state, a summary measure of total deviations is required. This can be achieved by calculating the sum of all the evaluated z-scores displayed in Figure II.18 plus those toward China and Albania (whose "pos-neg" evaluations are like those targets below the dotted line in Figure II.18). Thus in Table II.6 we see each country's z-scores arranged by target and listed under the sign indicating its positive or negative evaluation from the alliance perspective, as described above. The sum obtained is the Index of Deviance (ID) for each WTO member and for Yugoslavia.

It could be argued against this procedure that by summing all the z-scores one in effect weights interactions with all targets equally, disregarding important differences in either 1) total amount of attention devoted to certain targets, or 2) relative importance of potential partners. This would suggest the need for some sort of weighting scheme to reflect the relative importance of deviations with regard to the various partners. However, any observer-judgement weighting scheme of targets would inject an undue amount of subjectiveness into the index. Different observers weighting differently the relative importance of deviations toward Albania or Yugoslavia, for example, would arrive at a totally different ID, thus impairing the validity of the measure. On the other hand, a measure based on amount of attention devoted to the potential target would produce nonsensical results. For example, all WTO interactions with China represent only .9% of all WTO interactions, while the members' interactions with Nonaligned states represent 12.9%.[58] But to weight the z-score with China at .009 and that of the Nonaligned at .129 distorts reality by greatly undervaluing the importance of

TABLE II.6
INDIVIDUAL COUNTRY
EVALUATED [1] Z-SCORES BY PARTNER

BULGARIA	+	–	
Alb		.559	
CPR	.845		
Oth Comm.		.669	
USSR		.512	
EE		.932	
Yug	.630		
Non		2.040	
Ar	.228		
Is	.553		
West		.286	
US	.576		
FRG	.158		
tot	2.990	–4.998	= –2.008

CZECH	+	–	
Alb	.783		
CPR		1.014	
Oth Comm.		1.172	
USSR	1.059		
EE		.113	
Yug		.905	
Non	.395		
Ar		.094	
Is	.553		
West	.268		
US	.068		
FRG	.625		
tot	3.751	–3.298	= +.453

GDR	+	–	
Alb		1.678	
CPR	1.351		
Oth Comm.	1.339		
USSR	.966		
EE	1.628		
Yug	.534		
Non	.444		
Ar	1.837		
Is	.553		
West	1.667		
US	1.425		
FRG	.392		
tot	12.136	–1.678=	+10.458

HUNGARY	+	–	
Alb	.783		
CPR	.676		
Oth Comm.		.418	
USSR	.597		
EE	.695		
Yug		1.129	
Non	.379		
Ar		.961	
Is	.553		
West	.330		
US	.068		
FRG	.801		
tot	4.882	–2.508=	+2.374

1. See p. 41.

TABLE II.6
(continued)

POLAND	+	—
Alb		.112
CPR		.845
Oth Comm.	1.088	
USSR		.882
EE		.344
Yug	1.398	
Non	.444	
Ar		.243
Is		.293
West		1.027
US		.696
FRG		.018
tot	2.930	−4.460= −1.530

ROMANIA	+	—
Alb	.783	
CPR		1.014
Oth Comm.		.167
USSR		1.199
EE		.932
Yug		.521
Non	.379	
Ar		.763
Is		1.921
West		.951
US		1.459
FRG		1.946
tot	1.162	−10.873= −9.711

YUG	+	—
Alb		.559
CPR	.676	
Oth		2.678
USSR		1.172
EE	1.177	
Non		1.524
Ar	.475	
Is	.553	
West	.353	
US		3.579
FRG	.275	
tot	3.509	−9.512= −6.003 [1]

1. One less category is included in this total, the Yugo-Yugo dyad being impossible.

deviations toward the People's Republic. Therefore summing the evaluated z-scores unweighted seems the best solution. It in effect does heighten the importance of the z-scores for, e.g. the CPR, Albania, Israel, the US, and FRG relative to, say, the USSR, EE and the Nonaligned by considering them equally even though their mathematical proportion of overall relations is small. This would seem to better reflect reality while avoiding an overly subjective effect on the measure.

Summing the evaluated deviation scores for all targets, we see in Table II.8 the resulting spectrum of deviance for the East European states. Clearly evident now is the substantial overall deviations of Romania to the negative or nonaligned side, and of the GDR toward the alliance. The former even exceeds Yugoslavia and while it is true that Yugoslavia has one less partner category, the difference between their scores (3.708) is still substantial.

Especially interesting is the negative deviant position of Bulgaria. While the overall magnitude of this deviance is not great, general views of Bulgaria's position in the bloc would lead one to expect a positive, even greatly positive, deviance if any.[59] Instead it is Czechoslovakia and Hungary who back up East Germany as the alliance's most loyal member in terms of international interactions.

TABLE II.8
EAST EUROPEAN INDICES OF DEVIANCE (ID) [1]

East European States						
Romania	(Yugo) [2]	Bulgaria	Poland	Czech.	Hungary	GDR
−9.711	−6.003	−2.008	−1.530	+.453	+2.374	+10.458

1. ID = $\sum$ evaluated z-scores; see Table II.6.
2. It should be recalled that Yugoslavia's ID is based on one less category, the Yugo-Yugo Dyad being impossible.

HYPOTHESES TESTING

Using the Index of Deviance we can now seek some answers to the questions posed earlier regarding phenomena associated with deviance. Table II.9 ranks the countries according to the degree of negative deviance, highest to lowest (most positively deviant). In the same table the countries are ranked according to GNP per capita, highest to lowest, as an indicator of economic development. By looking for an association between them we are testing the hypothesis that negative deviance is related to level of economic development. This is done by computing a Spearman r correlation for the two ranks. Doing so we find a correlation of −.857 (significant below the .01 level). This means that those nations whose international behavior deviates from the norm in a negative way are those nations with a *low level* of economic development.

TABLE II.9
RANKINGS OF EAST EUROPEAN COUNTRIES

Country	Negative Deviance Interactions	GNP Per Cap. [1]	Trade Independence [2]	Military Position [3]
Romania	1	5	3	4
Yugoslavia	2	7	1	1
Bulgaria	3	6	7	2
Poland	4	4	2	5
Czech.	5	2	5.5	3
Hungary	6	3	5.5	6
GDR	7	1	4	7

1. Source: International Bank for Reconstruction and Development, *World Bank Atlas* (Washington: I.B.R.D., 1970). Ranking based on 1968 figures.
2. Source: Kintner and Klaiber *op. cit.*, pp. 226-34 . For items making up this index see note 60 below.
3. Source: Kintner and Klaiber *op. cit.*, pp. 234-36 . Country ranked first is in the most favorable military position vis-a-vis the USSR. Albania, which Kintner and Klaiber included with a ranking of 1, has not been included here and other countries have been moved up one rank to maintain a 1-7 ordering. For items making up this index see note 61 below.

In a similar manner we can determine the existence of a relationship, if any, between negative deviance and trade dependence. Table II.9 shows the ranking for the countries according to the trade dependence indicator developed by Kintner and Klaiber, Yugoslavia being the least dependent, Bulgaria, the most.[60] The degree of association between these ranking tests the hypothesis that greater trade independence is related to greater negative deviance. The Spearman r for the two rankings is +.384 which is not statistically significant. Hence, our conclusion contradicts that of Kintner and Klaiber who found a positive relationship between trade dependence and their "Index of Conformity."[61] We find that there is no significant relationship between trade independence and negative deviance.[62]

Finally, Table II.9 shows the ranking of the countries according to an index of military position vis-a-vis the USSR utilized by Kintner and Klaiber.[63] Measuring the association between these two rankings tests the hypothesis that there is some direct relationship between a favorable military position vis-a-vis the USSR and degree of negative deviance. This hypothesis is confirmed by a Spearman r of +.715 (significant below the .05 level).[64]

What do these results mean? The fact that the most economically developed states are the least negatively deviant suggests that the desire for increasing ties with non-Soviet countries that undoubtedly accompanies such development is more than counter-balanced by inhibitions against jeopardizing the smooth functioning of a more advanced, interdependent and hence vulnerable national economic structure. Therefore, courses of action that might provoke the use of sanctions by the Soviet Union are less frequently followed.

The fact that overall trade dependence is not significantly related to level of negative deviance indicates that such sanctions are or can be utilized in a very specialized and subtle manner against deviating members.[65] This finding also supports the notion that it is the anticipation or fear of this kind of pressure that restrains the more developed members of the alliance, a fear rendered more real by the use of such pressure against such nonadvanced but deviating states as Yugoslavia, China and Albania.[66] Of course the lack of a significant correlation between overall trade dependence and negative deviance reflects the failure of the economic levers to hold all allies in line, Yugoslavia and Romania being examples of states which have deviated despite the actual or potential use of economic pressures.[67] This may suggest, further, than an important national attribute to be considered in a discussion of deviance is that of resource base; the relatively good position, for example, of Romania on this score contrasts with that of, say, Hungary.[68]

The fact that it is the less developed states who have deviated to the negative side suggests that one source for foreign policy deviance may be a states' desire to improve its economic situation rapidly and/or broadly, combined with the recognition that this may not be possible within the existing alliance structure or plans. Such states "fear and suspicion" of their more advanced allies' plans for them causes them to seek ties and contacts elsewhere, while their relatively less developed economies make them less vulnerable to economic counterpressure.

The third result (strong positive correlation between military position and deviance) suggests an additional condition of negative deviance, i.e. a relatively secure military position vis-a-vis the USSR. This conforms to a *prima facie* observation of post-World War II East Europe where it has been Hungary and Czechoslovakia, militarily and geographically vulnerable, who have felt the Soviet iron heel directly, while Yugoslavia and Albania, more distant and secure, have not. Given this result, the behavior of Romania appears all the more remarkable. (For a discussion of Romania's military/geographic position, see pp. 196-7 below.)

However suggestive these results might be, we must reserve furtther judgement or discussion until the second half of the story is told, that of verbal or attitudinal foreign policy deviance.

CHAPTER THREE

INTERNATIONAL ATTITUDES:

CONCEPTS AND METHODS

In order to more fully compare the foreign policy behavior of the East European states during the period under investigation, it is necessary to supplement the measure of interactional behavior with measures of the states' international perspectives, that is, their attitudes with respect to key aspects of the international environment. This is necessary for several reasons. First, and most simply, nations act in a verbal as well as non-verbal manner. Virtually all definitions of foreign policy and analyses thereof, their disparity notwithstanding, include both verbal and physical items as part of a nation's foreign relations behavior.[1] Kegley has demonstrated empirically that the two principal dimensions of foreign policy behavior are affect, i.e. positive and negative behavior, and participation, passive or active involvement in international politics.[2] In the present study we can tap affect by examining the foreign policy attitudes of the states under consideration, while participation has been investigated through interaction data.

Second, analysis of international perspectives provides information regarding the nature of a government's "foreign policy doctrine," which Brodin defines as,

> a system of normative and empirical beliefs about the international system and the role of one's own country in that system, as declared in public by the official decision-makers of that country.[3]

This investigation of the international attitudes of the East European states is thus concerned with exploring for comparative purposes key aspects of their foreign policy doctrine. That this is necessary from a purely substantive point of view is indicated by the very small number of comparative studies of East European foreign policies which focus explicitly on the attitudes of these states in a precise, rigorous, systematic manner.[4]

The underlying model implicit in this analysis is that of "mediated stimulus response." As adumbrated by Holsti, Brody and North, events in the environment are considered stimuli to the actors. Their responses are mediated by the decision-makers' perceptions, i.e. their definition of the situation, and these same decision-makers' expressions of intent with regard to their own actions.[5] Were this a decision-making analysis we would concentrate on this mediation process in order to determine the nature of the decision-maker's belief system and the effect of that system upon the nation's foreign policy:[6] or we would concentrate on the decision-making process itself and its effect on policy.[7]

However we are concerned with foreign policy behavior as output; therefore our analysis focuses upon the responses as, "an action of an actor, without respect to his intent or how either he or other actors may perceive it."[8] For use as our dependent variable it is only necessary to add the qualification that the act be verbal, as we have already focused our attention on nonverbal interaction.

The method employed herein is that of content analysis. This refers to "any technique for making inferences by objectively and systematically identifying specified characteristics of messages."[9] While content analysis has been used for a variety of purposes in the study of international relations,[10] there are at least two serious criticisms of the method which must be considered as they affect not only the techniques but the over-all approach utilized.

First, content analysis does not produce "hard" data in the sense of observed aggregations of measurable physical resources, such as troop movements, diplomatic visits or GNP per capita. Data generated by content analysis stems from the perceptual system of the decision-makers and thus varies, to some degree, with their idiosyncracies of perception and expression. Furthermore, it is assessed equally subjectively, using observer judgement methods for identifying, coding and analyzing the content of expressed attitudes.[11] Causal inference from content data must be limited, it is pointed out,[12] since 1) the decision-maker's expression of attitude may at times be disingenuous,[13] or ignorant- simply incorrect- with regard to the motives for their behavior;[14] and 2) inference is overly dependent upon the technique of analysis. Frequency count analyses are particularly vulnerable to the latter criticism, as Mitchell notes.[15]

While objective systematic techniques for content analysis, such as the semantic differential technique,[16] can minimize the distortion of analytic methods, those utilizing content analysis have generally recognized the need to confirm causal inferences through independent validation.[17] In the present study the attitude comparison is part of a comparative study of the foreign behavior of states. The attitudinal output to be content analyzed is the dependent variable. The positions expressed by the decision-makers are to be explored for degrees of similarity and dissimilarity, with each other not as solitary proof of underlying motivations.[18] The problem distortion due to the technique will be minimized through the use of a simple thematic judgment described below.

An additional criticism of the content analysis method also relates to the analytical techniques. It is argued by Mitchell that expressions of attitudes are of limited use because in most instances they are not specific responses to "structured stimuli uniformly presented so as to elicit information on a number of predetermined dimensions."[19] They are not, in short, answers to questionnaires. Therefore such output as occurs can not truly be seen as a comparable stimulus-response situation. It is true that some studies peruse material for expression of attitudes on subjects with no clear indication of what the precise stimuli are for their data, if any.[20]

In order to specify the stimuli as precisely as possible and thus render the responses more strictly comparable, the present study is constructed so that the attitudes of the East European states are indicated by their responses to key events or issues which occurred during the time period studied. Thus, for example, their view of the United States is operationalized as their specific response to U.S. actions in Southeast Asia and to statements and trips by U.S. Presidents. The effect of this approach is to tighten the analytical situation structually at the loss of some amount of generality.[21] Issues and events are substituted from broader but looser country-specific attitudes. International perspectives are thus operationalized as the decision-maker's intended, expressed position or attitude in response to an event or issue occurring in international politics. This operationalization, in addition to adhering more truly to the s-r model, retains a concept of foreign policy behavior congruent with the requirements presented earlier.[22]

Furthermore, the method of comparison is designed to insure as truly comparable an s-r situation as possible by analyzing the foreign policy response in the most similar context possible. Axline, drawing upon Teune and Przeworski, explains that,

> . . . categorizing foreign policies according [to] the situations which occasion them, that is, the particular situation to which they respond, would provide a degree of similarity against which differences could be studied, both in terms of *form* and substance. And the greater similarity the context in which the occasion, or stimulus to foreign policy occurs the more comparative rigor will result.[23]

We will thus try to control the number of extraneous effects upon the foreign policy response behavior in order to be able to more clearly link the dependent and independent variables.

The specific comparative approach is based upon the co-orientation model development by Newcomb for studying interpersonal communications[24] and employed by Hopmann for the study of international alliance cohesion.[25] Building upon the work of Heider[26] and Festinger,[27] Newcomb illustrates that two individuals, A and B, and a subject of their mutual perception, X, form a "co-orientation system," whose components are interdependent. This system, in its minimal form, includes: A's orientation toward X; A's orientation toward B; B's orientation toward X; and B's orientation toward A. It is illustrated as follows.[28]

FIGURE III.1
CO-ORIENTATION SYSTEM

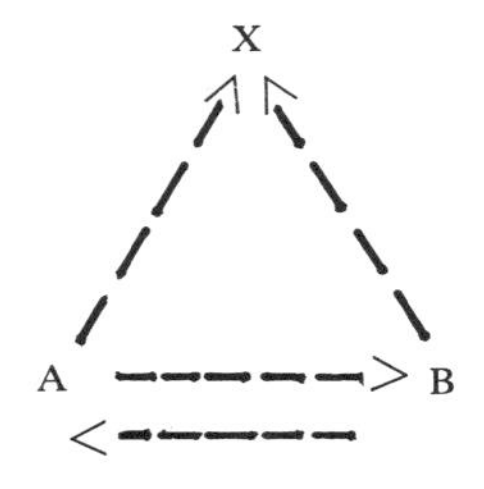

The interdependence of the orientations of the system means that neither one of A's orientations–toward B or toward X–exists in isolation from the other. There is therefore a strain toward balance or symmetry in the system.[29] The system continually strains toward an equilibrium wherein the number of negative co-orientations is even or zero. Simply put, if A has a positive orientation toward X and toward B, but B has a negative orientation toward X, there will be a tension or strain induced to remove one aspect of this discrepancy; either by A's changing his orientation toward B or X, or persuading B to do so.[30] The implication of this model relevant for the present analysis is that one can judge the relationship between A and B by their co-orientation toward X. If A and B both have similar orientations toward X, then their mutual orientation will tend to be positive; if they differ with regard to X their relationship will tend to be negative. The amount of strain induced in the system depends upon other parameters,[31] but the clear implication of this model is that the nature and direction of the relationship between A and B can be determined in part by the degree of consensus between them with regard to X.

By focusing upon the degree of similarity of the East European states' orientations toward a number of issues in the international environment we are also assessing the nature of the relationship between the states themselves. Thus studying the international attitudes of the East European states adds a verbal dimension to the study of their foreign policies and, in addition, gives an indirect measure of their intra-alliance, inter-state and inter-party relationships.

DATA GATHERING

A study of this sort is only as good as its data and the measurement thereof; thus some pains will be taken to explain the nature of the data operation. The time period for the study is the same as mentioned previously, April, 1965–April, 1969. For this period a list of events was compiled which took place either across international borders or within countries but having international implications. It was expected that these events would produce reactions on the part of the East European states. In the process of searching for reactions, this list was amended and changed; certain events which were thought to be stimuli were dropped, and others which elicited reactions were added.

The term "events" actually embraces several different types of occurrences. The outbreak of the Arab-Israeli War in June, 1967 and the Sino-Soviet border clashes in March 1969 are perhaps the purest type of "event." But the eighteen-party call for a world communist conference, issued in Moscow in November, 1967, and the several meetings of the preparatory commission for this world conference also are events, as they evoked broadly comparable responses from the states. Frequently the states created their own events, as when the Political Consultative Committee of the Warsaw Treaty Organization met and issued statements on world events, or when a salient anniversary, such as that of the Comintern, caused reactions. In short, occurrences during this period which produced a minimum number (5) of comparable reactions by the East European states were coded as events and the reactions included.[32] The complete list of events is found in Table III.3.

It should be noted that, in accordance with the reasoning outlined earlier, the Soviet Union is included in this aspect of the study.[33]

SOURCES

Several sources of information were employed to find the attitudes, positions and reactions of the East European governments to these events. The "Daily Report" of the Foreign Broadcast Information Service (FBIS) was the main source of statements and positions, owing to its overall reliability and attractiveness as a uniform translating source. Moreover, as a reproducer of direct verbatim party/government radio and journal statements and articles, it was the purest form of primary data available for all countries. This was supplemented by *Deadline Data on World Affairs*, the *Press Survey* of Radio Free Europe, *Notes and Etudes Documentaires*, and the *Yearbook on International Communist Affairs*, all of which were useful for filling in gaps in the FBIS material, but whose typical source citation was somewhat briefer.[34] Finally, when these sources failed to produce a reaction or statement logically expected or whose existence was known by the author, three country-specific sources were employed: *Pravda* and *Izvestia* of the Soviet Union (through the *Current Digest of the Soviet Press*), *Scinteia* of Romania, and the *Yugoslav Survey* and *Review of International Affairs* for Yugoslavia. Still it frequently happened that no reaction could be found for a country or countries when there was every reason to expect one. This missing data problem imposes certain limitations on the analysis and its interpretation, which will be discussed below.

Within countries certain sources were considered more appropriate than others. One of the few advantages in gathering data for research on East Europe is that the press is, to say the least, controlled, and thus represents at all times some aspect of the official government position. However, some sources are more authoritative and useful than others for assessing foreign policy attitudes. Therefore a rank ordering was employed in gathering the data on those occasions when the sources produced more than one statement or article regarding an event. Official government statements, labelled as such, were primary; next were international and domestic news service broadcasts (TASS, Aegerpress, Prague Domestic Service, etc.).[35] Considered of equal value to these, and indeed often identical, were party dailies (*Pravda, Neues Deutschland, Rabotnichesko Delo*); government dailies ranked next (*Isvestia, Politika*) or, when appropriate, party monthly or theoretical journals (*Kommunist*) and foreign affairs weeklies (*Lumea, Horizont*). Last came political front organs (*Dziennik Ludowy, Bashkimi*), youth dailies (*Mlada Fronta*), trade union or military organs (*Glos Pracy, Narodna Armiya*), and regional sources (*Pravda* [Bratislava] , Radio Zagreb).

Usually entire articles or broadcasts were available and used; often more than one for a country if several of equal value were found. On the other hand, occasionally only parts of or briefer articles were available, and were this a frequency count study this would be a serious failing in the data. Since, however, the document coding was for thematic affect–that is, a comparison of the positive-negative attitudes expressed, rather than a precise attention study or word count—such irregular sized primary data could be included with little danger of serious data distortion.

LIMITATIONS

It is important to recognize the limitations of the analysis drawn from this data set, in order to keep the claims of the study in line with supporting evidence. First, certain events that were expected to produce a sufficient number of comparable reactions failed to do so. For example, the holding of West German federal elections in West Berlin (February, 1969), the thirteenth anniversary of the WTO (May, 1968), and much of the internal turbulence of the Chinese Cultural Revolution (1966-69) failed to produce either the requisite number, breadth or comparability of reactions to be included as a stimulus-event. This study is aimed at comparing a group of countries over a five-year period, and therefore a minimum cutoff point for inclusion is necessary to insure that the study retain its broad scope.

A second limitation concerns missing data. If an event did produce a sufficient quantity and quality of reactions, it usually was the case that reactions for all nine countries could not be found in the sources. It would be tempting to merely code such null cells as "no reaction," which then would be a reaction itself, especially in East Europe. But such a statement in the absence of an actual search of the relevant party and government journals themselves at the time of the event, in order to verify that no reaction was forthcoming, would be speculation.[36] This search *was* done in two important cases, that of the Soviet Union and Romania. For these two countries a "no reaction" score could be given when they held silent. For all other cases a "no value" missing data code is utilized.[37]

Third, it is recognized by close readers of *podtext* that, though the press is controlled by each of these governments to varying degrees, the separate organs within each country frequently reveal differing and even opposing lines on some subject, however obliquely. In the Soviet Union, for example, *Trud, Krasnaia Zvezda, Pravda* and *Izvestia* have in the past reported speeches or editorialized "differently" on some subject. While these differences are not exactly of the *New York Times* vs. *Cleveland Plain Dealer* variety, a one-country, one-issue study would find them valid and useful.[38] This study has no mechanisms for including these within-country nuances and differences. Instead the attempt is made to obtain the most authoritative, highest government/party level statement on an issue by utilizing the document rank ordering explained above.

Finally, this is not a study of the individual foreign policies of the East European countries, at least according to the various existing conceptual schemes of foreign policy.[39] Rather it measures, compares and offers explanatory hypotheses for the attitudinal component of the foreign policy behavior of nine East European countries, across fifty-seven specific issues over five years. It is, then, an assessment of the behavior in international politics of a group of countries allied by geographic region, ideology, economic structure, and, with two exceptions a formal treaty of alliance.

THE RAW DATA

On forty-six separate occasions events took place which caused the nations of East Europe to issue statements or reactions to one or more aspects of the event. One event, such as the Warsaw meeting in July, 1968 of five WTO members on developments in Czechoslovakia, frequently stimulated reactions on two separate but related dimensions. In this case, for example, both the country's reactions to the meeting itself and to the Czech developments which produced it were coded. Thus a total of fifty-seven responses represents the maximum number of cases for the time period for each country.[40]

For each case (referred to interchangeably as an event or issue) each country's reaction was rated by three independent, expert coders.[41] They were instructed to read each document and rate it according to the following scale:

TABLE III.1

CONTENT ANALYSIS CODING SCHEME

Score	Description
+3	Strongly positive; strongly support; praise
+2	Positive; support
+1	Mildly positive; tend to support
0	Neutral; ambiguous
−1	Mildly negative; tend to criticize
−2	Negative; criticize; discourage
−3	Strongly negative; strongly oppose; condemn

In addition, particular descriptive phrases were added when necessary for certain situations.[42]

Intercoder reliability was assessed with various measures. Average Pearson r correlation—used to test the overall association of the three coders across all countries—was .875. Average rank-order correlation (Kendall tau) was .715. Moreover the three coders matched exactly in codings an average 41% of the time.[43]

Being thus convinced that the coding scheme is reliable and that the coders' assessments are both strongly associated and highly correlated in rank ordering, it is possible to utilize the average of the three coders' scores for each case as the raw data to be examined further. The scores computed for each country in this way are listed in Table III.2. The event area within which each case falls is listed in the leftmost column and they are defined in Table III.3. Each discrete event has a unique number, listed in the second column. The scores for that event for each country follow horizontally.

FINDINGS, I: DESCRIPTION AND ILLUSTRATIVE STATEMENTS

What, then, were the attitudes of the East European states toward the key issues that developed in the second half of the sixties? How closely grouped were they to each other; that is, how similar were they? Where are the divergences and who were the deviants? It will be useful in beginning to answer these questions to discuss briefly each issue represented in Table III.2 and present illustrative sections of representative statements which exemplify the countries scores on the coding scale.

Arab-Israeli War

In June, 1967, war erupted between Israel and her Arab neighbors. This event (AIS 01 in Table III.2) produced a strongly negative reaction on the part of the USSR toward Israel, as the score indicates. The tone of the statement can be gathered from the following excerpts:

> The Soviet Government has condemned Israeli aggression and demanded that the Israeli Government should, as the first urgent step to ending the military conflict, stop immediately and unconditionally its military actions against the UAR, Syria, Jordan, and other Arab countries and pull back its troops beyond the truce line. . . .
>
> Thus a military conflict has flared up in the Middle East because of the adventurism of the rulers of one country, Israel, which was encouraged by covert and overt actions of certain imperialist circles. . . . By launching aggression against neighboring Arab states the Israeli Government has trampled underfoot the U.N. Charter and elementary standards of international law.[44]

TABLE III.2
REACTION SCORES BY EVENT

Event	No.	Sov	Bul	Cze	GDR	Hun	Pol	Rom	Yug	Alb	WTO[1]
AIS	01	−3.0	−2.7	−1.7	9.0	−2.7	−2.7	+0.3	−2.3	9.0[2]	−2.18
AIS	02	−3.0	−3.0	−3.0	−3.0	−3.0	−3.0	7.0[3]	−3.0	9.0	−3.0
AIS	03	−1.3	−1.3	−1.3	−1.3	−1.3	−1.3	−1.3	−1.3	9.0	−1.3
AIS	04	−2.7	−2.7	−2.7	−2.7	−2.7	−2.7	−2.7	−2.7	9.0	−2.7
CPR	11	−2.0	−2.7	−2.0	−3.0	−2.3	−0.7	7.0	9.0	9.0	−2.117
CPR	12	+3.0	+2.0	+1.3	+2.0	+2.0	+2.0	0.0	0.0	−3.0	+1.757
CPR	13	−2.7	−3.0	9.0	9.0	−3.0	−2.5	9.0	9.0	9.0	−2.8
CPR	14	−2.7	9.0	9.0	−2.7	−0.7	−1.0	+0.7	0.0	9.0	−1.28
CZE	21	−0.3	−0.3	−0.3	−0.3	−0.3	−0.3	+3.0	+2.7	9.0	+0.171
CZE	22	−3.0	9.0	+3.0	−2.3	−0.7	−2.5	+1.3	+2.3	9.0	−0.7
CZE	23	+2.7	9.0	−2.7	9.0	9.0	+2.0	9.0	−2.0	9.0	+0.667
CZE	24	−2.3	−2.3	7.0	−2.3	−2.3	−2.3	7.0	7.0	7.0	−2.3
CZE	25	−2.3	−2.3	+2.3	−2.3	−2.3	−2.0	+3.0	+2.3	9.0	−0.843
CZE	26	+2.7	+3.0	−1.0	+3.0	+2.7	+2.3	9.0	−2.3	9.0	+2.117
CZE	27	−3.0	−3.0	+3.0	9.0	−2.0	−2.0	+2.7	+2.0	9.0	−0.716
CZE	28	+1.7	9.0	+0.7	9.0	+2.0	+1.0	0.0	−0.7	9.0	+1.08
CZE	29	−3.0	9.0	+2.7	−2.0	0.0	9.0	+2.0	+2.0	9.0	−0.06
CZE	30	+3.0	+2.7	+1.0	+3.0	+2.3	+2.3	−0.7	+0.7	9.0	+1.943
CZE	31	+3.0	+3.0	−2.7	+3.0	+3.0	+2.7	−2.7	−2.7	−3.0	+1.329
CZE	32	+3.0	+3.0	−2.7	+3.0	+2.7	+2.3	−3.0	−2.7	9.0	+1.186
CZE	33	+2.3	+3.0	−0.3	+2.7	+1.3	+1.7	0.0	−2.0	−3.0	+1.529

TABLE III.2 (continued)

Event	No.	Sov	Bul	Cze	GDR	Hun	Pol	Rom	Yug	Alb	WTO
EUR	41	+2.7	9.0	+2.7	+3.0	+2.7	+2.7	+2.3	9.0	9.0	+2.683
EUR	42	+2.7	+2.7	+2.7	+2.7	+2.7	+2.7	7.0	−1.5	7.0	+2.7
EUR	43	+2.7	+2.7	+2.3	+3.0	+2.7	+2.7	+2.7	+1.3	9.0	+2.686
EUR	44	+3.0	9.0	+2.7	9.0	+3.0	9.0	+0.3	−1.3	9.0	+2.25
EUR	45	+3.0	+3.0	+1.7	+2.7	+2.7	+2.0	+1.3	0.0	9.0	+2.343
EUR	46	−2.7	9.0	−2.3	−2.7	9.0	−3.0	9.0	−1.7	9.0	−2.675
EUR	47	+2.3	+2.3	+2.3	+2.3	+2.3	+2.3	−1.0	−1.3	−2.7	+1.829
WCM	51	+3.0	9.0	+2.7	9.0	9.0	+2.7	−2.0	−1.7	−3.0	+1.6
WCM	52	+2.7	+2.0	+1.7	9.0	+2.0	+1.7	7.0	−2.7	9.0	+2.02
WCM	53	+1.7	+2.3	+2.3	+1.3	+2.0	+2.7	+1.0	−2.0	9.0	+1.9
WCM	54	+3.0	+2.7	+0.3	9.0	+1.3	+2.0	−1.0	−1.7	9.0	+1.383
WCM	55	+3.0	+2.5	+1.3	9.0	+1.0	+3.0	−2.5	−2.0	9.0	+1.383
WCM	56	7.0	9.0	0.0	9.0	−2.0	9.0	+3.0	+0.7	9.0	+0.333
WCM	57	+2.7	9.0	−0.7	9.0	+2.7	9.0	−1.7	−2.3	9.0	+0.75
WCM	58	+2.7	+2.3	+1.7	9.0	9.0	9.0	+0.7	−1.3	−3.0	+1.85
WCM	59	+2.3	9.0	+1.7	+2.3	+2.7	+1.3	0.0	−1.0	9.0	+2.06
WCM	60	+2.7	+2.7	9.0	+2.3	+1.7	9.0	−1.7	−1.7	9.0	+1.54
USV	71	−2.7	−2.3	−2.3	−3.0	−3.0	−2.0	−2.3	−1.0	−2.7	−2.514
USV	72	−2.7	−3.0	−2.7	−2.7	−2.3	−2.7	−3.0	−2.7	−3.0	−2.729
USV	73	−3.0	−3.0	−3.0	−3.0	−2.7	−2.3	−3.0	−2.3	9.0	−2.857
USV	74	+3.0	9.0	+3.0	+3.0	+2.7	+2.7	+2.7	9.0	9.0	+2.85

TABLE III.2 (continued)

Event	No.	Sov	Bul	Cze	GDR	Hun	Pol	Rom	Yug	Alb	WTO
USV	75	−3.0	−2.7	−3.0	9.0	−2.7	−3.0	−3.0	−1.7	−3.0	−2.9
USV	76	−3.0	−3.0	−3.0	−3.0	−3.0	−3.0	7.0	7.0	7.0	−3.0
USV	77	+3.0	9.0	+0.7	+2.7	+2.7	+2.3	+2.0	+1.7	+3.0	+2.233
USV	78	−2.7	−2.7	−2.7	−2.7	−2.7	−2.7	−2.7	7.0	7.0	−2.7
USV	79	−2.0	9.0	+1.3	−2.5	−1.7	−0.3	−1.0	−0.3	9.0	−1.033
USV	80	+2.0	+1.0	+1.3	+1.0	+1.0	9.0	+2.0	+2.7	−1.0	+1.383
USL	81	−2.0	−1.7	−1.3	−2.7	−2.0	−1.7	−0.7	0.0	9.0	−1.729
USL	82	−2.7	9.0	−0.3	−3.0	−1.0	−1.3	−1.3	−1.7	9.0	−1.6
USL	83	−2.0	−2.3	−2.0	9.0	−1.3	−2.0	−1.3	−1.0	9.0	−1.817
USL	84	−2.0	9.0	−1.3	9.0	−0.3	−1.3	9.0	−0.7	−3.0	−1.225
USL	85	−1.0	9.0	+1.3	−1.7	−1.0	0.0	+0.3	−0.3	9.0	−0.35
USP	86	−2.3	−2.3	−2.0	−2.7	−1.3	−2.0	−2.3	−2.0	−2.7	−2.129
USN	91	+0.7	9.0	0.0	9.0	+0.3	+0.3	−0.7	+0.3	−2.7	+0.12
USN	92	−1.5	−0.7	+1.3	−2.3	−0.3	−1.0	0.0	+0.3	9.0	−0.643
USN	93	−1.3	−2.0	+1.0	9.0	−0.3	+0.3	+1.0	+2.3	9.0	−2.167

1. WTO=Mean score for reporting WTO countries
2. 9.0= No reaction found (missing data)
3. 7.0=No reaction issued or state not part of conference or group issuing statement

N=57

TABLE III.3
EVENTS (STIMULI)

Event	No.	Description	Place	Date
AIS	01	Outbreak of war in Middle East		6/6/67
AIS	02	Statement of EE states on Middle East ***	Moscow	6/9/67
AIS	03	Statement of EE Dep. Prems. on assistance to Arab states ***	Belgrade	9/4-6/67
AIS	04	Statement of EE For. Mins. on Middle East ***	Warsaw	12/19-21/67
CPR	11	11th Plenum of CC of CCP; communique on CPSU	Peking	8/66
CPR	12	Sino-Soviet border clashes	Ussuri River	3/69
CPR	13	9th National Congress of CCP (pre-)*	Peking	4/69
CPR	14	9th National Congress of CCP (post)**	Peking	4/69
CZE	21	Meeting of 6 WTO states on developments in Czechoslovakia+	Dresden	3/23/68
CZE	22	Meeting of 5 WTO states on developments in Czechoslovakia (events)++ (pre-)*	Warsaw	7/14-15/68
CZE	23	Meeting of 5 WTO states on developments in Czechoslovakia (meeting) a (pre-)*	Warsaw	7/14-15/68
CZE	24	Meeting of 5 WTO states on developments in Czech. Letter to Czech CP ***	Warsaw	7/14-15/68
CZE	25	Meeting of 5 WTO states on developments in Czech. (events)++ (post)**,†	Warsaw	7/14-15/68
CZE	26	Meeting of 5 WTO states on developments in Czech. (meeting) a (post)**	Warsaw	7/14-15/68

TABLE III.3 (continued)

Event	No.	Description	Place	Date
CZE	27	Meeting between CPSU Politburo and Czech Presidium (events)++,$	Cierna (Czech)	7/29-8/1/68
CZE	28	Meeting between CPSU Politburo and Czech Presidium (meeting) a	nad Tisou	
CZE	29	Meeting of 5 party leaders with Czech CP (events) ++	Bratislava	8/1/68
CZE	30	Meeting of 5 party leaders with Czech CP (meeting) a	Bratislava	8/1/68
CZE	31	Soviet invasion of Czechoslovakia %		8/21/68
CZE	32	Soviet invasion of Czechoslovakia @		8/21/68
CZE	33	Talks between Czech and Soviet party and govt leaders	Moscow	8/27/68
EUR	41	Statement of WTO PCC meeting on European peace and security (Bucharest Declaration)	Bucharest	7/5-9/66
EUR	42	Statement of Karlovy Vary conference of European CP's on European peace and security ***,c	Karlovy Vary	4/24-26/67
EUR	43	Appeal on Europe of WTO PCC meeting	Budapest	3/17/69
EUR	44	11th Anniversary of Warsaw Pact		5/14/66
EUR	45	WTO PCC meeting††	Budapest	3/17/69
EUR	46	FRG Note on peace, disarmament and world tension		3/25/66
EUR	47	Statement of WTO PCC meeting on Nonproliferation Treaty∞	Sofia	3/8/68
WCM	51	Conference of European communist and workers' parties	Karlovy Vary	4/24-26/67
WCM	52	18-party call for holding of consultative meeting on holding of world conference of communist and workers' parties ⇳	Moscow	11/24/67
WCM	53	Consultative conference on world conference of parties (pre-)*	Budapest	2/28-3/4/68
WCM	54	Consultative conference on world conference of parties ⇳	Budapest	2/28-3/4/68
WCM	55	Consultative conference on world conference of parties (post)**	Budapest	2/28-3/4/68

TABLE III.3 (continued)

Event	No.	Description	Place	Date
WCM	56	Romanian walkout from consultative conference	Budapest	3/1/68
WCM	57	Meeting of preparatory commission for world conference of parties ⇧	Budapest	4/24-29/68
WCM	58	Meeting of working group and prep. comm. for world conference of parties ⇧	Budapest	9/27-10/1/68 and 11/18-21/68
WCM	59	Meeting of prep. comm. for world conference of parties ⇧	Moscow	3/18-22/69
WCM	60	50th Anniversary of the Comintern		3/4/69
USV	71	President Johnson announces increases in troop strength in Vietnam and in draft calls	Washington	7/28/65
USV	72	United States resumes bombing of North Vietnam after 37-day lull		1/31/66
USV	73	United States bombs suburbs of Hanoi and Haiphong		6/29/66
USV	74	Statement of WTO PCC meeting on Vietnam	Bucharest	7/5-9/66
USV	75	United States bombs Hanoi		12/13-14/66
USV	76	Appeal on Vietnam of Karlovy Vary conf. of European CP's ***	Karlovy Vary	4/24-26/67
USV	77	Tet offensive by NLF in Vietnam		1/31-2/1/68
USV	78	Statement of WTO PCC meeting on Vietnam ***	Sofia	3/8/68
USV	79	Pres. Johnson announces limitation of bombing in Vietnam	Washington	3/31/68
USV	80	Pres. Johnson halts all bombing of North Vietnam		11/1/68
USL	81	State of Union message of Pres. Johnson	Washington	1/13/66
USL	82	State of Union message of Pres. Johnson	Washington	1/13/67

TABLE III.3 (continued)

Event	No.	Description	Place	Date
USL	83	State of Union message of Pres. Johnson	Washington	1/18/68
USL	84	State of Union message of Pres. Johnson	Washington	1/15/69
USL	85	Pres. Johnson announces that he will not be a candidate for reelection in November	Washington	3/31/68
USP	86	Seizure of USS *Pueblo* by North Korea		1/25/68
USN	91	First press conference of Pres. Nixon	Washington	1/27/69
USN	92	Tour by Pres. Nixon of major West European capitals (pre-)*		2/24-3/3/69
USN	93	Tour by Pres. Nixon of major West European capitals (post)**		2/24-3/3/69

* Statements and reactions issued just prior to event
** Statements and reactions issued during and/or after the event
+ communique of meeting scored for those attending; separate reactions for Romania and Yugoslavia
++ Reactions are to the events taking place in Czechoslovakia
a Reactions are to the holding of the meeting itself
† For Czechoslovakia the reaction scored is the "Reply" to the Warsaw letter
$ Reactions issued between the time of the announcement of the Cierna meeting (7/22/68) and the end of the meeting itself (8/1/68)
% Reactions are the initial announcements and reporting of the invasion
@ Reactions are the immediate post-invasion explanation and/or reactive statements
*** Statement itself scored
†† Reactions are to the Pact itself
∞ Reactions are the Statement itself for all states except Romania, Yugoslavia, and Albania, for which they are separate reactions to the NPT
⇧ Reactions are to the holding of a world conference of parties
c Separate Yugoslav raction scored

Similar statements were issued by the governments of Bulgaria, Hungary, Poland and Yugoslavia. The Czechoslovak government's reaction was also negative, though somewhat more mild:

> . . . the Czechoslovak Government expresses profound indignation over Israel's attitude toward efforts of the United Nations and peace-loving countries to restore peace in this area, and emphatically demands that Israel immediately suspend military operations and comply with the Security Council resolution.[45]

Contrast these statements, however, with the "appeal of the government of the Socialist Republic of Romania for the cessation of hostilities in the Near East,"

> Today in the afternoon First Deputy Foreign Minister George Macovescu invited Mr. Muhammad Fahmi Hamad, the ambassador of the UAR to Bucharest, to the Foreign Ministry and communicated to him in the name of the government of the Socialist Republic of Romania the anxiety of the Romanian people and government regarding the start of hostilities in the Near East between the UAR and Israel. Taking into account the danger of the continuation of war to peace in the Near East and to peace in the world, the government of the Socialist Republic of Romania addresses an appeal to the government of the UAR for an immediate stop to the hostilities, and it would welcome with particular satisfaction the solving by means of negotiations of divergencies, the reaching of fair and rational agreements which would take into account the legitimate rights of the interested peoples.[46]

The Romanian government statement goes on to report an identical appeal to the government of Israel through its ambassador in Bucharest.[47]

Events AIS 02 through AIS 04 are joint statements issued by the East European states on the Middle East in the six months following the war. The statement of June 9, signed in Moscow, is clear and condemnatory of Israeli action.

> The states participating in this meeting demand that Israel immediately stop military actions against the neighboring Arab

countries and withdraw all its troops from their territories behind the truce line. It is the duty of the United Nations to condemn the aggressor. If the Security Council does not take the proper measures, grave responsibility will rest with these states which failed to fulfill their duty as members of the Security Council. . . . If the Government of Israel does not stop the aggression and withdraw its troops behind the truce line, the socialist states which signed this statement will do everything necessary to help the peoples of Arab countries to administer a resolute rebuff to the aggressor, to protect their lawful rights, and to extinguish the hotbed of war in the Middle East and restore peace in that area.[48]

While Yugoslavia saw fit to join the Warsaw Pact meeting in this action- President Tito even travelled to Moscow himself to sign- the Romanians were conspicuously absent. That their absence reflects a policy position is clear when one considers the next statement the states issued on the Middle East, in Belgrade in September. The Romanians did attend this meeting, which was at the Deputy Premier level, and did sign the statement. The statement, however, was substantially milder on the Arab-Israeli issue, eschewing the formula condemnation of Israeli aggression and flouting of the U.N. Instead it said,

The participants in the meeting expressed their readiness to consider, together with the Arab countries, on a bilateral basis or in some other suitable form, practical measures for the further expansion of economic cooperation, and, at the same time, informed one another of their intentions and concrete moves made with a view to extending assistance to, and expanding economic cooperation with, these countries.

The participants in the meeting are convinced that this cooperation will help the Arab countries in their efforts to overcome the existing difficulties and to strengthen their national and economic independence, and that it will represent a further expression of the solidarity of the socialist countries with the Arab peoples in their just stuggle against imperialism and neocolonialism.[49]

The statement issued by the meeting of foreign ministers in Warsaw in December did return to a somewhat tougher line on the issues in the area.

> The parties to the meeting unanimously emphasized that the withdrawal of the Israeli forces from all occupied territories of the Arab states to the positions held prior to 5 June this year, is the main and indispensable condition for the restoration and consolidation of peace in the Middle East. In this context, the ministers pointed to the great importance of fulfilling the resolution of the UN Security Council of 22 November 1967, and the immediate withdrawal of the Israeli Armed Forces from all occupied territories of Arab states and emphasized the impermissibility of acquiring territories through war. Any interpretations designed to weaken this fundamental element of the Security Council resolution run counter to its letter and spirit.[50]

There was, however this notable new inclusion:

> The ministers stressed at the same time the need for the recognition by all UN member nations in the area of the fact that each of them has the right to exist as an independent national state and live in peace and security. Israel's actions directed at keeping any part of the occupied Arab territories hamper the solution of other problems of the are on the basis of the above-mentioned principles as well as the principles of noninterference in internal affairs, territorial integrity, and such actions must be condemned.[51]

In sum, scores AIS 01 through AIS 04 represent the fact that Romania held clearly and distinctly diverging views on the Middle East issue, though by the end of 1967, it did see fit to sign a statement closer to the position of the Soviets and the other East European states than to its own original position of the summer.

China

Within the issue area relating to China (scores CPR 11 through CPR 14 in Table III.2), the first event which produced broad comparable reactions was the Eleventh Plenum of the Central Committee of the Chinese Communist Party, held in August, 1966. This plenum, considered the official inauguration of the Cultural Revolution, ended in a communique harshly condemning the CPSU.[52]

> The new leading group of the Communist Party of the Soviet Union has inherited Khrushchev's mantle and is practicing Khrushchev

> revisionism without Khrushchev. Their line is one of safeguarding imperialist and colonialist domination in the capitalist world and restoring capitalism in the socialist world. The leading group of the CPSU has betrayed Marxism-Leninism, betrayed the great Lenin, betrayed the road of the Great October Revolution, betrayed proletarian internationalism; betrayed the revolutionary cause of the international proletariat and of the oppressed peoples and oppressed nations, and betrayed the interests of the great Soviet people and the people of the socialist countries. They revile the Communist Party of China as being "dogmatist" "sectarian" and "left adventurist." In fact, what they are attacking is Marxism-Leninism itself.[53]

All of the Pact members reacted negatively to the plenum and to this attack, though in varying degrees. In addition, they took it as a focus for commenting upon the Cultural Revolution idea in general. The East Germans were the most hostile:

> The communique of this 11th plenum stresses explicitly that the leadership of the CCP fully persists in the anti-Marxist and anti-Leninist policy which it has pursued for years to the detriment of the international communist movement. . . . This divisive policy has encouraged the aggressive forces of U.S. imperialism. It has promoted the ultraleftist and adventurist elements and has done grave harm to the communist party and the entire national-democratic movement in countries like Indonesia. By continuing this disastrous line in spite of all serious hints from the communist and workers' parties, the CCP leaders have gone into self-isolation. The documents of the 11th plenum of the CCP Central Committee show that the Chinese leaders regrettably intend to continue to persist in this disastrous line . . . . The CCP leaders took new steps at the plenum to step up the anti-Soviet campaign. The SED sharply condemns the base slanders against the Leninist policy of the CPSU and considers them an attack on all fraternal parties. The unfounded associations of the Chinese leaders are obviously meant to hide the fact that, by their attitude, by their rejection of joint actions, the plans of the imperialists to split and weaken the anti-imperialist forces are receiving new impetus.[54]

Poland's reaction was the mildest.

> . . . It appears from the facts mentioned above that in the developments connected with the "cultural revolution" much space is given to slogans directed against the international unity of the communist and working class movement. These developments are being watched with keen attention and anxiety by the communist movement and other progressive forces.[55]

Once again the Romanians stood apart by their silence. *Scinteia,* the RCP daily, published no reaction whatsoever, reflecting the Romanian attempt to remain neutral or even mediate the Sino-Soviet split.[56]

In March, 1969, serious border clases broke out between Chinese and Soviet forces along the Usurri River (event CPR 12). *Pravda* editorialized the Soviet view of the conflict, and the Soviet score (+3) represents that position; the Albanian score (−3) reflects support of the Chinese. Most of the WTO members supported the Soviets. The view of the Hungarian party was typical:

> . . . the provocations committed on March 2 and 15, repeatedly bringing fatal sacrifices, clarified without any doubt left that the leaders of China are aiming at the methodical smashing of interstate relations between the Soviet Union and China. The border provocations, the armed assaults planned well in advance, and the senseless massacres have become everyday policy of the Peking leaders.[57]

Romania, and to a lesser degree Czechoslovakia, deviated from this type of position. Prague's reproach was more measured and balanced.

> As a socialist country and as a [communist] party we cannot but express our disagreement with the Chinese shooting into Soviet territory and the present Chinese campaign . . . . Even today we see in the Chinese Republic the fruit of the anti-colonial and anti-imperialist revolution. We consider her to be a part of the socialist community, from which it should not be expelled, but from which she should not even expel herself . . . . It is not good for this country to draw attention to itself with facts such as border incidents with another big socialist power or territorial claims (which should have no place at all in relations between socialist countries) . . .[58]

Bucharest took the remarkable step of publishing the dispatches of both TASS and the New China News Agencies on the clash, and the respective governments' protest notes over the incidents.[59] It was a scrupulous study in neutrality. In addition, on this issue the Yugoslavs did not see fit to join the Pact group majority and issued a similarly neutral report.[60]

Finally, the Ninth National Congress of the Chinese Communist Party, held April 1 through 24, 1969, elicited comment from the East European states either before, during or after the events (scores CPR 13 and/or CPR 14). Comment was uniformly negative and strong before the congress, but during and just after the Congress, while the Soviets and Germans continued to be strongly critical, Poland and Hungary merely chided the CCP for the secrecy of the Congress and its deification of Mao Tse-tung. Again the Yugoslavs stuck to a neutral reportage of the Congress, even reprinting part of a *Jen Min Jih Pao* editorial.[61] The Romanian response was once more exceptional, as they reprinted entire New China News Agency dispatches on both the opening and closing of the Congress, and included their own mildly favorable reports in between.[62]

Events in Czechoslovakia

One of the most significant developments in world communism and certainly in East Europe since World War II was the "Prague Spring," which can be roughly dated from Alexander Dubcek's ascension as Party First Secretary in January 1968 until the Soviet invasion of August that year.[63] The scores in group CZE 21 through CZE 33 indicate the reactions of the East European states, expressed in various ways, to both the developments occurring in that country and the efforts of the Warsaw Pact to deal with those events.

The first concerted reaction of the Warsaw Treaty states to what was happening in Czechoslovakia took place at a meeting of six of the states (Romania was not invited) in Dresden, on March 23 (CZE 21). The communique ending this meeting made only oblique reference to "the progress in the enactment of the decisions of the January plenum of the Central Committee of the Communist Party of Czechoslovakia aimed at realizing the line mapped out by the 13th congress of the party."[64] Its content and tone, however, give sharp contrast to a Romanian article which appeared on *Scinteia's* front page less than a week after the Dresden meeting. The article hailed the "Intense Efforts of the Communist Party and People of Czechoslovakia for the Perfection of Social Life on the Road to Socialism."

> The political effervescence, generated by the plena of December 1967 and January 1968, continues as a process sustained in the direction of the democratic and progressive development of a friendly country. This process of innovation, initiated and led by the Communist Party of Czechoslovakia, has included the great mass of working people, public organizations, different sectors of Czechoslovak society. It finds expression in the measures with political and social character adopted recently, in the active participation of the communists, of different categories of citizens in public life, in the large discussion in the press, in meetings and sessions of party and state organs, of mass organizations. In the framework of these discussions are formulated proposals of great significance for the progressive development of the Czechoslovak S.R. along the road to socialism, for the welfare and happiness of the Czech and Slovak peoples.[65]

In addition, the Romanian paper quoted Dubcek at length on Czech domestic and foreign policy, and reprinted the statement of *Mlada Fronta*, the organ of the Czech youth organization, that "the problems of the internal evolution of our country are our internal matters and that it is an exclusive affair of our party and society in which manner these problems will be resolved."[66] Yugoslavia, too, boosted the Czechs and indicated that developments there "testify to the existence of a wide variety of ways of socialist development and speak for the never-ending progressive development of the socialist world, thus revealing the vitality of the socialist system and the rising consciousness of the socialist forces."[67]

A second meeting occurred in Warsaw July 14-15, and both before and after the meeting most of the countries commented directly on occurrences in Czechoslovakia (items CZE 22 and CZE 25). By this time the unofficial Czech manifesto of socialist humanism, entitled "Two Thousand Words," had been published and the allies were concerned for the existence and future of socialism in the neighboring section of the socialist camp.[68] The Soviet polemicists led the attack.

> It has now become more obvious than ever that the appearance of "The 2,000 Words" is not an isolated phenomenon, but evidence of the increasing activity in Czechoslovakia of rightist and overtly counterrevolutionary forces obviously linked with imperialist reaction. They have gone on to make fierce attacks against the

> foundations of socialist statehood. Evidently, forces hostile to the Czechoslovak people are hastening to take advantage of the unstable situation in the country in order to achieve their counterrevolutionary goals. The support that these forces have found among imperialists in the West is playing a considerable role in all this, as can be distinctly seen in connection with the publication of "The 2,000 Words."[69]

Ominously, *Pravda* recalled earlier events it considered similar.

> Such tactics are not new. They were resorted to by the counterrevolutionary elements in Hungary that in 1956 sought to undermine the socialist achievements of the Hungarian people.[70]

The GDR and Poland followed suit, attacking "the reactionary forces and imperialist maneuvers" that both detected.[71] The Hungarian reaction was much more mild, and though still warning of "anti-Socialist and hostile forces," Janos Gosztonnyi, editor-in-chief of *Nepszabadsag,* stated flatly, "There is definitely no parallel between the Czechoslovak events and those in Hungary in 1956—in the sense of a counterrevolution in Czechoslovakia."[72]

The Czechs themselves were critical of the Warsaw meeting (CZE 23 and CZE 26), as it was "held without us,"[73] and at the same time affirmed their own road to socialism. They reiterated this position in their reply to the letter that the five parties, meeting in Warsaw, had addressed to the Czech party. The stern, apprehensive tone of this letter expressed the participants' view of events in Czechoslovakia.

> The development of events in your country arouses deep anxiety among us. The onslaught of reaction against your party and the foundations of the social order of the Czechoslovak Socialist Republic, supported by imperialism, threatens—we are profoundly convinced of this—to push your country off the road to socialism, and consequently, it threatens the interests of the whole socialist system . . . .
>
> Recently, political organizations and clubs formed outside the framework of the National Front have to all intents and purposes become chiefs of staff of the reactionary forces. The social democrats

> stubbornly strive to from a party of their own; they organize underground committees, seek to bring about a split within the worker movement in Czechoslovakia; they want to organize the country's leadership in order to restore the bourgeois regime. The antisocialist and revisionist forces hostile to socialism have grabbed the press, radio, and television and turned these media into rostrums for attacking the Communist Party, for deceiving the working class and all workers . . . .
> In your country a whole series of events in recent months indicates that counterrevolutionary forces supported by imperialist centers have launched attacks on a broad front against the socialist system. On the other hand, the necessary resistance is lacking on the part of the party and the popular power.[74]

On the eve of the Warsaw meeting, the Romanian party published a long article emphasizing both the leadership role and independence of individual communist parties. In contrast to their worried allies the Romanians confidently asserted that it was the Czech Communist Party that was playing the leading role in the "rejuvenation" process in that country.[75] In the course of affirming their support of the Czechs, the Romanian party thus tried to assuage Soviet fears of loss of control in Czechoslovakia and to reiterate the legitimacy of events there by showing their origins to be, quite properly, the party itself.[76]

The Central Committee of the League of Communists of Yugoslavia were at least as forceful in their support of the Prague Spring, both before and after the Warsaw meeting, and, moreover, criticized the Warsaw meeting explicitly, which the Romanians did not do.[77] All of the other parties reacted positively to the meeting, praising it and the letter to the Czech party as an "act of solidarity" and "socialist internationalism."[78]

At the end of July another meeting, this time an extraordinary session between the entire CPSU Politburo and the entire Czech Presidium, took place at Cierna nad Tisou, in Czechoslovakia near the Soviet border. Between the time of the announcement of this meeting, June 22, and its conclusion July 31, reactions of most of the parties to further developments in Czechoslovakia were evident (item CZE 27). The Soviet and Bulgarian parties were unremitting in their hositility to Czech developments. *Pravda* of July 28 added a new element–urgency.

> The danger hanging over the socialist achievements of the Czechoslovak workers is increasing in intensity. Time does not wait!–thus

> their class brothers warn the communists and workers of Czechoslovakia. Like all fighters for the cause of socialism and peace, the Soviet people are exhorting the working class and all workers of Czechoslovakia—led by the communists—to bar the path to counterrevolution and protect their historic and revolutionary gains.[79]

Poland and Hungary were both also negative toward Czech developments and positive toward the meeting at Cierna (CZE 28).[80] While the Hungarian reaction is interesting for its mildness in comparison to the shrill shrieking of the Soviet polemecists, it is also clear that by this time Budapest has reversed itself on the analogy with 1956. Claiming that the situation in Czechoslovakia had changed, *Nepszabadsag* warned on July 25,

> What we saw and experienced in the period before the counterrevolution, is being repeated all over again in Czechoslovakia. The same writers, journalists and other persons, who formerly used slogans about democracy and freedom of criticism, have, after they seized power, suppressed criticism by every available means and silenced everybody who disagreed with them. The process is analogous, even to the point that we find in the forefront those writers and journalists who are troubled by their bad conscience, and who were the worst sectarians just a few years ago and did their utmost to glorify the former regime. But today they are overdoing it in the opposite direction, posing as champions of justice and dealing blows to "guilty" officials, as if they had never served the former regime as well (or even better).
>
> We have experienced all this, and were pushed into tragedy by it. When we raise our voice, it is not to give anyone a lecture, or to interfere in Czechoslovak internal affairs, but to cry out a warning: comrades, don't let the same thing happen (in Czechoslovakia) that led to open counterrevolution in Hungary. Do not repeat the mistakes we made . . . .[81]

The Czechs themselves, of course, spoke out in support of their own activities, but this time they were not negative on the meeting with the CPSU, since it was bilateral talks they preferred all along.[82] Nor did the Romanians condemn the meeting, drawing instead upon their most skilled Aesopian linguists.

> Public opinion in Romania is following with special attention the development of discussions which have begun between the Politburo of the Central Committee of the Communist Party of the Soviet Union and the Presidium of the Central Committee of the Communist Party of Czechoslovakia. It harbors the conviction that these discussions are proceeding in the spirit of the norms of relations between fraternal parties, in the spirit of mutual trust and respect, with the desire to contribute to the achievement of understanding, to the removal of differences of view which have appeared.[83]

The Yugoslavs were somewhat more skeptical of the Cierna meeting, though they did not speak out as harshly as they had on the Warsaw gathering. They did, however, reiterate their support for the Czechs, as did the Romanians.[84]

Immediately following the Cierna bilateral meeting (the next day in fact), a meeting was held in Bratislava between the Czech party leaders and the leaders of five other fraternal parties (event CZE 30). The five parties were unanimous in their praise of this meeting. The Hungarian *Magyar Hirlap* was typical.

> . . . The Bratislava declaration of the six parties . . . strengthened the unity and cohesion of forces of the signatories and serves the unity of the entire international workers movement. It represents an important step on the road leading to the great international consultation of our revolutionary workers' parties.[85]

Czech party leader Dubcek was more circumspect, but generally positive.

> For the road on which we have embarked it is of vital importance to insure that we can concentrate on our work in peace and in the knowledge of international support. The tasks which we set for ourselves in January we can successfully attain only in the community of socialist countries with which our future is linked forever. Otherwise, economic expansion cannot be insured, and even less so can the security and independence of our country be safeguarded, which is our primary duty in the divided and turbulent world of today. The meeting of the six communist parties has contributed to all these tasks. Its results are creating an area for the realization

> of our socialist and humanist aims. At the meeting we drew up and approved one document only- the joint statement which was published in the daily press today. No other conclusions were adopted.[86]

The Yugoslav party, departing from its past practice of criticizing such meetings, instead held a hopeful but cautious attitude about the meeting's results.

> Regardless of the extent to which the disentanglement has taken a positive path, it still stands that the Bratislava meeting has been a kind of compromise, that many other questions remain unsettled, and that relations among the communist parties will have to be discussed much longer.[87]

It remained for the Romanians this time to take a negative tack toward the meeting. The RCP criticized it because, "the Bratislava declaration of certain communist and workers parties deals with a number of problems of great political importance directly affecting other socialist countries which are not invited and did not attend this conference."[88] At this time the Soviet and German parties reacted toward the Czech events (CZE 29) as they had in the past, but the Hungarians changed tone again to mute their criticisms.[89]

Events CZE 31 and CZE 32 tap the states' initial announcement and report of the Soviet invasion of Czechoslovakia and their immediate post-invasion reactions. The result is clear. The five participating parties lauded the act as fraternal assistance rendered to a socialist brother state threatened by counterrevolution. Bulgaria, for example, in addition to reporting in full the TASS announcement of the invasion, explained,

> The Soviet Government and the governments of the allied countries–the Bulgarian People's Republic, the Hungarian People's Republic, the GDR, and the Polish People's Republic–proceeding from the principles of the indissoluble friendship and cooperation and in accordance with the existing treaty obligations, have decided to respond to the aforementioned request for the necessary help to the fraternal Czechoslovak people. This decision is in full accordance with the right of the states for individual and collective self-defense as stipulated in the allied treaties signed

> by the fraternal socialist countries. This decision is also in harmony with the vital interests of our countries in connection with the protection of European peace, against the forces of militarism, aggression, and revanchism, which have repeatedly involved the European nations in wars.[90]

On the other side, the Yugoslvas, the Romanians, needless to say the Czechs, and not to be left out, the Albanians, condemned this action outright.[91] In Bucharest,

> The Central Committee, the State Council, and the Council of Ministers have unanimously expressed their profound concern in connection with this act, stressing that it represents a flagrant violation of the national sovereignty of a fraternal socialist free and independent state, of the principles on which the relations between socialist countries are based, of the unanimously recognized norms of international law.[92]

The communique of the RCP Central Committee went on,

> Nothing can justify this armed action—the occupation of Czechoslovakia by the troops of these countries. The interference in the internal affairs of the Czechoslovak people and of their Communist Party, the armed intervention against Czechoslovakia, represents a grave blow for the interests of the unity of the world socialist system, for the international communist and workers movement, for the prestige of socialism throughout the world, and for the cause of peace.[93]

Though these governments condemned the invasion, the Romanian reaction was the most forceful, perhaps reflecting a fear that they would be next.[94] Virtually all Romanian government and party organizations drafted declarations condemning the Soviet action. Further, in an unusual move, party General Secretary and State President Nicolae Ceausescu spoke to a huge crowd from the balcony of the Central Committee headquarters, blasting the invasion and issuing a call for "worker-peasant guards, defenders of the independence of our socialist fatherland." "We want our people," Ceausescu continued, "to have their armed units in order to defend their revolutionary achievements and in order to insure their peaceful work, and the independence and the sovereignty of our socialist fatherland."[95] Though the chances of these forces being able to repel the Red Army would have been certainly no better than those of the students at Wenceclas Square, this statement does illustrate the level of divergence of views reached during this period.

Lastly, reactions were forthcoming from all the states on the occasion of Czech-Soviet talks which followed soon after the invasion. Unable to produce a Kadar-type faction to replace Dubcek and the country's popular President Ludvik Svoboda, the Kremlin was forced to deal, albeit from a position of some strength, with the existing leadership. Dubcek and the others were flown to Moscow where tense discussions ensued.[96] The Soviets and their closest comrades saw these talks as a validation of their recent action.

> . . . It [the joint communique ending the talks] records an agreement between the communist parties and governments of our countries on those most important measures aimed at strengthening the development of unity and cooperation between them. The measures worked out during the talks will promote liquidation of the threat to socialism in the Czechoslovak Socialist Republic coming from internal and external reaction.[97]

Milder, though still supportive comment was drawn from Budapest and Warsaw, the latter offering,

> A common statement has been tensely awaited for the last few days. This statement calms all those to whom the achievements of socialism and the prospects of its development are dear. The communique confirms that the Czechoslovak people will continue to develop on the road of socialism in accordance with the resolutions of the January and May plenums of the Czechoslovak Communist Party.[98]

While Yugoslavia criticized the talks, the Albanians in their assessment reached new heights of mandibular gymnastics.

> According to the text of the communique, the Soviet revisionists aggressors and the Czechoslovak capitulating revisionists have reached a dirty compromise to save their positions at the expense

> of the Czechoslovak people and the cause of the freedom of the peoples of socialism.[99]

The Czechoslovak and Romanian parties were substantially more measured in their reactions, the former attempting to reassure their stunned people, and the latter merely "taking cognizance of the agreement concluded following the talks," eschewing additional criticism of Soviet action.[100]

Europe

Events EUR 41 through EUR 46 refer to reactions or statements issued by the East European countries on the situation in, or the future of, Europe.

In July, 1966, an historic meeting of the Political Consultative Committee of the Warsaw Treaty Organization took place in Bucharest.[101] A "Declaration on European Peace and Security" was issued (referred to as the Bucharest Declaration [event EUR 41]). The statement lamented the lack of a permanent stabilizing peace treaty after World War II, pointed out current threats to European security, which included "US aggressive circles," "reactionary forces of West Europe," especially "militarist and revanchist forces in West Germany;" condemned the proposed Multilateral Nuclear Force which envisioned West German access to nuclear weapons; and demanded the recognition of the GDR and "the actual state of affairs in Europe." At the same time, it called for an extension of economic, scientific and cultural relations between all the states of Europe, and "measures for the strengthening of security in Europe ," which included the simultaneous liquidation of NATO and the WTO, recognition of existing frontiers, reduction and withdrawal of all foreign troops, establishment of a nuclear-free zone, and a German peace settlement. Finally, a general European conference was called for to work out plans for insuring general European security.[102]

All of the reporting states reacted positively to this Declaration Poland's *Zolnierz Wolnosci* was typical in its enthusiasm.

> The documents adopted at the Bucharest conference of the Political Consultative Committee of the Warsaw Pact member-states both because of their political importance and the problems raised go far beyond the limits of our geographic region. One may say

> that they constitute not only the common current credo in foreign policy of the socialist countries represented in Bucharest, but are at the same time a creative proposal [for a] constructive program of action for all nations and states of our continent, both the socialist and capitalist states, the neutral, and those which are members of the Atlantic bloc.[103]

No significant dissidence was found from this view.

The next event which produced common reactions regarding the situation in Europe was the European Conference of Communist Parties, held in Karlovy Vary, Czechoslovakia, April 24-26, 1967. The parties attending the conference adopted a "Statement on Europe" which reiterated and supported the general line and proposals put forth in the Bucharest Declaration of the previous year. The scores on event EUR 42 indicate this Statement's approval of that policy. Notably absent on this score is Romania, whose party was not represented at this conference. The RCP explained that ". . . in the course of the exchange of views and consultations which took place a common accord was not reached beforehand regarding the character, aim, and manner of the holding of the conference," a reference to the Romanian party's unwillingness to participate in any conference which might try to excoriate China.[104] In addition, by this time the Romanian views of European security in general were beginning to diverge from the bloc majority line.[105] At this point a vote of "absent" significantly separates the Romanians from their Pact allies.

The Yugoslav party also did not attend the European conference, owing to its ardent adherence to the idea of individual communist party independence. But in addition, the LCY issued a letter on April 18 which explained that it saw developments in the West as being more favorable; and that it questioned whether the Karlovy Vary Conference and its issuances were

> . . . the most suitable means of effective mobilization in bringing together the expanded progressive forces in Europe and in the world, and are they in keeping with the new conditions which make it necessary to seek out new forms of cooperation.[106]

The Yugoslavs instead saw individual party independence and cultivation of influence of the European working classes as the key to the further extension of general European cooperation, rather than the adoption of a general line.[107]

In March, 1969, another meeting of the Warsaw Pact Political Consultative Committee, taking place in Budapest reaffirmed the proposals of the Bucharest Declaration, including especially the call for a general European conference on peace and security (item EUR 43). This time reactions, including that of the Romanians, were generally more positive. Speaking to a joint session of the State Council and Council of Ministers, State Council Chairman Nicolae Ceausescu stated,

> With regard to the appeal made in Budapest for security in Europe, it is of special importance because it comes at a time when, in our opinion, there are favorable conditions for taking steps forward in this direction. The way in which this appeal has been received by public opinion and by many leading circles in Europe shows the correctness of this [approach]. It shows that the appeal is welcome and that it is our duty to act firmly in order to contribute to the achievement of the goals it contains.[108]

Even the Yugoslavs added a positive note, welcoming the Budapest Appeal, but noting quickly the need to align deed with word.

> . . . Thus, what we are faced with here is the realization of the only practical foundations of mutual cooperation between European countries, foundations which, to be sure, were not discovered at the Budapest meeting but which were gravely shaken by political and other actions in 1968.[109]

Events EUR 44 and EUR 45 tap the countries' specific reactions to the Warsaw Pact itself. The eleventh anniversary of the Pact, coming on May 14, 1966, elicited strongly positive editorial reactions from the Soviet Union, Hungary and Czechoslovakia. Calling the organization an "Indestructible Combat Fraternity," *Pravda* explained its increasingly vital function.

> The 11 years that have passed since the signing of this historic document in Warsaw have conclusively shown that the decision of the fraternal countries to establish close cooperation in the

> military sphere was completely correct. In our time imperialist adventurism and its danger to the peoples, to the cause of peace and social progress have grown still more. More and more often it seeks a way out through military provocations, various types of conspiracies and outright military intervention.[110]

Romania did not mark the date as enthusiastically. Writing in *Lumea*, a foreign affairs weekly, N. Patrascu acknowledged the need for the Pact "signed under complex historical circumstances," and the continued need for vigilance "in order to thwart the designs of the reactionary aggressive circles." But Patrascu reminded his readers that,

> In the interval that has elapsed since the setting up of the Warsaw Treaty, the balance of power in the international arena has increasingly changed in favor of socialism, democracy and peace. Growing social forces are opposed to the aggressive imperialist policy. The present crisis in NATO, SEATO, and CENTO is accounted for precisely by the widescale rejection of the policy which led to the creation of these military blocs, dangerous to peace. Acknowledgement of the anachronism they represent in the new international context is manifested in increasingly wide and diversified circles. Even some member-states of the Western blocs are reevaluating their stand in relation to these blocs, and call for changes that should take into account the new international realities.[111]

Yugoslav commentator Bozidar Kicevic was less subtle. He declared flatly that both blocs "in the West and the East" were cold war anachronisms to be disbanded. Though he reserved his harsher criticisms for NATO, it was still clear that a policy of "active coexistence," "nonalignment and independence" should replace both alliances.[112]

On the occasion of the Budapest "Appeal" of March, 1969, the states expressed their opinion also on the Pact. The Soviet Union, Bulgaria, East Germany, Hungary and to a lesser degree Poland, set the positive line.

> We have often written that the Warsaw Pact is an organization of peace. It was about a decade and a half ago in May, 1955- 6 full years after the aggressive military coalition, NATO, was formed—that the Warsaw Pact declared its aim to be the defense

> of peace and the consolidation of the socialist states' security. That the military organization stands in the service of peace cannot even for a moment suggest contradictions. The strength of the Warsaw Pact has insured peace on our continent. By the appeal mentioned above the Budapest session demonstrated even more spectacularly the peaceful tasks of the Warsaw Pact.[113]

Czech reaction was slightly more reserved, with Party First Secretary Dubcek, nearing the end of this tenure, emphasizing the measures taken at Budapest for improving the command structure of the organization.[114] These aspects also pleased the Romanians, eliciting from them a more positive reaction to the Pact on this occasion.[115] Yugoslavia merely restated its opposition to a Europe divided into blocs, in the context of assessing the Budapest Appeal.[116]

Another proposal on Europe which produced some reactions by the East European states originated in West Germany. On March 25, 1966, Bonn sent a note to all countries with which it had relations, to the governments of East Europe (except Albania), and to the Arab States, outlining its views on European peace and disarmament, and containing several specific proposals in these areas (event EUR 46). After reviewing official German policy and views on Europe in general and relations with Czechoslovakia, Poland and the Soviet Union specifically, the German note proposed *inter alia*: a) European-wide renunciation of production or transfer of nuclear weapons and a reduction of their current numbers; b) formal declarations between East and West renouncing the use of force; c) bilateral agreements on exchanging observers for military maneuvers; and d) an international disarmament conference.[117]

All of the reporting states reacted negatively to this West German note. The Polish reply, sent on April 29, was especially harsh in tone, focusing on the specifics of the issues between the two states.

> The Note . . . contains the absurd argument that Germany continues to exist within her 1937 frontiers. The Government of the Polish People's Republic has repeatedly pointed out that such a position is devoid of any foundation, and once more rejects it categorically.
>
> The frontier on the Oder and Lusatian Neisse is final.[118]

Warsaw demanded that the FRG realign its view of the past as well as the present, as a basis for peace,

> The renunciation of territorial claims against Poland; recognition of the frontier on the Oder and Neisse; recognition of the Munich Agreement as null and void, not because Hitler invalidated it by his perfidy and crimes, but because it sanctioned the rape of Czechoslovakia and was a stage in the policy of conquest; recognition of the existence of the German Democratic Republic as an equal German State and a partner for the unification of Germany; a clear answer to the question [of] what that unified Germany should be like; the renunciation of armaments and, in particular, the abandonment of the intention to gain access to nuclear weapons; a constructive approach to even partial and gradual solutions aimed at relaxation in Europe—that is the real and only road toward the consolidation of security and peace. . .[119]

The East German, Soviet and Czech replies were similary negative, criticizing in addition Bonn's Hallstein doctrine (by which the FRG severed relations with any state recognizing the GDR) and its proposed participation in a NATO nuclear sharing force. These last two aspects were central to Yugoslavia's negative reply issued on July 28. In addition, Belgrade took a different tack on the German problem from both the FRG and its East European neighbors.

> The Government of the Socialist Federal Republic of Yugoslavia, however, is convinced that the solution of the German problem can only be brought about as a result of the process of promoting inter-European cooperation and of the overcoming of the existing division and mistrust in Europe, and certainly not as the precondition of the beginning of this same process.[120]

In addition to these European-centered events, included also with this group is the WTO position on the proposed Nuclear Nonproliferation Treaty (NPT), outlined by the Political Consultative Committee at Sofia in March, 1968 (EUR 47). The Statement, adopted and signed by all the Pact members except Romania, declared the signatories' support for a U.S.-Soviet draft treaty submitted to the Eighteen-nation Disarmament Committee the previous January and about to be discussed at that Committee's session beginning March 15. The signers urged a prompt signing and coming into force of a nuclear nonproliferation treaty.[121] Romania demurred and on March 13 *Scinteia* explained the Romanian position. While acknowledging the January draft as a step forward, Bucharest urged stronger security guarantees for states renouncing nuclear weapons and, in addition, inclusion of a more direct tie to nuclear disarmament.

> By putting no brakes whatsoever on an unhampered continuation of the arms race, such a treaty would cause its perpetuation and even its growth and implicitly enhance the atomic war danger. That is why Rumania proposes that the treaty not limit itself merely to laying down the intention of conducting negotiations for nuclear disarmament, but stipulate the concrete, unequivocal pledge of the nuclear powers to undertake measures of nuclear disarmament, because only such a clause would have the nature of producing legal effects.[122]

Further, Romania insisted on inclusion of additional control measures aimed at preventing access to nuclear weapons through the "foreign military bases" on one's territory, and measures for periodic conferences and treaty withdrawal.[123] Yugoslavia also criticized the Soviet draft and indicated its desire for a more precise statement of obligations by the powers on general nuclear disarmament, on effective international control and nondiscriminatory sharing of peaceful nuclear material.[124] The Albanians denounced the whole treaty as part of a sinister plot directed against China.[125]

World Communist Movement

Events WCM 51 through WCM 60 refer to occasions on which the states issued reactions on issues of world communism, most often the actual or potential holding of international conferences of communist parties. When most of the European parties met at Karlovy Vary in April, 1967 (WCM 51), *Pravda* applauded.

> . . . The Karlovy Vary conference, the Politburo of the CPSU Central Committee notes in its approval of the work of our party delegation, was new proof of the loyalty of European Marxist-Leninist parties to the ideas of proletarian internationalism, proof of their striving to strengthen the alliance of revolutionary forces of our times in the struggle against the aggressive and reactionary policy of imperialism, for the cause of peace, democracy, national independence, and socialism. . . . The success of the Karlovy Vary

> conference is yet further proof of the vital need for joint actions by Marxist-Leninist parties and the usefulness of collective meetings and conferences.[126]

Poland and Czechoslovakia echoed these sentiments.[127] But the Romanian party criticized the principle of this conference and refused to take part. *Scinteia* emphasized each party's "exclusive right to determine its strategy and tactics, paths, methods, and forms of struggle. . . ."[128] Party leader Ceausescu made it clear that his party's nonparticipation was based on its unwillingness to accept "old practices which make relations between parties depend on the acceptance or non-acceptance of certain points of view, and which permit the exercise of certain pressures under one form or other." The Romanian party was unwilling to acknowledge "an international coordinating centre" and reaffirmed "the legitimate right to take part in an international meeting if it considers it necessary and useful, as well as the legitimate right not to take part."[129]

The Yugoslav party similarly stressed independence for individual parties and lamented that even after bilateral meetings with some other parties,

> . . . . the LCY arrived at the conviction that a change of any essential importance whatsoever could hardly be expected in the positions on which agreement had already been reached at the meeting of the Drafting Commission in Warsaw.[130]

The "change" the Yugoslav party particulary desired was a renunciation of the declaration of the international party conferences of 1957 and 1960, which support the general line of the CPSU on building socialism, "left" and "right" dangers to socialism, and the world ideological situation in general. The LCY opposed the concept of one leading general line on communism and its development, urging a broader inclusion of "progressive" forces into the anti-imperialist camp. It had signed neither document, and pushed for all future conferences to invalidate them.[131]

Finally, the Albanian party put the conference in perspective:

> [the Karlovy Vary meeting] exposed before the world public the depths to which the Khrushchev revisionists have degenerated, their complete capitulation to the bourgeoisie and other reactionary

> forces, and their eventual turning into a handful of social reformists serving capitalism and counterrevolution heart and soul.[132]

In November, 1967, eighteen parties issued a call from Moscow for a consultative meeting to be convened to discuss the holding of a new international conference of communist and workers' parties (event WCM 52). The CPSU fully endorsed the proposal.

> The idea of holding an international conference of representatives of communist and workers parties fully corresponds to the principles of proletarian internationalism. The documents of the Moscow conferences of 1957 and 1960 which determined the general line of the international communist movement, stressed the great importance of this form for strengthening the military unity of the fraternal parties. The international conference of communist and workers parties, as is pointed out in the statement of 1960, is an effective form for the mutual exchange of opinions and experience, the enrichment by collective efforts of Marxist-Leninist theory, and for the elaboration of united positions in the struggle for common aims. . . . A consultative meeting will undoubtedly be a forum which will thoroughly discuss the question of the nature and aims of a new international conference, the coordination of the unity of revolutionary forces, and the strengthening of a united anti-imperialist front—a genuine requirement at the present stage of world development. A necessary prerequisite for the achievement of this high aim is the cohesion of the Marxist-Leninist parties.[133]

The Czech view represents a more restrained but still positive reaction.

> The increasing activities of U.S. imperialism and the intricate methods of U.S. global strategy, aimed against the socialist countries and against the national liberation movement, require a joint coordinated procedure which would take into account the conditions of forces, and possibilities of individual parties and their opinions. Such a procedure necessarily rules out all doubts that it would lead to the assertion of hegemony by one party.[134]

While Yugoslavia reacted negatively to the proposed meeting—Radio Zagreb calling the conference a means of exerting pressure on individual parties—Romania issued no reaction at all.[135]

As the Consultative Conference approached, the various parties issued their views and positions on it, including the Romanians who announced their intention to attend two weeks before the conference was scheduled to begin. Romanian acceptance was carefully circumscribed, indicating that the RCP saw the meeting as an exchange of views regarding the holding of an international conference—a contingency not taken for granted by Bucharest. The conference, in its view, should be open to all communist and workers' parties and be

> . . . carried out on the basis of strict respect for the norms of relations between the communist parties proceeding from the understanding and recognition that the unity and internationalist solidarity which the communist movement needs so much can be built only on the foundations of independence and the equality of all the parties.[136]

Further, the Romanian vision of the tasks and workings of the conference is notable for its differences from that expressed by *Pravda*.

> The Romanian Communist Party Central Committee is of the opinion that an international conference, to serve the cause of unity, rapprochment, and understanding between the parties, must be of the nature of a meeting for a broad and fruitful exchange of opinions regarding the common problems of the struggle against imperialism, in order to achieve a general consensus without adopting directive and program documents for the activity of the communist parties, which are the only ones with the right and in a position to work out their internal and international policy.
>
> The Romanian Communist Party considers that it would be useful at the consultative meeting for there to be an exchange of opinions regarding the agenda of the conference, and it shares the opinion that this should be restricted to the problem of the concrete tasks of the current struggle against imperialism.[137]

Finally, the RCP listed this specific injunction:

> . . . the Budapest meeting, as well as a future international conference, must not discuss and criticize the internal or international political life of any fraternal party which is or is not present

> at these meetings, and the practice of blaming or condemning other parties must not be resorted to in any form—a practice which has severely damaged the communist movement. The participation or nonparticipation of a party at an international conference—a problem which can be decided only by the respective party—must in no case be reason for epithets or labelling, and must not affect the relations of comradely collaboration between the communist and workers parties.[138]

The Romanian attitude is in sharp contrast with the hard line of the Poles:

> . . . The strengthening of the unity of communist and workers' parties, and of their cooperation in the world arena, is today an indispensable condition of further successful struggle against the forces of imperialism, reaction and war, and for the sacred cause of the freedom of nations, socialism and peace. In the present situation the attitude of a party toward the question of a new conference is the criterion of its attitude toward the unity of the entire movement, the criterion of its faithfulness to the principles of proletarian internationalism.[139]

The reaction of the other parties, all positive except the Yugoslav, ranged in between. The Soviet view, as expressed in *Kommunist* and broadcast by TASS is worthy of particular note. In a restrained tone, clearly directed toward mollifying its critics to the south, the CPSU stated that

> . . . all preparation for another international meeting of communist and workers parties and its proceedings [must] be concentrated on the positive tasks of the communist movement. The meeting must not be a disciplinary court or an ecumenical council authorized to 'excommunicate' someone from the communist movement. Its purpose must be cohesion for the sake of the supreme interests of the working class—for struggle against imperialism and the threat of war, for defense of the interests of the working people.[140]

Next, the Soviets offered an olive branch to Belgrade.

> The jointly worked out general line of the international communist movement retains all its significance. However, after the 1960

> meeting, many new problems accumulated which call for a solution. . . . For the CPSU the recognition of the complete equality of all parties, large and small, in discussing . . . the current problems of the communist movement is one of the key points of its program. What is characteristic of the CPSU's policy is the attention to and respect for other fraternal parties, their opinions, views, and experience. Any hegemonistic tendencies . . . are alien to Soviet communism.[141]

If the Romanians were persuaded to give the consultative meeting a try, the Yugoslavs were not. The earlier Declarations were still the sticking point.[142]

The meeting itself produced three different types of reactions by the parties (events WCM 54, WCM 55 and WCM 56). First, most parties explained their stands on the holding of a world communist conference. The Soviets, Bulgarians and Poles were most favorable to such a conference. A plenum of the Bulgarian Communist Party Central Committee declared:

> The conference, which will discuss the tasks connected with the struggle against imperialism at the present stage and the unity of the communist and workers parties and of all anti-imperialist forces, will reflect the striving for further rallying the ranks of the international communist and workers movement. This will furnish new prerequisites for an upsurge in the struggle of the revolutionary and democratic forces and also for national and social liberation, peace, democracy, and socialism.[143]

The Romanians and Yugoslavs remained skeptical, however, and viewed the prospect of an international conference negatively.[144]

And well might Bucharest feel concerned about the nature of an international conference of parties. For at the consultative meeting itself, the Romanian party was bitterly attacked by the head of the Syrian Communist Party delegation, Khalid Bakdash. After the conference refused to disavow the Syrian attack and after a "virgorous statement of protest," the Romanian delegation left the meeting.[145]

Apparently chagrined by the Syrian-Romanian incident, the Hungarian and Czechoslovak parties presented views more restrained than before, but still positive toward the upcoming conference. The latter, in an article in *Prace,* took the unusual step of relating each area of controversy over the upcoming conference, and the opposing views represented therein.[146]

As to the consultative meeting itself and the Romanian incident (events WCM 55 and WCM 56), the same parties which urged the conference also praised the February meeting; similarly, the Czechs and Hungarians were restrained in their positive attitudes; and needless to say, Bucharest and Belgrade had little good to say on the consultative meeting. The Soviets, Bulgarians and Poles did not mention the Romanian walkout at all. Hungary, however, was openly critical of the Romanian action. After asserting that the consultative meeting would be an "important positive influence on the further evolution of the anti-imperialist struggle," *Nepszabadsag* stated,

> This fact cannot be changed by the event that, as it is clear from the text published by the consultative meeting, the delegation of the Romanian Communist Party under the pretext of its dispute with the Syrian Communist Party regretably withdrew from the meeting. Nobody supported the demand by representatives of the Romanian fraternal party that the meeting participants should collectively convict the Syrian delegation. They also regarded as unfounded the Romanian demands that the parties should be limited in their rights to freely express their opinions. In the patiently led and thorough debate, each delegation explained: They saw no reason to meet the demands that violate the democracy of the meeting and the sovereignty of parties . . . . It is the inalienable right of every party to decide what steps are necessary and when. The interests of unity in any case desire that the Romanian Communist Party revise its standpoint and return to the common work.[147]

The Czech press, though reporting the walkout, refused to condemn it: [148] nor did the Yugoslavs, not surprisingly.[149] The Romanians explained their official position at Budapest, and a special plenary session of the Central Committee of the RCP issued a long supportive Communique.[150]

In 1968 and early 1969 there were three meetings in Budapest and one in Moscow of the working group and preparatory commission for the holding of the international conference, made up of the representatives of sixty-seven parties.[151] Items WCM 57, WCM 58 and WCM 59 represent the views expressed at these times by the various parties on the holding

of the international conference.[152] The supporting group is fairly consistently made up of the Soviet Union, Bulgaria, the GDR, Hungary and Poland. The Czech and Romanian scores on this issue, it will be noted, move from negative to either positive or neutral. The effect of the intervening Soviet invasion on expressed policy seems clear. Contrast, for example, these aspects of the Czech view of the conference:
In April, 1968, the Czech party stated,

> As regards the proposed agenda for the conference . . . in no case should there be a verbal declaration of war against imperialism; but each party, on the basis of a thorough and sober analysis of the real situation in the world and in its country, should consider concrete possibilities for contributing to this struggle and toward increasing the attractiveness of the ideas of socialism, and should come forward with appropriate proposals.[153]

In March, 1969, *Rude Pravo* wrote,

> . . . preparations—as is obvious from their organization—are taking place with the wide and unlimited participation and therefore most democratically. This on the other hand does not mean that the draft which is now being distributed can be taken to be a mere collection of opinions and standpoints expressed during its preparations. The experience of similar meetings shows that not all reminders can be taken into consideration. Whether the central committees will withdraw them or whether they continue to insist on them still remains to be seen.[154]

Whereas the Czech party had the previous year criticized the preparatory commission's workings, urging that "there should be a really all-around and open discussion (and not a series of monologs)," and calling for the inclusion of the Romanian and Yugoslav parties, in March of 1969, according to *Rude Pravo,* the preparations were taking place "with wide and unlimited participation and therefore most democratically." No further mention was made in the Czech statements of November, 1968 or of March, 1969 of the Romanians (who had by this time joined the preparatory commission) or of the Yugoslavs (who had not), nor of the Moscow Declaration of 1960, which had been labelled "unjust" previously.

While the Yugoslavs maintained their negative view of the conference, though modifying their degree of criticism, the results of the Czech events was apparently not lost on Bucharest. Ceausescu had stated in a speech to the Bucharest party *aktiv* in April, 1968, that,

> . . . in order for us to envision participation in the conference, in the work of the preparatory commission and in the preparation of the conference in general, it is necessary that a public agreement should be reached before-hand to the effect that in the preparatory commission as well as the conference other fraternal parties will not be attacked and criticized, whether they participate or not in the proceedings of the commission or in the conference. . . . if we jointly organize an international conference it is absolutely natural that we must reach a previous understanding, a mutually acceptable agreement on the nature of the meeting, on the spirit in which the questions will be put and debated.[155]

But by the fall of that year the Romanians evidently felt able to not only support an international conference but attend, for the first time, a preparatory commission meeting in Budapest, which, they reported, was held "in an atmosphere of an open exchange of opinions and in the spirit of comradely cooperation."[156]

Not all parties saw fit to reassess their positions. On October 8, the Albanian party paper published a long editorial on the "Scandalous Failure of the General Revisionist Conference."[157]

The last event in this issue group is the fiftieth anniversary of the Comintern (WCM 60), marked on March 4, 1969 with, among other things, celebrations in Moscow. The Soviets, Bulgarians and East Germans praised the organization in speeches at the celebration, and Moscow *Kommunist* published a long laudatory article entitled, "The Living Heritage of the Comintern."[158] The Yugoslavs also felt the heritage of the Comintern was alive, but saw that as reason for criticism.

> Views on the Comintern are used to express the attitudes for the forthcoming conference of communist parties. Everything that was said about the Comintern in today's report, [of a speech by Mikhail Suslov praising the Comintern] as well as what has been repeated countless times during the last year, is just for that purpose. In other words, the legacy of the dead Comintern is the attitude which sets the solidarity with the Soviet Union as the criterion of some-

> body's Marxism . . . . Solidarity, yes—but the formation should be turned around. Today, too, the solidarity of the Soviet party with the other parties is just as necessary as the solidarity of the other parties with the Soviet party. Everything else is nationalism and hegemony. The right of primogeniture does not exist. Solidarity must be mutual. There is no monopoly of correctness of policy. There is no guarantee that the Soviet [policy] is always correct, and that the policies of others can be correct, but need not be.[159]

The Romanian Communist Party published a similar article under the collective pseudonym V. Iliescu, noting in a review of the history of both the Comintern and Cominform, the presence of certain "negative effects of some practices and methods,"

> . . . such as the establishment by the Comintern—without thorough knowledge of prevailing realities in each country—of the political line and specific tasks of the communist parties, prescription of their forms and methods of activity, the enforcing of recipes considered to be universally valid, and immixture in the activity of the parties. Life itself has proved that owing to the great diversity of conditions and concrete tasks of the communist parties, their being directed from a center is both practically and objectively impossible.[160]

Both the Romanian commemoration and, less subtly, that of the Yugoslav party, were directed against any contemporary analogs of past Comintern practices, in whatever form they might arise.[161]

United States

The last major group of events listed in Table III.2 relates to the United States, its actions and/or leaders. Events USV 71, USV 72, USV 73, and USV 75 are direct United States' actions increasing the level of the Vietnam war effort. (The exact nature of each event can be determined from Table III.3.) The countries' reactions to President Johnson's initial large increase in troop strength in 1965 and concomitant increase in draft calls (USV 71), for example, met with unanimous disapproval.[162] Reactions ranged from irritated confusion on the part of Yugoslavia at the American President's escalation of the war simultaneously with offering peace negotiations [163] to East Germany's fierce denunciation of Johnson as a war criminal dispatching "murderers" to Vietnam.[164]

The bombing of North Vietnam (events USV 72, USV 73 and USV 75) in 1966 produced similar strongly negative reactions. Calling the bombing an "unprecedented barbarity," TASS spoke thus:

> The American raid on the suburbs of Haiphong and Hanoi shows that Washington has taken another step, and probably several, in the notorious "excalation" of its aggression in Vietnam. Despite the resolute protests of the peoples, despite the growing opposition to the extension of the war in the U.S. proper, the aggressors have taken this extremely dangerous step . . . . The increased bombings of the DRV are a gesture of despair on the part of the American aggressors, who have encountered in Vietnam a protracted and prospectless war which has frustrated all their notions about the "local" and "special" nature of this venture.[165]

On the December bombing of Hanoi only Yugoslavia was somewhat less negative.

> The acts of U.S. bombing raids on residential quarters in the city of Hanoi must be considered a grave political offence and amount to a war adventure . . . the U.S. air strikes against apartment blocks in Hanoi that took place on 13 and 14 December were not only at variance with the peaceloving intentions of the United States, so frequently proclaimed, but also with the latest U.S. hint of willingness to get the Christmas ceasefire extended on the battlefields of Vietnam.[166]

In addition, three statements adopted by most of the states in conferences clearly condemn U.S. activities in Vietnam. The same meeting of the Political Consultative Committee of the WTO which produced the Bucharest Declaration (July 5-9, 1966) also adopted a stern "Statement on Vietnam," which condemned the war in Vietnam as "the most cynical manifestation of the aggressive policy of American imperialism." "It is an outrage," the statement went on, "upon international law and international agreements and a gross breach of the U.N. Charter."[167] The statement, setting the tone for future Pact positions on Vietnam warned the U.S. of the dangers of a wider war, blasted the "incompatibility" of acts of war with talk of peace, and demanded that the U.S. stop the war, withdraw its forces and "respect the basic national rights of the Vietnamese people." Finally the Pact members offered their support, material and economic assistance, and even volunteers to the people of Vietnam.[168] All of the reporting states reacted positively to this statement in supplementary editorials in their party and state organs (event USV 74).

A meeting of the WTO Consultative Committee which took place in Sofia in March, 1968, adopted a similar statement. This statement is coded itself as event USV 78.

Finally, the delegates attending the Karlovy Vary Conference of European communist and workers' parties issued a Vietnam "Appeal" (USV 76) which said *inter alia*,

> The delegates in the Karlovy Vary conference strictly condemn in their appeal the aggressive war in Vietnam and express their solidarity with the heroic struggle of the Vietnamese people. They also stress their firm determination to work actively for the isolation and defeat of the aggressive policy of U.S. imperialism, to fight for the withdrawal of all foreign forces from Vietnam and for the recognition of the right of the Vietnamese people independently to solve their internal problems.[169]

Neither Romania nor Yugoslavia were signatories to this Appeal as neither attended this conference.

On the other hand, one development in Vietnam to which the East European states did react in a generally positive way was the Tet offensive of January-February, 1968 (USV 77). Hailing the country-wide attacks by the Vietcong and North Vietnamese as "the most damaging failure for U.S. interventionist policy," *Pravda* in an article entitled "One Blow After Another," exuded,

> In the atmosphere of a powerful upsurge of the national liberation movement in South Vietnam, a new broad front of the patriotic forces is coming into being which, alongside the NFLSV, comprises other democratic and national organizations working for the liberation and genuine independence of South Vietnam.
>
> The latest events in South Vietnam have yet again forcefully

> demonstrated to the whole world the incontestible fact that the broadly representative NFLSV is the only logical representative of the people of South Vietnam, and not the miserable handful of traitors that are grovelling to Washington.[170]

Hungary called it "[e]qual in value to Dien Bien Phu,"[171] and a GDR commentator exclaimed, "What a success!"[172] The Yugoslavs and particularly the Czechs were more restrained in their reactions.

> From all this, no hasty conclusions should be drawn for the time being. We can merely point out what this could possibly mean. The strength and skill of the Liberation forces' attack confirm that any optimistic prospects of the U.S. generals for a victory are groundless.[173]

USV 79 and USV 80 are the two actions by the United States in the reverse direction, that is, toward winding down the war. In March, 1968 President Johnson announced a limitation of the bombing north of the DMZ, i.e. in North Vietnam. At the same time, he announced that he would not be a candidate for reelection in November. Reaction to the bombing limitation was varied. The East Germans retained their previous level of hostility. *Berliner Zeitung* called the bombing limitation

> . . . nothing but a blackmailing maneuver, for it demands that in return for this unbounded kindness the Vietnamese accept the U.S. conditions, and that they no longer defend themselves, in other words that they capitulate. However, peace will only come about when U.S. bombs no longer fall on even a square inch of Vietnamese soil and when the last American has been withdrawn. Whoever refuses to learn this lesson will end up like Johnson—an unsuccessful murderer.[174]

Moscow reacted in a similar tone, and claimed that the United States did not even respect its own limitation.

> But even now, in giving the order to stop bombing the DRV, Washington is not sincere. First, the United States admits that 10 percent of the DRV territory near the Demilitarized Zone

> continues to be under fire. Second, yesterday [April 1] the Americans dealt a blow to an area 300 kilometers from the Demilitarized Zone.[175]

Yugoslavia and Poland, while still critical, especially of the fact that the bombing limitation was not complete, were less negative in their reactions. Both saw cause for possible hope, if through sardonic circumstances. The Belgrade government daily, *Politika*, said people

> . . . will sense the encouragement stemming from the belief that the more broadminded and dignified America, led by the elite minds, has won a victory through the irresistible pressure of its high moral principles to which even the man in the White House has finally yielded.[176]

And Warsaw even speculated that

> . . . Johnson hopes that his announced step may now lead to negotiations, it could be that even before the Democratic Party convention, which nominates its choice for the presidency, Johnson might be recognized by a large segment of public opinion as the person who has finally brought about an alleviation of the Vietnam conflict and possibly even peace.[177]

Czechoslovakia's reaction went further in this direction, with Prague's analyst stating,

> . . . I would like to add that President Johnson, by his calm and matter-of-fact analysis yesterday, has won over the American public, which is confident that he regards his proposals sincerely and that this is not a political game. This assessment is primarily related to his proposal regarding a peaceful solution to the Vietnam conflict, a partial cessation of the bombing and the limitation of military actions.[178]

Reaction was almost totally favorable, if not effusive, on the total bombing halt (of North Vietnam) ordered by President Johnson in November, 1968, and announced in conjunction with an agreement to begin quadrapartite negotiations. *Scinteia* called the action a "Step toward the solution of the Vietnam problem" and wrote:

> The public opinion of the whole world took note with satisfaction of the news of the unconditional halt of the American bombing of the territory of the Democratic Republic of Vietnam. The act of ceasing the bombing, which for four years has ravaged the earth, has struck towns and peaceful villages, economic and cultural objectives of the DRV, constitutes a positive event in international political life.[179]

But, the organ of the RCP noted,

> The cessation of the bombing is only a beginning for assuring peace in Vietnam. What is incumbent now is to move further along, to assure the withdrawal of all foreign troops from Vietnam, to recognize the inalienable right of the Vietnamese people to live in liberty, without foreign occupation.[180]

The Soviet government, in a similar statement, was careful to lay the burden for future movement on the road to peace on the United States.

> The progress in the forthcoming talks between the sides will depend on whether the U.S. Government displays a realistic and serious approach to the position of the DRV and the NFLSV, which reflects the aspirations of the Vietnamese people and accords with the Geneva agreements of 1954.[181]

These reactions were typical of all the states, save, of course, Albania, who managed to detect that

> . . . this decision of the American administration is the result of the many secret bargains and the known plot of American imperialism with the leaders of the Soviet revisionist clique to impose, at all costs, upon the Vietnamese people the devilish plan for peace talks . . .[182]

The groups of events USL 81 through USL 85 scores the states' reactions to U.S. President Lyndon B. Johnson specifically, and U.S.

foreign policy in general. Events USL 81 through USL 84 are the four successive "State of the Union" messages delivered by Johnson from 1966 through 1969. When the states react to a U.S. President's State of the Union message they are reacting to the United States in general, especially its foreign policy behavior. For this reason the event is particularly useful as a stimulus.

The Soviet reaction over the four years was consistently negative, criticizing in particular the war in Vietnam, and relating it to U.S. economic and social difficulties. Moscow summed up Johnson's administration on the occasion of the last message.

> The violence lying at the foundation of U.S. foreign policy has served as a stimulus for an unprecedented epidemic of violence within America itself. As it departs, the Johnson administration bequests to the whole world the memory of unprecedented racist programs and brutal suppression of Negro unrest. It leaves the memory of a wave of political terror that has engulfed America, claiming as its victims Negro leader Martin Luther King and Senator Robert Kennedy . . . . the departing government has been incapable of drawing the appropriate conclusions from its mistakes and President Johnson repeated once more that the way to peace in Vietnam lies through fierce fighting, and thereby gave it clearly to be understood that he sees the solution of the Vietnamese problem not in the immediate cessation of American aggression, but in the intensification of military pressure. He called for strengthening unity in Europe, thus emphasizing that Washington should, as before, count on strengthening the aggressive North Atlantic Alliance, which represents a serious threat to the peace and security of European peoples.[183]

The Soviet line was matched by the Bulgarians, exceeded by the Germans, and softened somewhat by Poland. While Warsaw echoed the critical Soviet assessment of the foreign policy of the Johnson administration, it did grudgingly point out that,

> In the big problems essential for world peace, only the recent period, beginning approximately last spring, in the face of the fiasco of the Vietnam escalation and other failures, brought a gradual abandoning of a collision course. The result was the

> signature on the agreement on nonproliferation of nuclear arms, and then the decision on a halt to the bombing of democratic Vietnam as well as the opening of the way for peace talks in Paris. Johnson also initiated the preliminary contacts for opening the American-Soviet discussion limiting armaments in missiles and anti-missiles.[184]

The reactions of Hungary and Yugoslavia to Johnson's final message were substantially muted, with the latter concerning itself more with the prospects offered by the President's successor.[185]

Over the four years Romania and Czechoslovakia varied somewhat, within a range slightly less negative than the Soviets; but only one reported reaction, that of Yugoslavia to the 1966 speech, even approached neutrality, much less positiveness. On that occasion *Borba* wrote:

> Although President Johnson's State of the Union Message, on the whole, shows the U.S. Government's determination to continue the war in Vietnam, it nevertheless contains some elements which, given a good deal of benevolence, might be interpreted in the sense that the President seriously nurtures the idea of a possible peaceful settlement of the Vietnam problem, or at least negotiation. . . . Johnson's message is characterized by extraordinary caution and a very careful choice of words and terms, so that an undesired interpretation of what has been done, and hints on what is intended, might be avoided . . . the U.S. President has failed to state further specific prospects of the further course of the war in Vietnam.[186]

The most specific reactions to President Johnson himself came when he announced his withdrawal as a candidate for reelection, in March 1968 (event USL 85). Reactions to this aspect of Johnson's speech were in general more measured, with that of East Berlin being the harshest.

> The fact that, under pressure of the international and domestic protests against a policy of aggression, a U.S. President has been compelled to give up is a yardstick for the present-day balance of forces in the United States itself and in the whole world . . . . He has now been compelled to admit ignominiously what has long been known on all continents: Johnson's policy of the "Great

> Society" has failed miserably. The great, powerful, and rich United States has been steered by him and the mighty monopolists behind him into a blind alley, indeed into a catastrophic situation. The pitcher which is taken to the well too often gets broken at last. Even the U.S. President cannot indefinitely trample on mankind's will.[187]

The Soviet response, as well as that of Hungary, Romania and Yugoslavia, was to reprimand the departing President for his Vietnam policy and draw the conclusion that his resignation represented the failure of that policy.[188] All of these mentioned the suspicion that Johnson's withdrawal might be a political ploy and that he might indeed show up again for the Democratic Party's candidacy. But Czechoslovakia, in a generally more positive response, noted that, "It is clear that it will be extremely hard for him now to go back on his word," and called his action a "virtue born of necessity."[189]

Event USP 86 codes the states' reaction to the seizure of the United States ship *Pueblo* by the North Koreans off their coast in January, 1968. Little comment is required here except to say that U.S. action in this situation was roundly blasted by all the states as a provocation and extension of United States' aggression.

The last three events relate to the next U.S. President, Richard Nixon. On January 27, 1969 Nixon held his first Presidential press conference and the reporting states reacted with studied blandness (USN 91). The Soviet Union issued no major editorial on the Nixon statements and merely reported the issues he touched on, taking particular note of what it took to be a mellowing in Nixon's arms race rhetoric.

> . . . During his election campaign he spoke about the United States having to [aim] for military superiority over the Soviet Union and to talk with the Soviet Union from a so-called position of strength. Now Richard Nixon mentioned neither the position of strength nor military superiority. He stated that instead of military superiority one should now only talk about sufficient military power. This, some observers believe, also manifests a more realistic approach to questions of foreign policy.[190]

Most of the other responses were similar with each state noting, for better or worse, items of particular interest in the President's comments. The Romanians, for example, took care to label Nixon's position opposing Chinese membership in the U.N. as "obstructionist."[191] The Albanians, however, did not shrink from attacking such a "diehard counterrevolutionary as Nixon," and put their lexicographers to work overtime.

> He admits as a vital necessity, in the first place for the very supreme interests of the United States and of the ruling class, the imperialist-revisionist counterrevolutionary hegemonist collaboration, as a very much necessary means to cope with and oppose the invincible socialist great China of Mao Tse-tung, to defeat and bring down to their knees the revolutionary peoples of the world, to halt and annihilate the liberation and anti-imperialist movements in the world.[192]

President Nixon's first major diplomatic venture was a trip to Western Europe, a tour of the major NATO capitals there, lasting from February 24 to March 3, 1969. The reactions to this trip (items USN 92 and USN 93) were mixed, but again not severe in either direction. *Izvestia* warned against "Atlanticism"—strengthening NATO—on the eve of the trip and upon Nixon's return reassured itself and others that NATO unity, along with American dominance, had departed Europe.

> . . . America's political strategists who brought NATO into being 20 years ago have proved to be shortsighted. They thought that the myth of a so-called Soviet menace would enable them to cement their aggressive military alliance and hold it firmly together permanently so that their country might exercise hegemony in Western Europe. But the myth burst, and on his arrival in Europe President Nixon found his closest allies torn by dissension. . . . The very foundations underlying the North Atlantic pact, aimed as it is at the Soviet Union and its allies, are toppling. At one time NATO kept going with American subsidies and the fear of nonexisting menace, but those were transient factors, good only for a time. They lost their bite a long time ago.[193]

Reactions from Sofia and East Berlin were similar. *Rabotnichesko Delo* wrote:

> Brushing aside all talk about 'a new attitude' of the U.S. to its allies, about the intention to act in the future in coordination with them and not to impose on them its will . . . there remains the naked truth which caused the long tour of the U.S. President. This is the striving of the U.S. to tighten NATO under its management so that it might be used in a yet more efficient manner in the future policies of the new administration.[194]

Romania and Yugoslavia, on the other hand, reacted positively to the Nixon trip, the latter praising the President as an "urbane, charming and even sincere guest," and moreover as having "displayed qualities of a statesman."[195] More interesting still was the positive reaction of the Czechs who, in contrast to the Soviets, saw the tour as a proper consultation trip on the part of a U.S. President, especially one anticipating an approaching summit with the Soviet Union.[196] Moreover, the Slovak paper *Pravda* commented, upon Nixon's return,

> The American President achieved practically all he set out to do in his tour of Europe . . . . His West European allies received him and his conceptions with understanding . . . . Nixon is returning with an idea of the complexity of the problems standing in the way of his diplomatic 'advance on Moscow' . . . . It must be said, however, that Nixon's European policy gives more attention to the needs and realities of the crisis-ridden European continent than did Johnson's.[197]

These, then, are the issues to which the states of East Europe reacted in greatest measure. The above sampling of responses is intended to provide a substantive overview of some of the positions taken by the various states on these issues. At the same time, by demonstrating with texts the range of these responses, it is hoped that a richer contextual framework has been sketched for the overall quantitative analysis which follows.

The impressions one receives from this review are that Romania and Yugoslavia rather consistently differed with the Soviet Union and the other Pact states in their international perspectives. The Bulgarians and East Germans seem the most ardent representatives of a tough, uniform Pact line on the various issues, at times even exceeding Moscow in their hostility toward enemies of socialism and socialist unity. Hungary and Poland appear to be the most centrist of the Pact states, taking generally Pact-supportive, but not overly exaggerated, positions. Finally, the Czech responses are the most mercurial, as one would expect given their internal history during the period.

To substaniate these impressions, effect a more precise comparison of the international attitudes of these states, and to facilitate objective hypotheses testing, the development of certain quantitative indicators of conformity and deviation is appropriate.

FINDINGS, II: PAINTING BY NUMBERS

Overall, what can be said about the responses of these countries to the major international issues that occurred between 1965 and 1969? Specifically, how close were the members of this international region to each other in their reactions to these events? Using the data listed in Table III.2 we can construct various measures of comparability, ranging from the more general to the more specific, to answer this question.

Table III.4 lists Pearson product-moment correlation coefficients for the scores of all nine countries with each other, across all cases in which reactions were found for both. We can see that, over the course of this quinquennium, the reactions of Bulgaria, the GDR, Hungary and Poland were very highly correlated with those of the Soviet Union. Less strongly associated, but still significant, are the reactions of Czechoslovakia and the Soviet Union. Finally, the table shows no statistically significant correlation between the reactions of the Soviet Union and those of either Romania, Yugoslavia or Albania. Row 7 indicates that the Romanian reactions were in fact most highly correlated with those of Yugoslavia, with the Romania-Czechoslovakia correlation next highest. (The correlation with Albania, although high, is less meaningful for interpretation due to the low number of cases for Albania.)

In Row 10 we see how highly correlated the individual states were with the Warsaw Pact as a whole. The variable WTO was computed for each issue by taking the mean score of all reporting Pact members. Not surprisingly, the Soviet Union, Bulgaria, the GDR, Hungary and Poland all show near perfect correlation; with Czechoslovakia lower, the Romanians lower still, and Yugoslavia and Albania showing no significant relationship.

TABLE III.4:
PEARSON CORRELATION COEFFICIENTS OF REACTION SCORES

	SOV	BUL	CZE	GDR	HUN	POL	ROM	YUG	ALB	WTO
SOV	1.0000 (0) S=0.001	0.9876 (38) S=0.001	0.4369 (52) S-0.001	0.9862 (39) S=0.001	0.9398 (52) S=0.001	0.9662 (50) S=0.001	0.1666 (47) S-0.132	-0.1379 (49) S=0.172	0.2830 (14) S=0.163	0.9319 (56) S=0.001
BUL	0.9876 (38) S=0.001	1.0000 (0) S=0.001	0.5300 (35) S=-.001	0.9863 (28) S=0.001	0.9688 (37) S=0.001	0.9655 (35) S=0.001	0.1874 (31) S=0.156	-0.0873 (33) S=0.315	0.0247 (10) S=0.473	0.9530 (38) S=0.001
CZE	0.4369 (52) S=0.001	0.5300 (35) S=0.001	1.0000 (0) S=0.001	0.4593 (36) S=0.002	0.5384 (49) S=0.001	0.5461 (47) S=0.001	0.6402 (46) S=0.001	0.5365 (48) S=0.001	0.2123 (14) S=0.233	0.6935 (53) S=0.001
GDR	0.9862 (39) S=0.001	0.9863 (28) S=0.001	0.4593 (36) S=0.002	1.0000 (0) S=0.001	0.9566 (38) S=0.001	0.9470 (36) S=0.001	0.3282 (33) S=0.031	0.0054 (33) S=0.488	0.2887 (9) S=0.226	0.9468 (39) S=0.001
HUN	0.9398 (52) S=0.001	0.9688 (37) S=0.001	0.5384 (49) S=0.001	0.9566 (38) S=0.001	1.0000 (0) S=0.001	0.9467 (47) S=0.001	0.2424 (45) S=0.054	-0.0418 (46) S=0.391	0.3735 (12) S=0.116	0.9461 (53) S=0.001
POL	0.9662 (50) S=0.001	0.9655 (35) S=0.001	0.5461 (47) S=0.001	0.9470 (36) S=0.001	0.9467 (47) S=0.001	1.0000 (0) S=0.001	0.2777 (41) S=0.039	-0.0342 (43) S=0.414	0.2701 (12) S=0.198	0.9557 (50) S=0.001
ROM	0.1666 (47) S=0.132	0.1874 (31) S=0.156	0.6402 (46) S=0.001	0.3282 (33) S=0.031	0.2424 (45) S=0.054	0.2777 (41) S=0.039	1.0000 (0) S=0.001	0.7412 (45) S=0.001	0.6344 (13) S=0.010	0.4543 (48) S=0.001
YUG	-0.1379 (49) S=0.172	-0.0873 (33) S=0.315	0.5365 (48) S=0.001	0.0054 (33) S=0.488	-0.0418 (46) S=0.391	-0.0342 (43) S=0.414	0.7412 (45) S=0.001	1.0000 (0) S=0.001	0.6750 (14) S=0.004	0.1218 (50) S=0.200
ALB	0.2830 (14) S=0.163	0.0247 (10) S=0.473	0.2123 (14) S=0.233	0.2887 (9) S=0.226	0.3735 (12) S=0.116	0.2701 (12) S=0.198	0.6344 (13) S=0.010	0.6750 (14) S=0.004	1.0000 (0) S=0.001	0.3326 (14) S=0.123
WTO	0.9319 (56) S=0.001	0.9530 (38) S=0.001	0.6935 (53) S=0.001	0.9468 (39) S=0.001	0.9461 (53) S=0.001	0.9557 (50) S=0.001	0.4543 (48) S=0.001	0.1218 (50) S=0.200	0.3326 (14) S=0.123	1.0000 (0) S=0.001
WTOS	0.8830 (56) S=0.001	0.9317 (38) S=0.001	0.7486 (53) S=0.001	0.9175 (39) S=0.001	0.9266 (53) S=0.001	0.9323 (50) S=0.001	0.5318 (48) S=0.001	0.2047 (50) S=0.077	0.3482 (14) S=0.111	0.9927 (57) S=0.001

Table III.4 Continued

	SOV	BUL	CZE	GDR	HUN	POL	ROM	YUG	ALB	WTO
WTCB	0.9171	0.9284	0.7194	0.9360	0.9359	0.9450	0.4989	0.1653	0.3533	0.9979
	(56)	(38)	(53)	(39)	(53)	(50)	(48)	(50)	(14)	(57)
	S=0.001	S=0.001	S=0.001	S=0.001	S=0.001	S=0.001	S=0.001	S=0.126	S=0.108	S=0.001
WTOC	0.9633	0.9740	0.5760	0.9740	0.9582	0.9675	0.3890	0.0317	0.3487	0.9885
	(56)	(38)	(53)	(39)	(53)	(50)	(48)	(50)	(14)	(57)
	S=0.001	S=0.001	S=0.001	S=0.001	S=0.001	S=0.001	S=0.003	S=0.413	S=0.111	S=0.001
WTOG	0.9115	0.9425	0.7297	0.9175	0.9367	0.9495	0.4912	0.1620	0.3388	0.9973
	(56)	(38)	(53)	(39)	(53)	(50)	(48)	(50)	(14)	(57)
	S=0.001	S=0.001	S=0.001	S=0.001	S=0.001	S=0.001	S=0.001	S=.130	S=0.118	S=0.001
WTOH	0.9201	0.9424	0.7114	0.9385	0.9065	0.9504	0.5080	0.1629	0.3333	0.9940
	(56)	(38)	(53)	(39)	(53)	(50)	(48)	(50)	(14)	(57)
	S=0.001	S=0.001	S=0.001	S=0.001	S=0.001	S=0.001	S=0.001	S=0.129	S=0.122	S=0.001
WTOP	0.9162	0.9424	0.7154	0.9405	0.9404	0.9335	0.4902	0.1541	0.3480	0.9977
	(56)	(38)	(53)	(39)	(53)	(50)	(48)	(50)	(14)	(57)
	S=0.001	S=0.001	S=0.001	S=0.001	S=0.001	S=0.001	S=0.001	S=0.143	S=0.111	S=0.001
WTOR	0.9645	0.9789	0.6344	0.9709	0.9708	0.9737	0.2765	-0.0053	0.2582	0.9832
	(56)	(38)	(53)	(39)	(53)	(50)	(48)	(50)	(14)	(57)
	S=0.001	S=0.001	S=0.001	S=0.001	S=0.001	S=0.001	S=0.029	S=0.485	S=0.186	S=0.001

A more precise manner of ascertaining the degree of individual Pact member correspondence with the rest, however, is to compute seven additional means for each issue: onw with each of the seven Pact countries left out.[198] The correlation of these new variables, WTOS through WTOR, with each country indicates the degree of association between the country and other Pact members as a group. Bulgaria, for example, correlates at the .928 level with the other Pact members over all issues; East Germany, .918; and Hungary, Poland and the Soviet Union have similarly high correlations with the rest of the Pact when their own scores are eliminated. When the Czech scores are removed from the WTO mean, however, its correlation with the rest drops to a greater degree (16%; from .694 to .576). The most dramatic, and for our purposes, significant variable is WTOR (the WTO without Romania). The association of Romania's reactions with those of the Pact drops by almost 40% (from .453 to .277) when the other members' scores are averaged separately as a group. Needless to say, the association of this new group (the Pact-without-Romania) is even higher with the Soviet Union and the other countries of the Pact.

The Pearson correlation coefficients are a useful overall measure of association of these countries' attitudes. However, it is only a measure of degree of association between variable score levels. It is theoretically possible, though very unlikely, that two countries' scores could be strongly associated, even positively so, acorss cases, but still be dissimilar. This could be the case, for example, if country A's reaction was negative and country B's positive, but both increased in positiveness together. In order to take into account this possible effect on our measure of association, we utilize the Kendall tau rank order correlation. In this case we are measuring the degree to which country A's scores, when rank ordered across cases, are similar to the rank ordered scores of country B. Thus overall similarity of reactions is measured by comparing the scores' rankings rather than their numerical value. The absolute values of the Kendall tau do tend to be lower due to the correction for ties built into its computation.[199]

The results, listed in Table III.5, are the same nevertheless, with strong correlations showing up between the Soviet Union and Hungary, the GDR, Poland and especially Bulgaria. There is much less similarity with Czechoslovakia, and no association at all with Romania and with the two non-Pact states, Yugoslavia and Albania. Again, the correlations are

TABLE III.5
KENDALL CORRELATION COEFFICIENTS
OF REACTION SCORES

Variable Pair	Variable Pair	Variable Pair	Variable Pair	Variable Pair	Variable Pair
SOV 0.8169 WITH N(38) BUL SIG.001	SOV 0.3309 WITH N(52) CZE SIG .001	SOV 0.7707 WITH N(39) GDR SIG .001	SOV 0.7318 WITH N(52) HUN .001	SOV 0.7976 WITH N(50) POL SIG .001	SOV 0.0961 WITH N(47) ROM SIG .186
SOV -0.0545 WITH N(49) YUG SIG. 303	SOV -0.0291 WITH N(14) ALB SIG .450	SOV 0.6934 WITH N(56) WTO SIG .001	SOV 0.6606 WITH N(56) WTOS SIG.001	SOV 0.6801 WITH N(56) WTOB SIG.001	SOV 0.7320 WITH N(56) WTOC SIG.001
SOV 0.6906 WITH N(56) WTOG SIG .001	SOV 0.6911 WITH N(56) WTOH SIG .001	SOV 0.6796 WITH N(56) WTOP SIG .001	SOV 0.7470 WITH N(56) WTOR SIG .001	BUL 0.3957 WITH N(35) CZE SIG .001	BUL 0.8588 WITH N(28) GDR SIG .001
BUL 0.8033 WITH N(37) HUN SIG .001	BUL 0.7376 WITH N(35) POL SIG .001	BUL 0.1827 WITH N(31) ROM SIG .085	BUL -0.0041 WITH N(33) YUG SIG .487	BUL -0.1188 WITH N(10) ALB SIG .338	BUL 0.7075 WITH N(38) WTO SIG .001
BUL 0.7075 WITH N(38) WTOS SIG .001	BUL 0.6787 WITH N(38) WTOB SIG .001	BUL 0.7739 WITH N(38) WTOC SIG .001	BUL 0.6846 WITH N(38) WTOG SIG .001	BUL 0.7005 WITH N(38) WTOH SIG .001	BUL 0.7174 WITH N(38) WTOP SIG .001
BUL 0.7492 WITH N(38) WTOR SIG .001	CZE 0.3970 WITH N(36) GDR SIG .001	CZE 0.4332 WITH N(49) HUN SIG .001	CZE 0.4708 WITH N(47) POL SIG .001	CZE 0.5099 WITH N(46) ROM SIG .001	CZE 0.3923 WITH N(48) YUG SIG .001
CZE 0.1685 WITH N(14) ALB SIG .228	CZE 0.5646 WITH N(53) WTO SIG .001	CZE 0.5842 WITH N(53) WTCS SIG .001	CZE 0.5762 WITH N(53) WTOB SIG .001	CZE 0.4641 WITH N(53) WTOC SIG .001	CZE 0.5803 WITH N(53) WTOG SIG .001
CZE 0.5650 WITH N(53)	CZE 0.5633 WITH N(53)	CZE 0.5653 WITH N(53)	GDR 0.8041 WITH N(38)	GDR 0.7023 WITH N(36)	GDR 0.3073 WITH N(33)

TABLE III.5 Continued

Variable Pair	Variable Pair	Variable Pair	Variable Pair	Varible Pair	Variable Pair
GDR 0.0748 WITH N (33) YUG SIG .281	GDR -0.0990 WITH N (9) ALB SIG .367	GDR 0.7580 WITH N (39) WTO SIG .001	GDR 0.7580 WITH N (39) WTOS SIG .001	GDR 0.7553 WITH N (39) WTOB SIG .001	GDR 0.8211 WITH N (39) WTOC SIG .001
GDR 0.7438 WITH N (39) WTOG SIG .001	GDR 0.7685 WITH N (39) WTOH SIG .001	GDR 0.7723 WITH N (39) WTOP SIG .001	GDR 0.78.27 WITH N (39) WTOR SIG .001	HUN 0.7811 WITH N (47) POL SIG .001	HUN 0.1949 WITH N (45) ROM SIG .036
HUN -0.0020 WITH N (46) YUG SIG .492	HUN 0.1284 WITH N (12) ALB SIG .299	HUN 0.7901 WITH N (53) WTO SIG .001	HUN 0.7783 WITH N (53) WTOS SIG .001	HUN 0.7763 WITH N (53) WTOB SIG .001	HUN 0.8168 WITH N (53) WTOC SIG .001
HUN 0.7747 WITH N (53) WTOG SIG .001	HUN 0.7545 WITH N (53) WTOH SIG .001	HUN 0.7953 WITH N (53) WTOP SIG .001	HUN 0.8156 WITH N (53) WTOR SIG .001	POL 0.2439 WITH N (41) ROM SIG .016	POL 0.0222 WITH N (43) YUG SIG .421
POL 0.0403 WITH N (12) ALB SIG .437	POL 0.7742 WITH N (50) WTO SIG .001	POL 0.7687 WITH N (50) WTOS SIG .001	POL 0.7685 WITH N (50) WTOB SIG .001	POL 0.7937 WITH N (50) WTOC SIG .001	POL 0.7805 WITH N (50) WTOG SIG .001
POL 0.7771 WITH N (50) WTOH SIG .001	POL 0.7418 WITH N (50) WTOP SIG .001	POL 0.8208 WITH N (50) WTOR SIG .001	ROM 0.5492 WITH N (45) YUG SIG .001	ROM 0.3581 WITH N (13) ALB SIG .066	ROM 0.3730 WITH N (48) WTO SIG .001
ROM 0.4014 WITH N (48) WTOS SIG .001	ROM 0.3816 WITH N (48) WTOB SIG .001	ROM 0.3581 WITH N (48) WTOC SIG .001	ROM 0.3749 WITH N (48) WTOG SIG .001	ROM 0.4024 WITH N (48) WTOH SIG .001	ROM 0.3868 WITH N (48) WTOP SIG .001
ROM 0.2478 WITH N (48) WTOR SIG .007	YUG 0.4970 WITH N (14) ALB SIG .015	YUG 0.1250 WITH N (50) WTO SIG .105	YUG 0.1478 WITH N (50) WTOS SIG .070	YUG 0.1343 WITH N (50) WTOB SIG .090	YUG 0.0599 WITH N (50) WTOC SIG .275

Table III.5 Continued

Variable Pair	Variable Pair	Variable Pair	Variable Pair	Variable Pair	Variable Pair
YUG 0.1334 WITH N (50) WTOG SIG .091	YUG 0.1319 WITH N (50) WTOH SIG .094	YUG 0.1343 WITH N (50) WTOP SIG .090	YUG 0.0948 WITH N (50) WTOR SIG .171	ALB 0.1527 WITH N (14) WTO SIG .247	ALB 0.2083 WITH N (14) WTOS SIG .176
ALB 0.2083 WITH N (14) WTOB SIG .176	ALB 0.0698 WITH N (14) WTOC SIG .378	ALB 0.1805 WITH N (14) WTOG SIG .210	ALB 0.1527 WITH N (14) WTOH SIG .247	ALB 0.1805 WITH N (14) WTOP SIG .210	ALB 0.0694 WITH N (14) WTOR SIG .378
WTO 0.9651 WITH N (57) WTOS SIG .001	WTO 0.9724 WITH N (57) WTOB SIG .001	WTO 0.9030 WITH N (57) WTOC SIG .001	WTO 0.9743 WITH N (57) WTOG SIG .001	WTO 0.9680 WITH N (57) WTOH SIG .001	WTO 0.9702 WITH N (57) WTOP SIG .001
WTO 0.8901 WITH N (57) WTOR SIG .001	WTOS 0.9595 WITH N (57) WTOB SIG .001	WTOS 0.8831 WITH N (57) WTOC SIG .001	WTOS 0.9576 WITH N (57) WTOG SIG .001	WTOS 0.9526 WITH N (57) WTOH SIG .001	WTOS 0.9585 WITH N (57) WTOP SIG .001
WTOS 0.8672 WITH N (57) WTOR SIG .001	WTOB 0.8865 WITH N (57) WTOC SIG .001	WTOE 0.9623 WITH N (57) WTOG SIG .001	WTOB 0.9566 WITH N (57) WTOH SIG .001	WTOB 0.9532 WITH N (57) WTOP SIG .001	WTOB 0.8769 WITH N (57) WTOR SIG .001
WTOC 0.8872 WITH N (57) WTOG SIG .001	WTOC 0.8859 WITH N (57) WTOH SIG .001	WTOC 0.8906 WITH N (57) WTOP SIG .001	WTOC 0.8362 WITH N (57) WTOR SIG .001	WTOG 0.9497 WITH N (57) WTOH SIG .001	WTOG 0.9513 WITH N (57) WTOP SIG .001
WTOG 0.8776 WITH N (57) WTOR SIG .001	WTOH 0.9469 WITH N (57) WTOP SIG .001	WTOH 0.8631 WITH N (57) WTOR SIG .001	WTOP 0.8747 WITH N (57) WTOE SIG .001		

strongest between the Warsaw Pact mean and the Soviet Union, Bulgaria, the GDR, Hungary and Poland, with all, in addition, retaining strong relationships with the Pact when their own scores are not part of the mean. The Czech scores are low for all members of the Pact—especially with the Soviet Union (.33)—and highest with Romania (.50). The Czech - WTO correlation, a respectable .56, drops 18% (to .46) when its own scores are excluded from the mean. The same is true for Romania, whose already low correlation with the Warsaw pact (.37) is reduced even further (to .25) when its own score is factored out. Romania's reactions are in general more similar to those of Yugoslavia than to any Pact member. A further interesting point is that the correlations of the five most strongly similar Pact members all improve even further when their reactions are tested against the variables WTOC (WTO-without-Czechoslovakia) and WTOR (WTO-without-Romania).

It is important to try to gain further detail in comparing the responses of these countries to world events. One would expect, for example, that the low Czech-Soviet correlation was created almost exclusively by their diverging reactions (to say the least) to the Czech events. In addition, it is important to see over what issue areas the Romanian-Soviet diversity occurred and in what areas their reactions were less divergent or similar. And even some of the staunchest allies may have varied somewhat in distinct issue areas, though their overall correlations were high. Thus at this point it is useful to separate the fifty-seven cases into distinct issue areas and statistically compare reactions in those areas.

Correlation would not be appropriate in this instance due to the relatively low number of cases which will be making up each issue group. Therefore, a similar, though less intuitively inferential measure will be utilized.

Figures III.2-15 plot the reactions of each of the countries, of the Pact as a whole (WTO) and of the Pact-minus-Romania (WTOR), for fourteen different issue groups or areas. The scores are the mean of the countries' reaction scores for the various discrete events making up the issue group. The construction of the issue groups is described below. Thus in Figure III.2 Bulgaria's score of −2.425 is the average of its four separate scores (−2.7, −3.0, −1.3, −2.7) on events AIS 01 through AIS 04. The same is done for the Pact (WTO) and the Pact-Without-Romania (WTOR). The clustering or dispersion of the states on the various issue

REACTIONS TO ARAB-ISRAEL ISSUE GROUP (AIS)

Fig. III.2

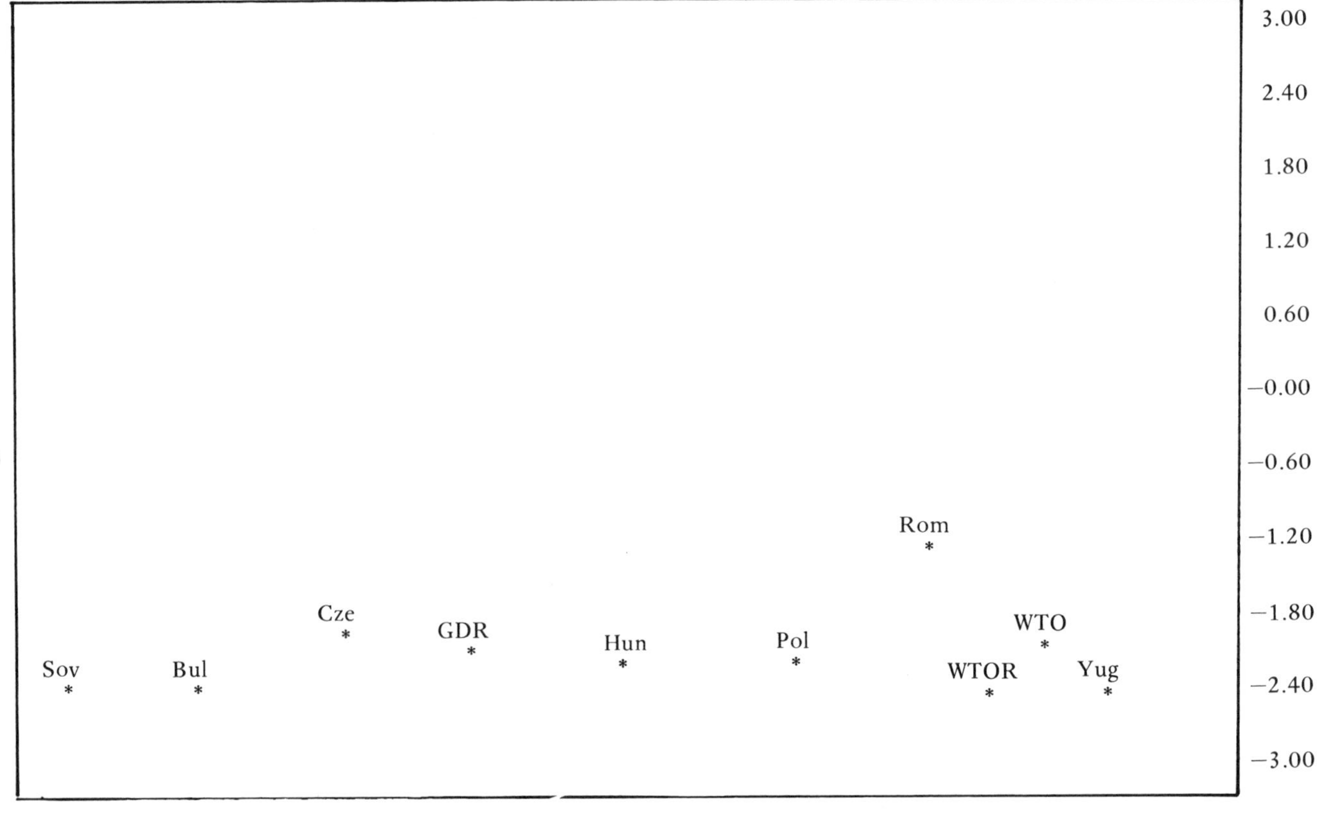

REACTIONS TO CCP CONFERENCES (CPRCO)

Fig. III.3

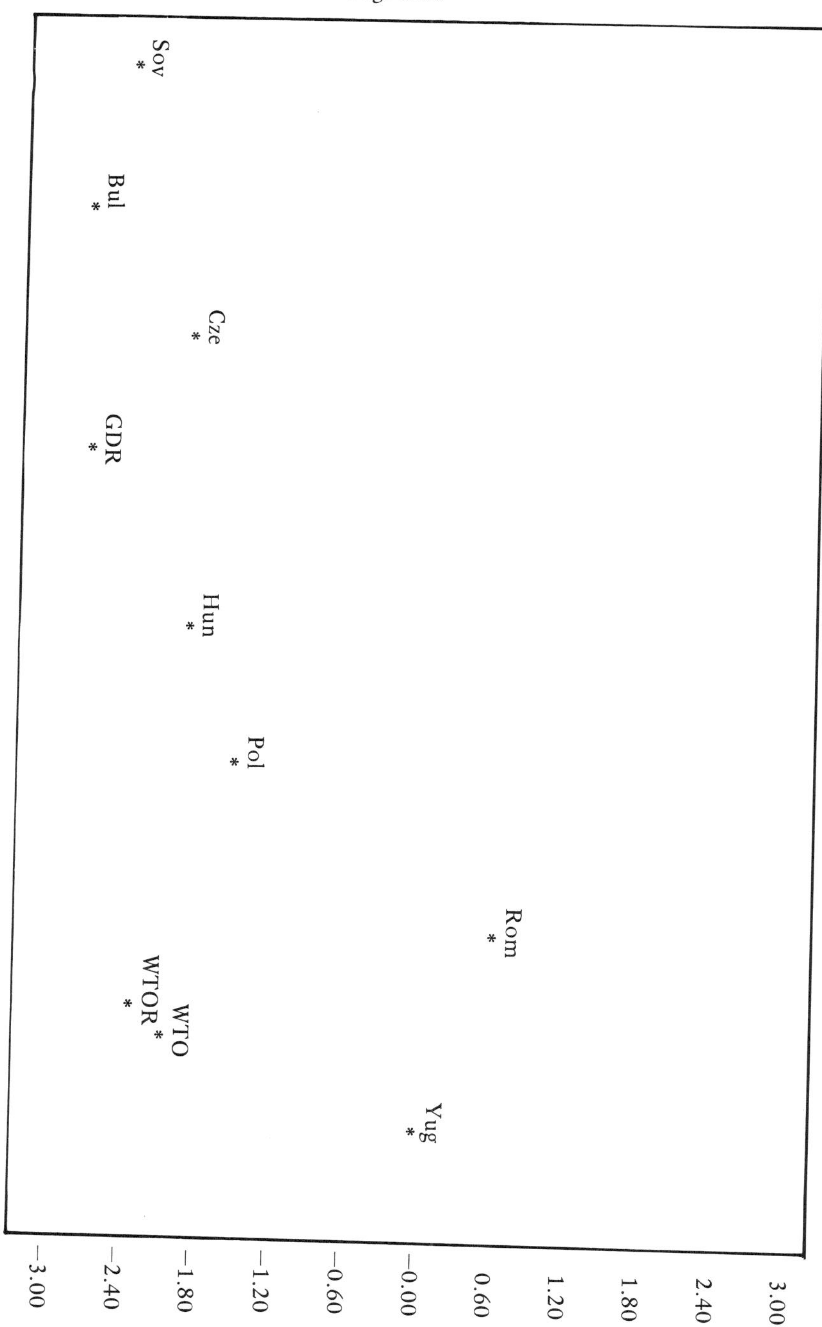

REACTIONS TO SINO-SOVIET BORDER CLASHES (CPRBR)

Fig. III.4

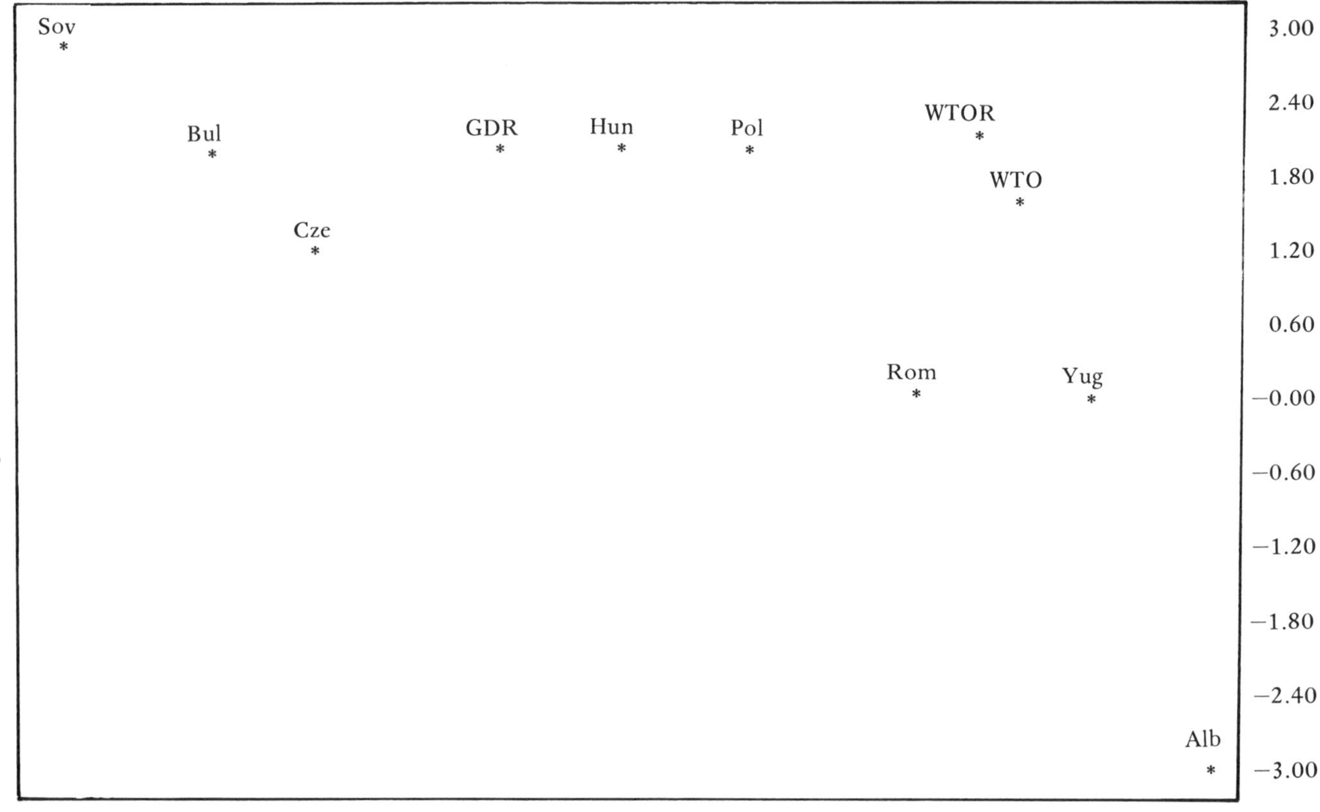

REACTIONS TO CZECH EVENTS (CZEEV)

Fig. III.5

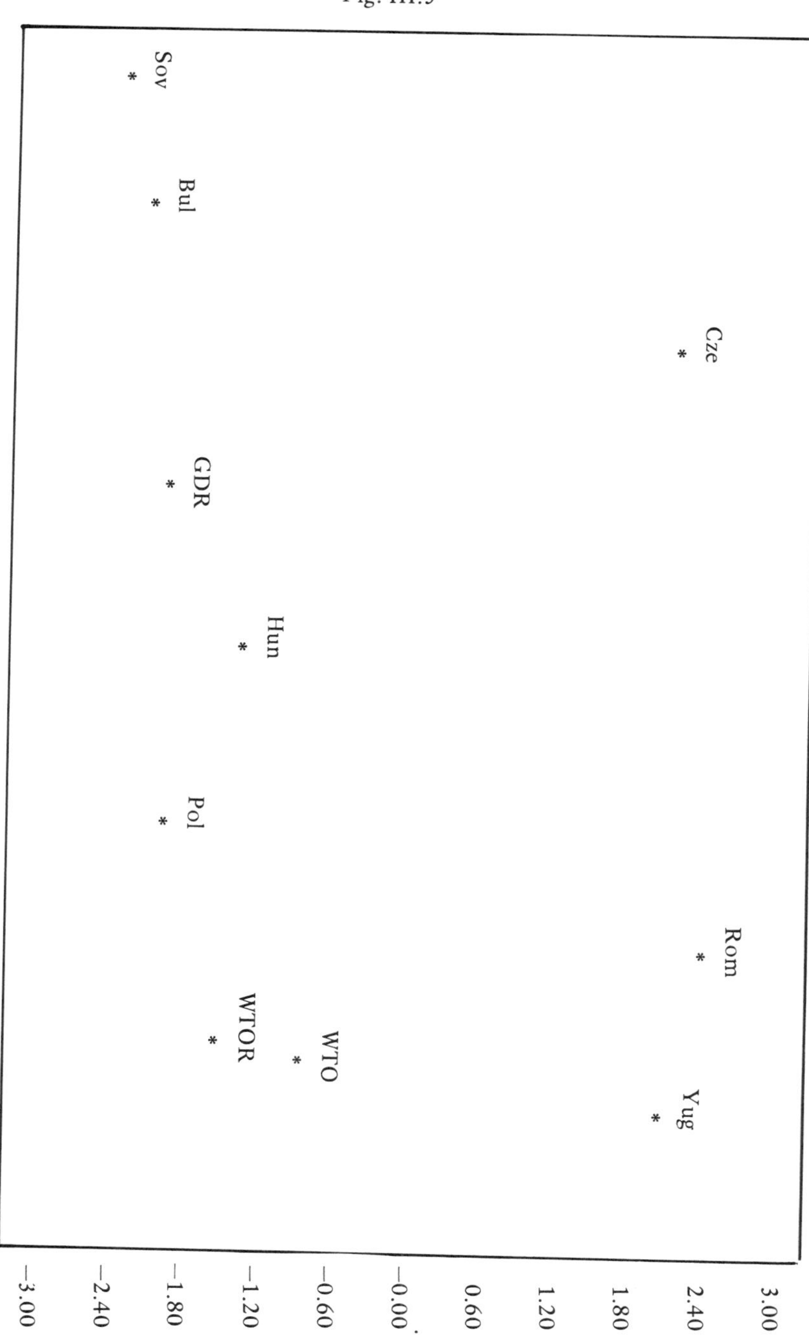

REACTIONS TO ACTIONS TAKEN RE: CZECHOSLOVAKIA (CZEAC)

Fig. III.6

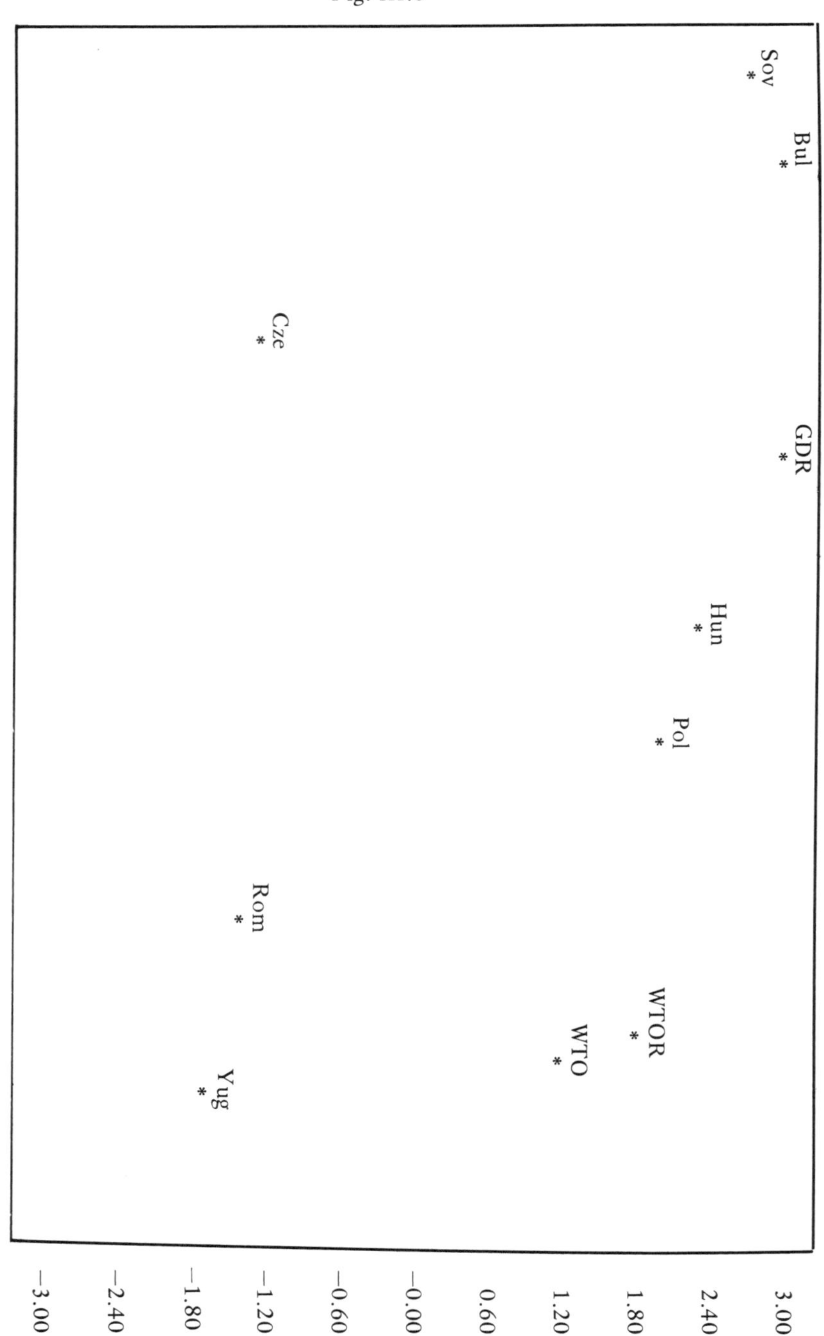

REACTIONS TO EUROPE ISSUE GROUP (EUR)

Fig. III.7

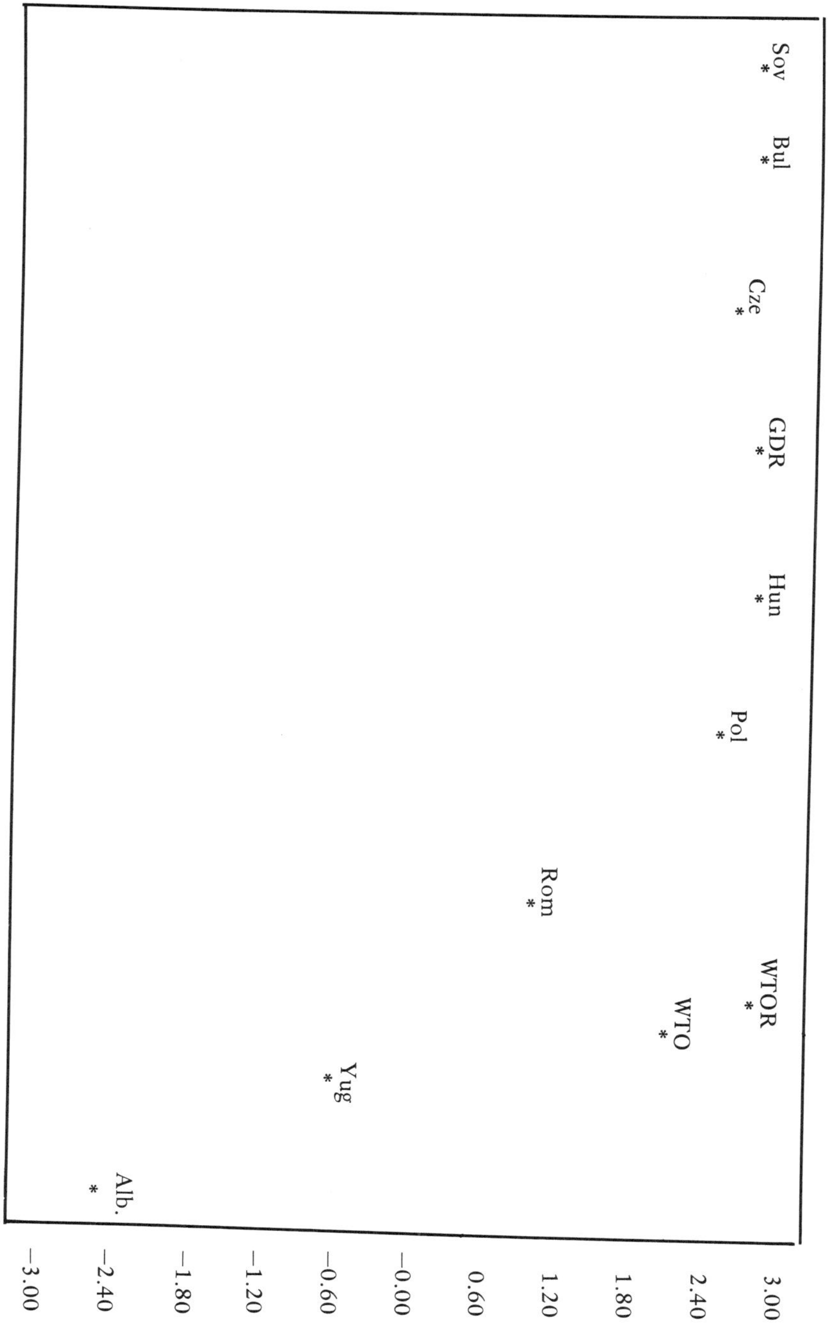

Sov *
Bul *
Cze *
GDR *
Hun *
Pol *
Rom *
WTOR *
WTO *
Yug *
Alb. *

3.00 2.40 1.80 1.20 0.60 –0.00 –0.60 –1.20 –1.80 –2.40 –3.00

REACTIONS TO FRG NOTE (FRG)

Fig. III.8

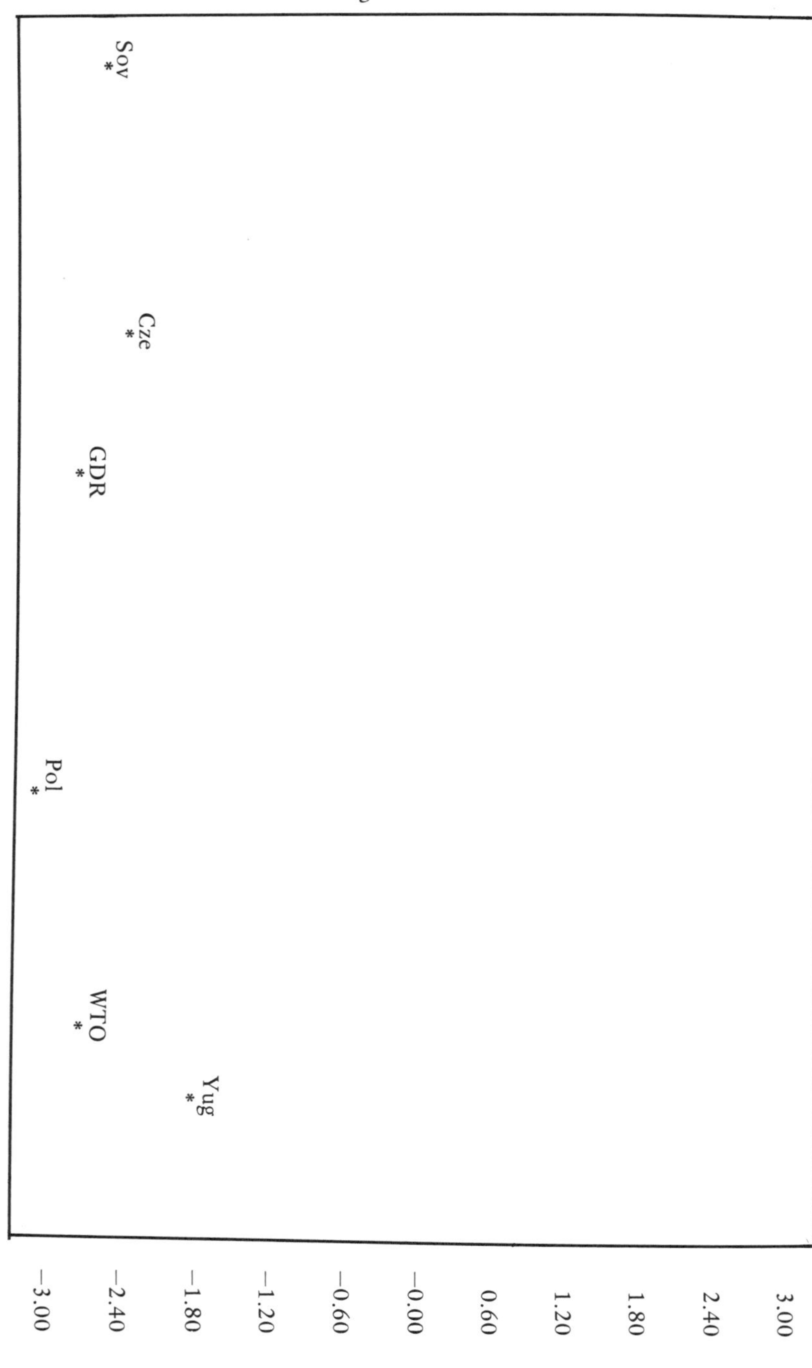

REACTIONS TO WORLD COMMUNIST MOVEMENT–CONFERENCES (WCMCF)

Fig. III.9

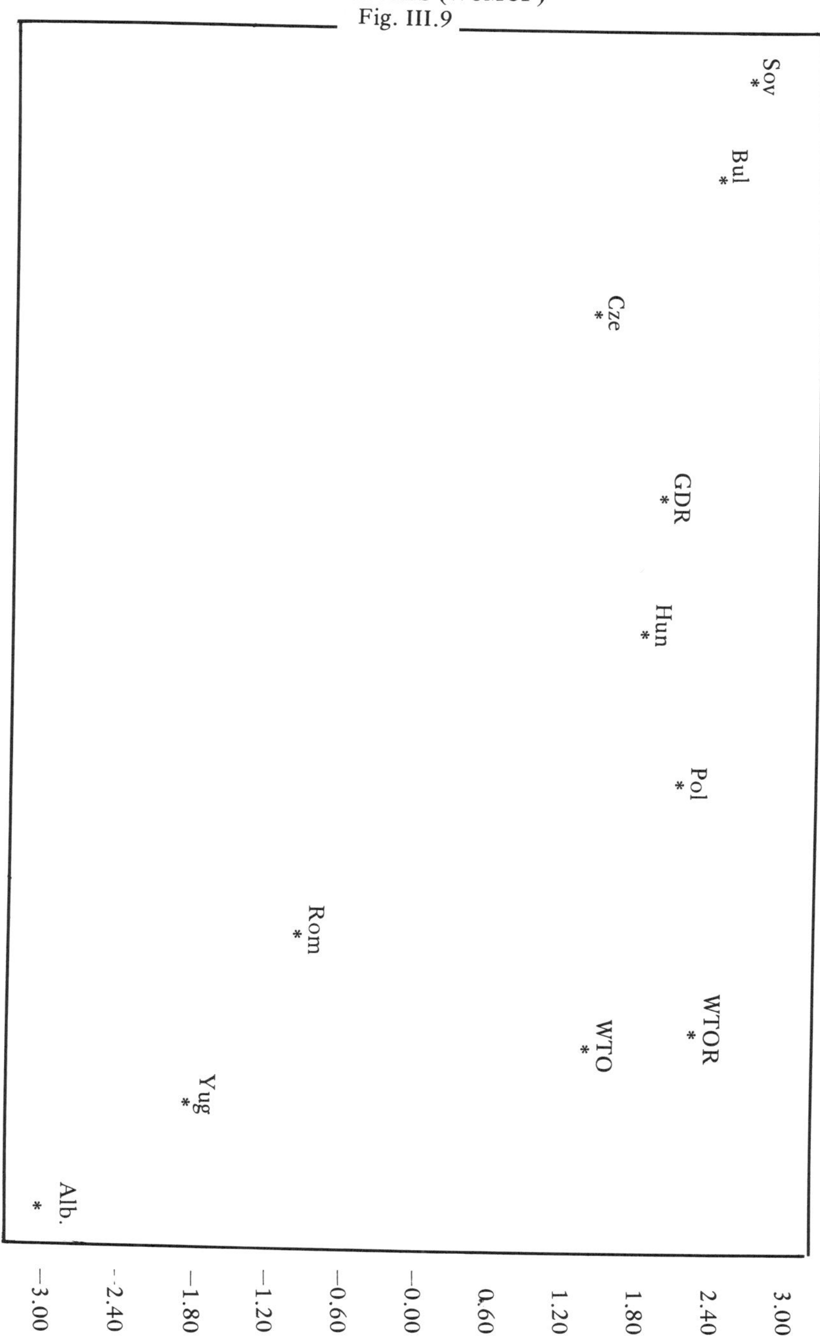

REACTIONS TO ROMANIAN WALKOUT FROM CONSULTATIVE CONFERENCE (WCMR)

Fig. III.10

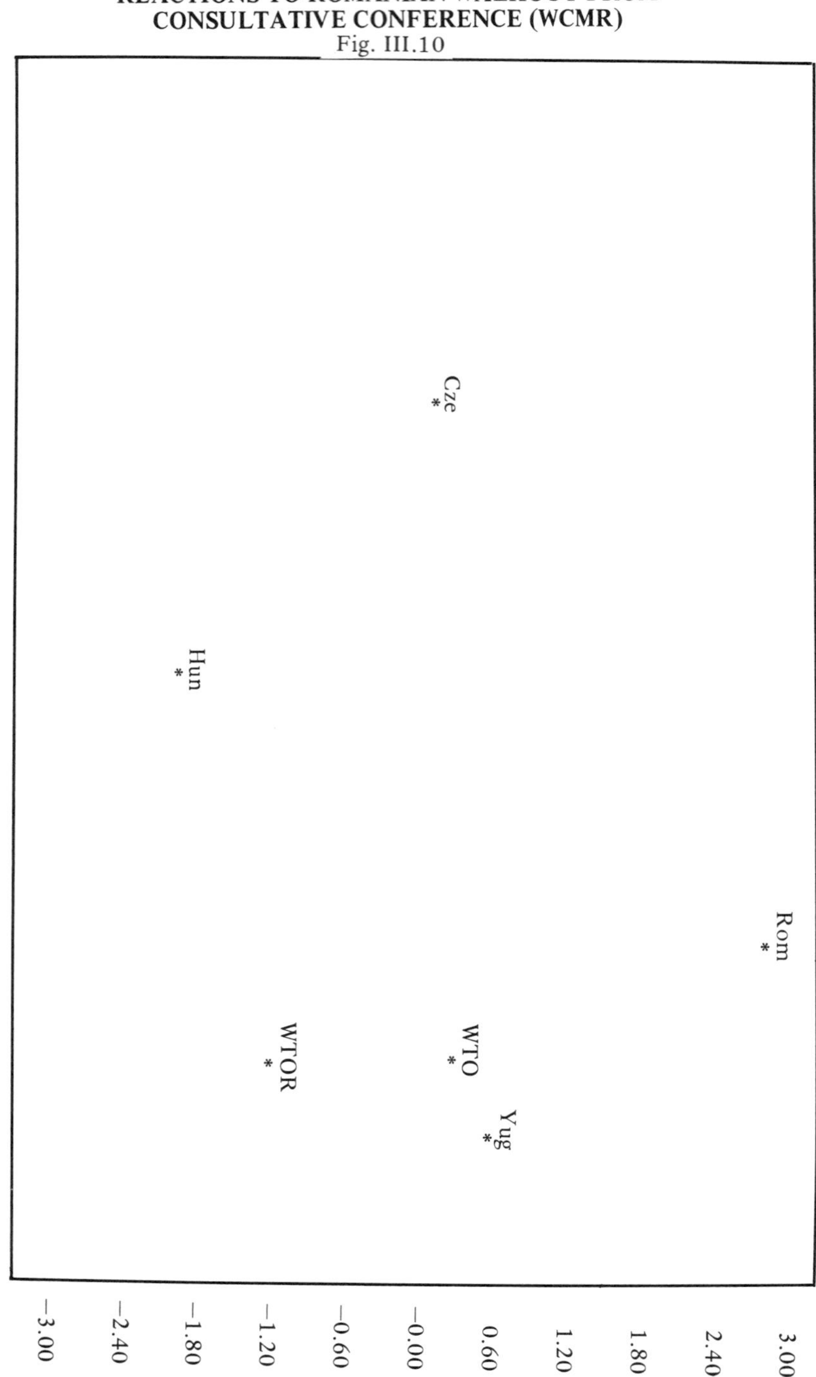

REACTIONS TO UNITED STATES' ACTIONS IN VIETNAM (USVAC)

Fig. III.11

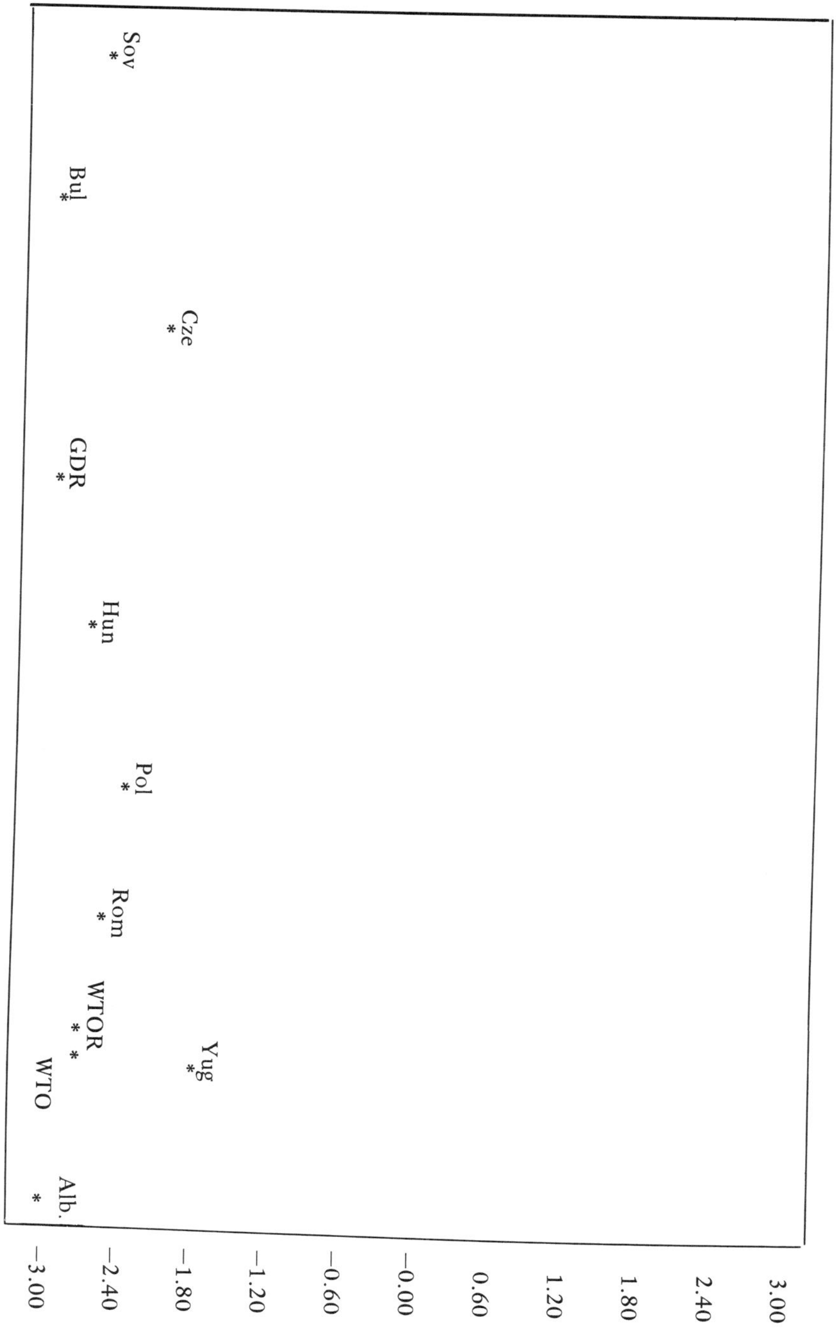

Sov
Bul
Cze
GDR
Hun
Pol
Rom
WTOR
Yug
WTO
Alb.

−3.00 −2.40 −1.80 −1.20 −0.60 −0.00 0.60 1.20 1.80 2.40 3.00

REACTIONS TO OTHER VIETNAM-RELATED EVENTS (USVO)

Fig. III.12

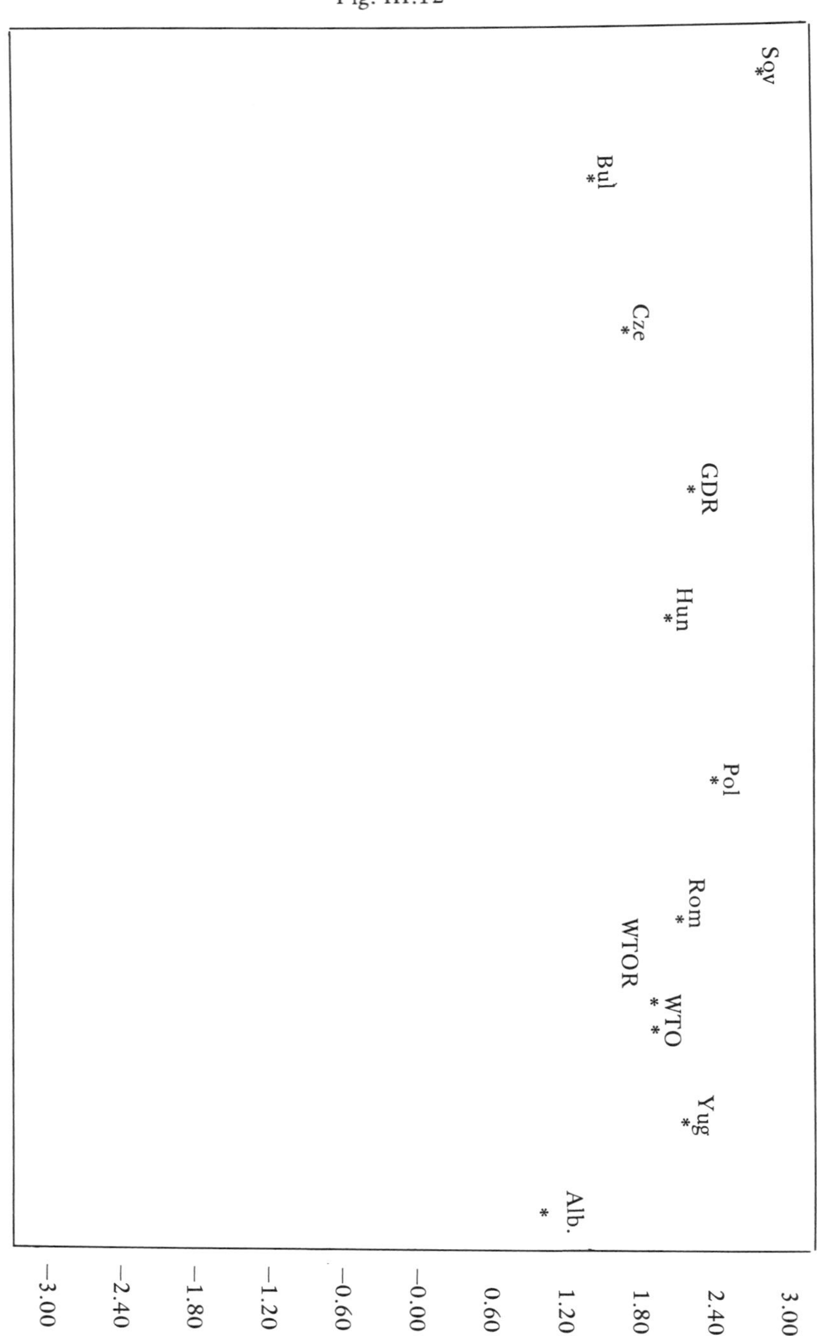

REACTIONS TO U.S. PRESIDENT JOHNSON (USLBJ)

Fig. III.13

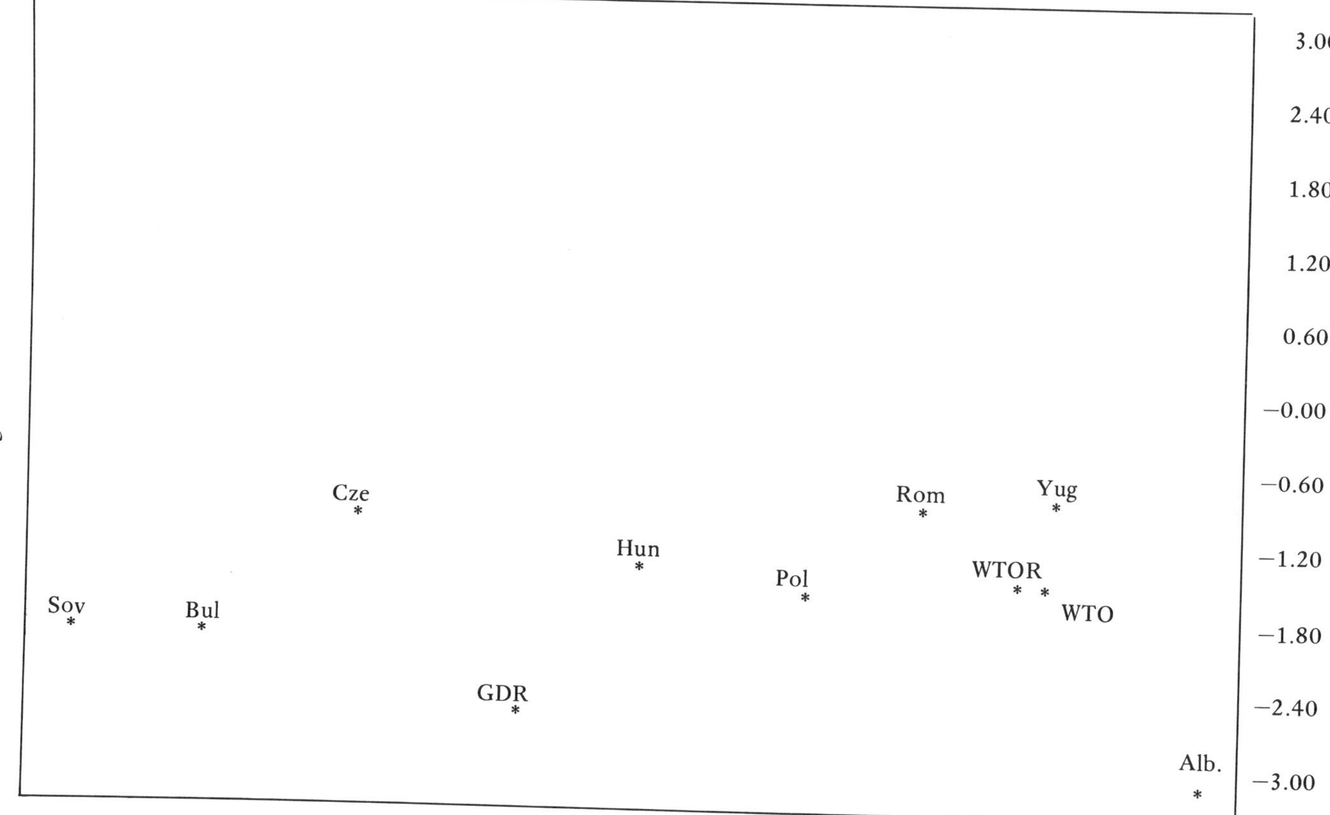

REACTIONS TO SEIZURE OF USS *PUEBLO* (USPUEB)

Fig. III.14

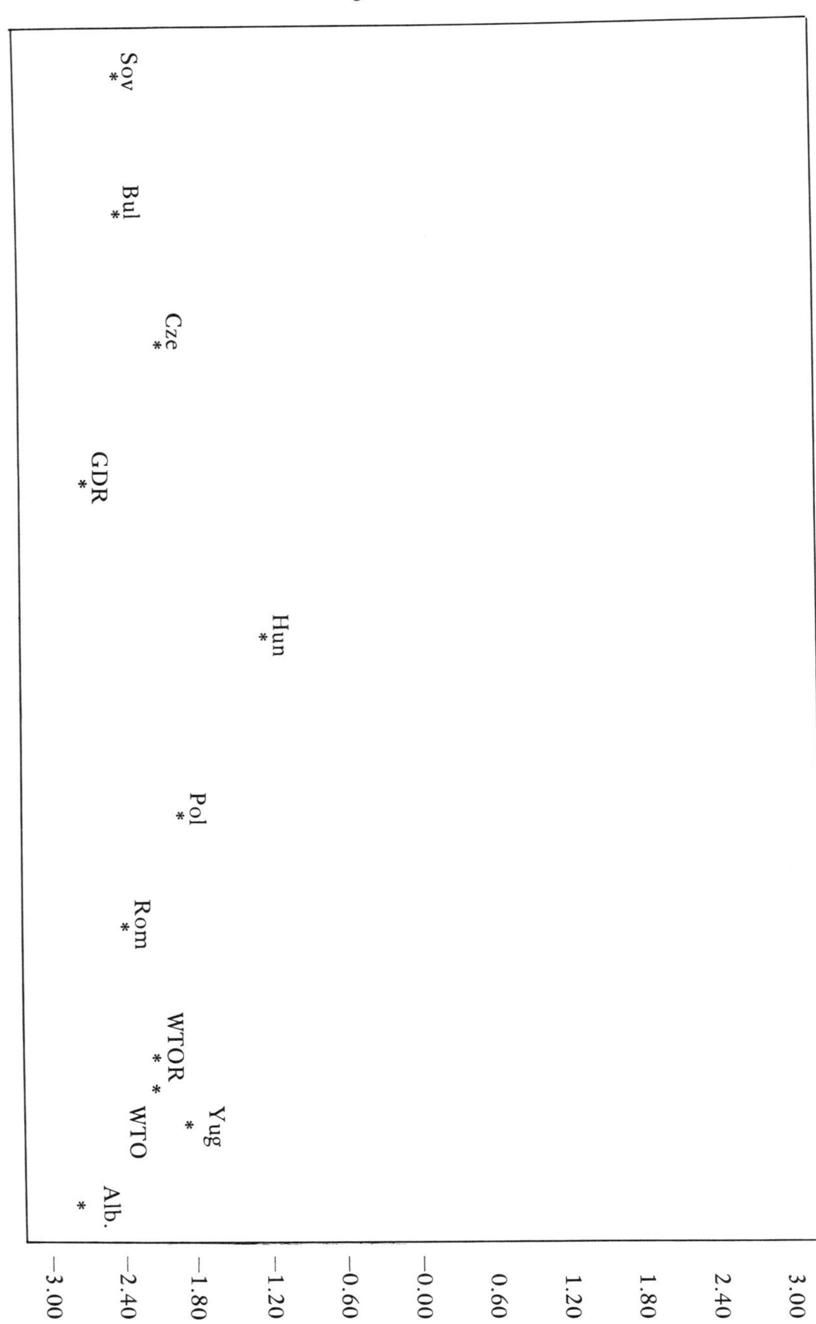

REACTIONS TO U.S. PRESIDENT NIXON (USNIX)

Fig. III.15

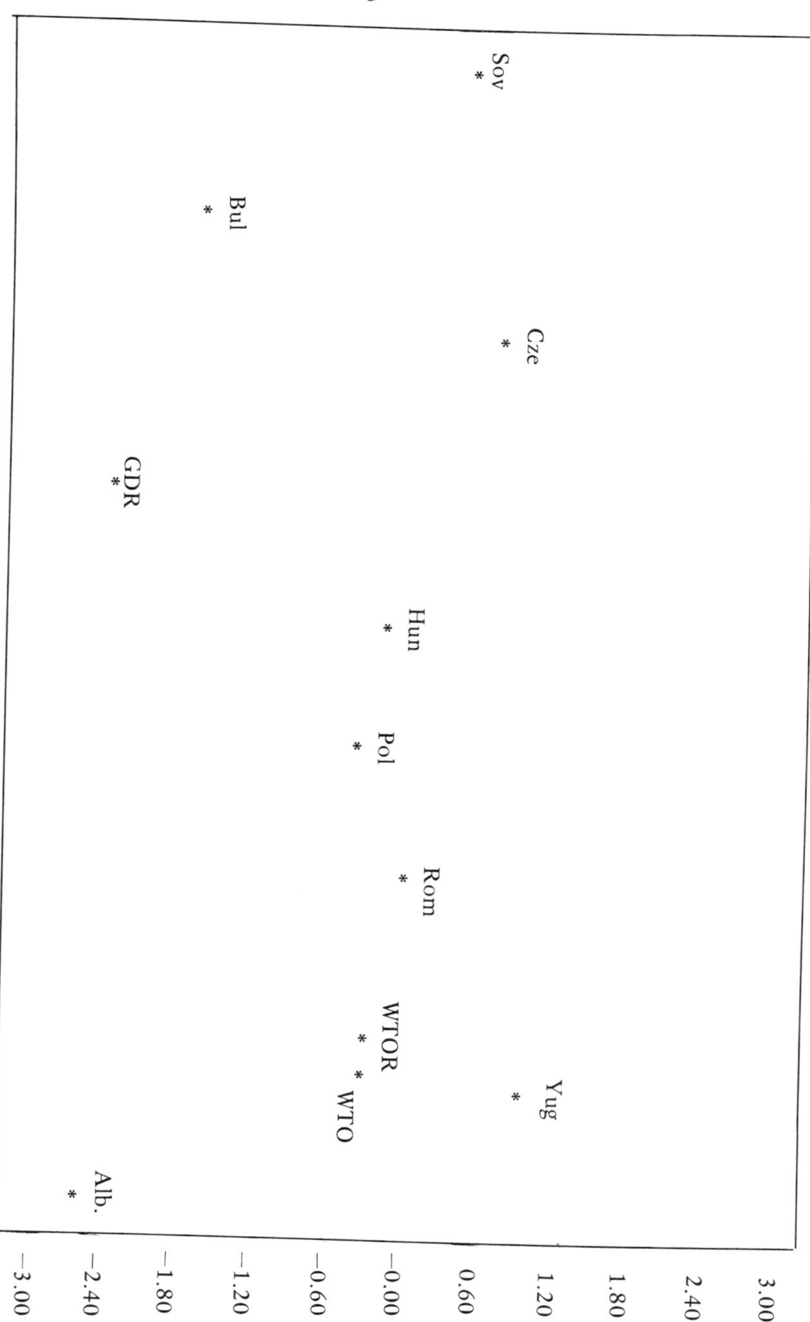

areas is readily evident from these graphs (and will be considered further below). At this point we wish to determine, for each issue group the distance of each country's score from the WTO mean score, and from the WTO with the respective country's score excluded.[200] For the Soviet Union, for example, these scores will be variables DEVSOV and DISSOV respectively, in Tables III.6-19 which list the scores computed for each issue group.[201]

The grouping of the issue areas from the discrete events is straight forward and seriatim in some cases, more complicated and selective in others. The nature of each will be explained as it is taken up.

Arab-Israel

The Arab-Israel issue area is made of events AIS 01 through AIS 04. The mean scores for all countries is listed in Table III.6. Notable is Romania's relatively mild −1.23 compared to the others and to the WTO mean of −2.27. When the Pact score is calculated without the Romanians this mean is −2.390. Romania has the greatest deviation from the Pact (.794) and the GDR the least (.000). With their own scores factored out, the latter's distance is still zero, and Romania's distance is even greater (.953).

China

Events CPR 11, CPR 13 and CPR 14 make up the first issue area for China, comparing reactions to the two CCP gatherings. Overall reactions, as Table III.7 shows, were harshest in Bulgaria and the GDR, with Yugoslavia and Romania being the mildest. The latter either published no reaction or roughly neutral ones. Consequently the WTO mean of −2.07 is made still harsher (to −2.33) when the Romanian score is excluded. The Czechs deviate the least from both of these means, with the Romanians, of course, the furthest away from both. The scores for Poland (DEVPOL = .666 and DISPOL = .817) reflect that country's relatively mild reaction to the Chinese activities (a mean of −1.40).

The second issue area for China is the border clashes of March, 1969 (CPR 12), scores for which range from +3, support for the Soviet Union to −3, support for China. (See Table III.8.) As this group is based only on the one case, interpretation must be cautious, but the Romanian distance is clear, as is the Soviet. This latter score reflects the fact that

TABLE III.6
REACTIONS FOR ISSUE GROUP AIS: ARAB-ISRAELI CONFLICT

Variable	Cases	Mean
SOV	4	–2.5000
BUL	4	–2.4250
CZE	4	–2.1750
GDR	3	–2.3333
HUN	4	–2.4250
POL	4	–2.4250
ROM	3	–1.2333
YUG	4	–2.3250
ALB	0	******
WTO	4	–2.2708
WTOS	4	–2.2250
WTOB	4	–2.2400
WTOC	4	–2.2900
WTOG	4	–2.2708
WTOH	4	–2.2400
WTOP	4	–2.2400
WTOR	4	–2.3900
DEVSOV	4	0.2292
DEVBUL	4	0.1542
DEVCZE	4	0.0958
DEVGDR	3	0.0000
DEVHUN	4	0.1542
DEVPOL	4	0.1542
DEVROM	3	0.7944
DEVYUG	4	+0.0542
DEVALB	0	******
DISSOV	4	+0.2750
DISBUL	4	+0.1850
DISCZE	4	–0.1150
DISGDR	3	0.0000
DISHUN	4	+0.1850
DISPOL	4	+0.1850
DISROM	3	–0.9533

TABLE III.7
REACTIONS FOR ISSUE GROUP CPRCO:CCP CONFERENCES

Variable	Cases	Mean
SOV	3	–2.4667
BUL	2	–2.8500
CZE	1	–2.0000
GDR	2	–2.8500
HUN	3	–2.0000
POL	3	–1.4000
ROM	1	0.7000
YUG	1	0.0000
ALB	0	******
WTO	3	–2.0656
WTOS	3	–1.9661
WTOB	3	–2.0044
WTOC	3	–2.0733
WTOG	3	–1.8883
WTOH	3	–2.0794
WTOP	3	–2.2167
WTOR	3	–2.2306
DEVSOV	3	0.5456
DEVBUL	2	0.3917
DEVCZE	1	0.1167
DEVGDR	2	1.1517
DEVHUN	3	0.3211
DEVPOL	3	0.6656
DEVROM	1	1.9800
DEVYUG	1	–1.2800
DEVALB	0	******
DISSOV	3	+0.6828
DISBUL	2	+0.4833
DISCZE	1	–0.1400
DISGDR	2	+1.4175
DISHUN	3	–0.4039
DISPOL	3	–0.8167
DISROM	1	–2.4750

TABLE III.8

REACTIONS TO SINO-SOVIET BORDER CLASHES (CPRBR)

Variable	Cases	Mean
SOV	1	3.0000
BUL	1	2.0000
CZE	1	1.3000
GDR	1	2.0000
HUN	1	2.0000
POL	1	2.0000
ROM	1	0.0000
YUG	1	0.0000
ALB	1	−3.0000
WTO	1	1.7571
WTOS	1	1.5500
WTOB	1	1.7167
WTOC	1	1.8333
WTOG	1	1.7167
WTOH	1	1.7167
WTOP	1	1.7167
WTOR	1	2.0500
DEVSOV	1	1.2429
DEVBUL	1	0.2429
DEVCZE	1	0.4571
DEVGDR	1	0.2429
DEVHUN	1	0.2429
DEVPOL	1	0.2429
DEVROM	1	1.7571
DEVYUG	1	−1.7571
DEVALB	1	4.7571
DISSOV	1	+1.4500
DISBUL	1	+0.2833
DISCZE	1	−0.5333
DISGDR	1	+0.2833
DISHUN	1	+0.2833
DISPOL	1	+0.2833
DISROM	1	−2.0500

TABLE III.9
REACTIONS FOR ISSUE GROUP CZEEV: CZECH EVENTS

Variable	Cases	Mean
SOV	6	−2.3167
BUL	4	−1.9750
CZE	5	2.1400
GDR	5	−1.8400
HUN	6	−1.2667
POL	5	−1.8200
ROM	5	2.4000
YUG	5	2.2600
ALB	0	******
WTO	6	−0.7413
WTOS	6	−0.4125
WTOB	6	−0.6117
WTOC	6	−1.1778
WTOG	6	−0.5536
WTOH	6	−0.6475
WTOP	6	−0.5933
WTOR	6	−1.1931
DEVSOV	6	1.5753
DEVBUL	4	1.0530
DEVCZE	5	2.7582
DEVGDR	5	1.0937
DEVHUN	6	0.5453
DEVPOL	5	0.9424
DEVROM	5	2.8296
DEVYUG	5	−2.6896
DEVALB	0	******
DISSOV	6	+1.9042
DISBUL	4	+1.2475
DISCZE	5	−3.3133
DISGDR	5	+1.3190
DISHUN	6	+0.6442
DISPOL	5	+1.1200
DISROM	5	−3.3717

TABLE III.10
REACTIONS FOR ISSUE GROUP CZEAC : ACTIONS TAKEN RE CZECHOSLOVAKIA

Variable	Cases	Mean
SOV	7	2.6286
BUL	5	2.9400
CZE	7	−1.1000
GDR	5	2.9400
HUN	6	2.3333
POL	7	2.0429
ROM	5	−1.2800
YUG	7	−1.6714
ALB	2	−3.0000
WTO	7	1.4019
WTOS	7	1.1607
WTOB	7	1.2590
WTOC	7	1.9226
WTOG	7	1.2590
WTOH	7	1.2905
WTOP	7	1.2833
WTOR	5	1.8799
DEVSOV	7	1.2120
DEVBUL	5	1.3229
DEVCZE	7	2.5165
DEVGDR	5	1.3229
DEVHUN	6	0.8819
DEVPOL	7	0.6492
DEVROM	5	2.4130
DEVYUG	7	−3.0880
DEVALB	2	4.4286
DISSOV	7	+1.4679
DISBUL	5	+1.5433
DISCZE	7	−3.0226
DISGDR	5	+1.5433
DISHUN	6	+1.0417
DISPOL	7	+0.7881
DISROM	5	−3.1599

TABLE III.11
REACTIONS FOR ISSUE GROUP EUR [1]: EUROPE

Variable	Cases	Mean
SOV	6	2.7333
BUL	4	2.6750
CZE	6	2.4000
GDR	5	2.7400
HUN	6	2.6833
POL	5	2.4800
ROM	5	1.1200
YUG	5	–0.5600
ALB	1	–2.7000
WTO	6	2.4151
WTOS	6	2.3411
WTOB	6	2.3833
WTOC	6	2.4050
WTOG	6	2.3728
WTOH	6	2.3494
WTOP	6	2.4106
WTOR	6	2.6433
DEVSOV	6	0.3183
DEVBUL	4	0.2857
DEVCZE	6	0.3278
DEVGDR	5	0.2919
DEVHUN	6	0.2683
DEVPOL	5	0.1690
DEVROM	5	1.2438
DEVYUG	5	–2.9214
DEVALB	1	4.5286
DISSOV	6	+0.3922
DISBUL	4	+0.3333
DISCZE	6	–0.3950
DISGDR	5	+0.3427
DISHUN	6	+0.3339
DISPOL	5	+0.1973
DISROM	5	–1.5187

1.Does not include reactions to FRG note (event EUR 46); see Table III.12

TABLE III.12
REACTIONS TO FRG NOTE
(FRG)

Variable	Cases	Mean
SOV	1	–2.7000
BUL	0	******
CZE	1	–2.3000
GDR	1	–2.7000
HUN	0	******
POL	1	–3.0000
ROM	0	******
YUG	1	–1.7000
ALB	0	******
WTO	1	–2.6750
WTOS	1	–2.6667
WTOB	1	–2.6750
WTOC	1	–2.8000
WTOG	1	–2.6667
WTOH	1	–2.6750
WTOP	1	–2.5667
WTOR	1	–2.6750
DEVSOV	1	0.0250
DEVBUL	0	******
DEVCZE	1	0.3750
DEVGDR	1	0.0250
DEVHUN	0	******
DEVPOL	1	0.3250
DEVROM	0	******
DEVYUG	1	0.9750
DEVALB	0	******
DISSOV	1	0.0333
DISBUL	0	******
DISCZE	1	0.5000
DISGDR	1	0.0333
DISHUN	0	******
DISPOL	1	0.4333
DISROM	0	******

TABLE III.13
REACTIONS FOR ISSUE GROUP WCMCF:
WORLD COMMUNIST MOVEMENT-CONFERENCES

Variable	Cases	Mean
SOV	9	2.6444
BUL	6	2.4167
CZE	8	1.3750
GDR	3	1.9667
HUN	7	1.9143
POL	6	2.2333
ROM	8	−0.9000
YUG	9	−1.8222
ALB	2	−3.0000
WTO	9	1.5715
WTOS	9	1.2837
WTOB	9	1.4617
WTOC	9	1.6178
WTOG	9	1.5485
WTOH	9	1.4820
WTOP	9	1.4844
WTOR	9	2.1222
DEVSOV	9	1.1174
DEVBUL	6	0.7439
DEVCZE	8	0.5754
DEVGDR	3	0.6478
DEVHUN	7	0.5257
DEVPOL	6	0.8117
DEVROM	8	2.4154
DEVYUG	9	−3.3937
DEVALB	2	4.7250
DISSOV	9	+1.4126
DISBUL	6	+0.9103
DISCZE	8	−0.7358
DISGDR	3	+0.7833
DISHUN	7	+0.6688
DISPOL	6	+0.9967
DISROM	8	−3.0350

TABLE III.14
REACTIONS TO ROMANIAN WALKOUT FROM CONSULTATIVE CONFERENCE (WCMCR)

Variable	Cases	Mean
SOV	0	******
BUL	0	******
CZE	1	0.0000
GDR	0	******
HUN	1	−2.0000
POL	0	******
ROM	1	3.0000
YUG	1	0.7000
ALB	0	******
WTO	1	0.3333
WTOS	1	0.3333
WTOB	1	0.3333
WTOC	1	0.5000
WTOG	1	0.3333
WTOH	1	1.5000
WTOP	1	0.3333
WTOR	1	−1.0000
DEVSOV	0	******
DEVBUL	0	******
DEVCZE	1	0.3333
DEVGDR	0	******
DEVHUN	1	2.3333
DEVPOL	0	******
DEVROM	1	2.6667
DEVYUG	1	0.3667
DEVALB	1	******
DISSOV	0	******
DISBUL	0	******
DISCZE	1	0.5000
DISGDR	0	******
DISHUN	1	3.5000
DISPOL	0	******
DISROM	1	4.0000

TABLE III.15
REACTIONS FOR ISSUE GROUP USVAC:
UNITED STATES ACTIONS IN VIETNAM

Variable	Cases	Mean
SOV	7	–2.7286
BUL	6	–2.7833
CZE	7	–2.2000
GDR	6	–2.8167
HUN	7	–2.5857
POL	7	–2.2857
ROM	6	–2.5000
YUG	5	–1.6000
ALB	3	–2.9000
WTO	7	–2.5333
WTOS	7	–2.4957
WTOB	7	–2.5343
WTOC	7	–2.5995
WTOG	7	–2.4771
WTOH	7	–2.5224
WTOP	7	–2.5776
WTOR	7	–2.5267
DEVSOV	7	0.2034
DEVBUL	6	0.1381
DEVCZE	7	0.4027
DEVGDR	6	0.3540
DEVHUN	7	0.2769
DEVPOL	7	0.2762
DEVROM	6	0.1270
DEVYUG	5	–0.8067
DEVALB	3	0.1857
DISSOV	7	+0.2424
DISBUL	6	+0.1622
DISCZE	7	–0.4814
DISGDR	6	+0.4211
DISHUN	7	+0.3271
DISPOL	7	–0.3262
DISROM	6	–0.1489

TABLE III.16
REACTION FOR ISSUE GROUP USVO:
OTHER VIETNAM-RELATED EVENTS

Variable	Cases	Mean
SOV	3	2.6667
BUL	1	1.0000
CZE	3	1.6667
GDR	3	2.2333
HUN	3	2.1333
POL	2	2.5000
ROM	3	2.2333
YUG	2	2.2000
ALB	2	1.0000
WTO	3	2.1556
WTOS	3	2.0533
WTOB	3	2.1811
WTOC	3	2.2533
WTOG	3	2.1400
WTOH	3	2.1600
WTOP	3	2.1611
WTOR	3	2.1400
DEVSOV	3	0.5111
DEVBUL	1	0.3833
DEVCZE	3	0.5889
DEVGDR	3	0.3333
DEVHUN	3	0.3333
DEVPOL	2	0.1083
DEVROM	3	0.3333
DEVYUG	2	+0.9250
DEVALB	2	1.5750
DISSOV	3	+0.6133
DISBUL	1	−0.4600
DISCZE	3	−0.7067
DISGDR	3	+0.4000
DISHUN	3	−0.4000
DISPOL	2	+0.1300
DISROM	3	+0.4000

TABLE III.17
REACTION FOR ISSUE GROUP USLBJ:
U.S. PRESIDENT LYDON JOHNSON

Variable	Cases	Mean
SOV	5	−1.9400
BUL	2	−2.0000
CZE	5	−0.7200
GDR	3	−2.4667
HUN	5	−1.1200
POL	5	−1.2600
ROM	4	−0.7500
YUG	5	−0.7400
ALB	1	−3.0000
WTO	5	−1.3440
WTOS	5	−1.2060
WTOB	5	−1.3257
WTOC	5	−1.4640
WTOG	5	−1.2017
WTOH	5	−1.4153
WTOP	5	−1.3587
WTOR	5	−1.4370
DEVSOV	5	0.5960
DEVBUL	2	0.2560
DEVCZE	5	0.7274
DEVGDR	3	1.2405
DEVHUN	5	0.5926
DEVPOL	5	0.1874
DEVROM	4	0.6238
DEVYUG	5	−0.6440
DEVALB	1	1.7750
DISSOV	5	+0.7340
DISBUL	2	+0.3067
DISCZE	5	−0.8720
DISGDR	3	+1.4778
DISHUN	5	−0.7340
DISPOL	5	−0.2267
DISROM	4	−0.7400

TABLE III.18
REACTIONS TO SEIZURE OF USS *PUEBLO* (USPUEB)

Variable	Cases	Mean
SOV	1	−2.3000
BUL	1	−2.3000
CZE	1	−2.0000
GDR	1	−2.7000
HUN	1	−1.3000
POL	1	−2.0000
ROM	1	−2.3000
YUG	1	−2.0000
ALB	1	−2.7000
WTO	1	−2.1286
WTOS	1	−2.1000
WTOB	1	−2.1000
WTOC	1	−2.1500
WTOG	1	−2.0333
WTOH	1	−2.2667
WTOP	1	−2.1500
WTOR	1	−2.1000
DEVSOV	1	0.1714
DEVBUL	1	0.1714
DEVCZE	1	0.1286
DEVGDR	1	0.5714
DEVHUN	1	0.8286
DEVPOL	1	0.1286
DEVROM	1	0.1714
DEVYUG	1	−0.1286
DEVALB	1	0.5714
DISSOV	1	+0.2000
DISBUL	1	+0.2000
DISCZE	1	−0.1500
DISGDR	1	+0.6667
DISHUN	1	−0.9667
DISPOL	1	−0.1500
DISROM	1	+0.2000

TABLE III.19
REACTIONS FOR ISSUE GROUP USNIX: U.S. PRESIDENT RICHARD NIXON

Variable	Cases	Mean
SOV	3	–0.7000
BUL	2	–1.3500
CZE	3	0.7667
GDR	1	–2.3000
HUN	3	–0.1000
POL	3	–0.1333
ROM	3	0.1000
YUG	3	0.9667
ALB	1	–2.7000
WTO	3	–0.2465
WTOS	3	–0.1750
WTOB	3	–0.1244
WTOC	3	–0.4256
WTOG	3	–0.1544
WTOH	3	–0.2750
WTOP	3	–0.2761
WTOR	3	–0.2950
DEVSOV	3	0.8402
DEVBUL	2	0.9202
DEVCZE	3	1.0932
DEVGDR	1	1.6571
DEVHUN	3	0.2021
DEVPOL	3	0.3513
DEVROM	3	0.8932
DEVYUG	3	–1.2132
DEVALB	1	2.8200
DISSOV	3	+1.0083
DISBUL	2	+1.1033
DISCZE	3	–1.2922
DISGDR	1	+1.9333
DISHUN	3	–0.2417
DISPOL	3	–0.4206
DISROM	3	–1.0783

the Soviet Union reacted, understandably, more negatively to the clashes than did any of its allies.

Czechoslovakia

Events CZE 21, CZE 22, CZE 24, CZE 25, CZE 27 and CZE 29 make up the "Czech Events" issue area. As seen in Table III.9 the Soviet reaction was on the whole the harshest; no other ally has an average score over −2.0. These scores clearly reveal Pact divergence of views on of the issue. The fears of Poland and East Germany of the events next door, as well as Bulgaria's slavish echoing of Soviet views, produce substantially negative scores. Hungary's somewhat more moderate approach is also represented (−1.267). The Czech score, of course, reflects their own support of the democratization process. That this score does not equal a pure +3 reflects the at times moderate tone of Czech supporting statements, which were an especial attempt to assure but not provoke the Soviets.[202] It is the Romanians and Yugoslavs who are most enthusiastic in support of the Czechs. The mildly negative WTO mean (−.7413) is thus a creation of the diversity of viewpoints on this issue, rather than a consistently and uniformly negative approach. This is borne out by the large deviations and distances for almost all the countries, with the Czechs and Romanians being the most deviant.

Actions taken by the Soviet Union and certain Pact members to deal with the Czech situation form another issue group, "Czech Action," made up of the reactions to events CZE 23, CZE 26, CZE 28 and CZE 30 through CZE 33. The diverging groups here are more clearly polarized. Two of the more interesting points on Table III.10 are the variables WTOC and WTOR. These scores are substantially more positive than the overall WTO score. The large distances of the Yugoslav, Romanian, Czech and Albanian scores from the others reflect the nature of their responses to Pact actions toward Czechoslovakia.

While in these two issue groups the Czechs certainly do differ from the Pact, the Romanians have differed in all five issue groups so far, and the non-Pact Yugoslavs in four. It will be interesting to see if the Czech deviation "spilled over" to its reaction to other world events.

Europe

Table III.11 displays the means scores for the reactions to events EUR 41 through EUR 45 and EUR 47, having to do with proposals for European security and disarmament. The "inner five," alike as usual in their reactions, are joined by the Czechs, and it is only the Romanians who are less enthusiastic about these proposals, as described previously.[203] Consequently, the Pact mean is quite positive, the Pact-without-Romania slightly more so, and the group in general is set quite significantly apart from Yugoslavia and Albania in their responses (see Figure III.7) The deviation scores measure this clustering.

Event EUR 46, the FRG note, produced fewer but all negative responses, with only Belgrade being somewhat less negative (Table III.12)

World Communist Movement

Events WCM 51 through WCM 55, and WCM 57 through WCM 60 measure the parties' reactions to the holding of regional and world communist party conferences. The range of reactions can be seen in Table III.13. The Soviets were overall most enthusiastic, followed by faithful Bulgaria. Somewhat less so, but still favorable were Poland, the GDR, and Hungary, followed by the less positive Czechoslovakia. On the negative side is the familiar trio of Romania, Albania and Yugoslavia. Though the Pact mean is favorable (+1.5715), it is even more so when Romania's scores are factored out (+2.1222). The large Romanian and sizeable Soviet distance from the respective means reflect their differences on the issue from the overall group, the one more negative, the other more positive.

Event WCM 56 (Table III.14) is the Romanian walkout from the February, 1968 consultative conference and the data, as can be seen, is sparse.

United States

The first issue group relating to the United States combines reactions to U.S. actions in Vietnam. Reactions to events USV 71 through USV 73, USV 75, USV 76, USV 78 and USV 79 are averaged and displayed in Table III.15. The group is almost uniformly negative with only the Yugoslavs and to a lesser extent the Czechs, somewhat less so. The Pact mean is a harsh −2.533 and does not change significantly regardless of which countries are excluded. All deviations and distances are low (save pershaps that of Yugoslavia), and for once the Pact seems to have reacted as a Pact.

Similar but not total uniformity is displayed in Table III.16, which combines some positive reactions to events relating to Vietnam: a WTO Appeal (event USV 74); the Tet offensive (USV 77); and the November 1968 bombing halt (USV 80). The only significant departure from general Pact approval of these events was a somewhat less positive Czech reaction (+1.667). (The Bulgarian score of +1.0 was based on the only one of these events for which a Bulgarian reaction was found, the bombing limitation, toward which Sofia was lukewarm.) Referring back to Table III.2 we can see that this is due mostly to a substantially less enthusiastic reaction by Prague to the February, 1968 Tet offensive (+.7 as compared to a WTO-without-Czechoslovakia mean of +2.5). It appears that in these two issue groups pertaining to the United States some movement is evident on the part of Czechoslovakia away from close association with the Soviet position. In both issue groups Czechoslovakia had the highest distances in the Pact and moreover, Prague's positions were substantively in the direction of a moderately less critical view of the United States.

Table III.17 charts the countries' views of President Johnson as scored by their reactions to his four State of the Union messages and his election withdrawal speech (USL 81 through USL 85). The range of reactions here is somewhat greater, though all the states, including Romania and Yugoslavia, yield a negative mean for the issue area. The GDR has the highest mean distance of the Pact members, (a very mean −2.4667). This issue is another instance of Czech deviation from the Pact; Prague has the second highest distance. Referring back to Table III.2 for a closer look, it can be seen that Prague was less negative than the Pact on two of Johnson's messages (1966) and 1967) and more negative on two others (1968 and 1969); however, on Johnson's withdrawal speech (March 1968), the Czech reaction was furthest of all from the Pact mean and was substantially more positive than the only other positive reaction, that of Romania.

Table III.17, reveals another curious phenomenon. Hungary and Poland both reacted overall more mildly, i.e. less negatively, to President Johnson than did either the Soviets or Bulgarians; their means are quite close (Poland, −1.26; Hungary, −1.12). Yet a glance at their distances shows a great difference (DISHUN = .7340, DISPOL = .2267). How could they both have similar means for the issue group and yet have differing mean distances from the Pact? The answer lies in their individual deviation patterns for the issue group. Referring back to Table III.2 we can compare both Hungary's and Poland's deviations for each event in the issue area. We see that while Poland's reactions scores tended to fall very close to the WTO mean on each event (the DEVPOL's are: .029, .3, .183, .075, .35), Hungary's scores were always over or under the WTO mean by larger amounts (DEVHUN: .271, .6, .517, .925, .65). Thus though the two countries' reactions overall averaged out similarly, and indeed they are roughly similar—Poland also scored a 0 once, on Johnson's withdrawal speech, but when the WTO mean was itself 0.35-on this issue Poland's reactions were more consistently close to those of its allies than were the reactions of Hungary.[204]

Table III.18 illustrates the nations' uniformly negative reaction to the USS *Pueblo* incident (USP 86), with only Hungary and the GDR standing out as somewhat less and more negative respectively.

Finally, the issue group relating to President Nixon is indicated in Table III.19, with the reactions ranging from the most negative, East Germany (−2.3), to the most positive, Yugoslavia (+.9667). The WTO average is only slightly negative, reflecting the fact that the Czechs and Romanians were mildly positive in their reactions and Hungary and Poland were only mildly negative. The deviations and distances in this issue group tend to be large, illustrating the dispersal of the group's reactions.

As this review indicates, the region demonstrated a degree of normative integration which varied across issue area. This finding can be related to some existing hypotheses and tests regarding conflict and intra-alliance cohesion and this will be done in the following chapter. For the present we will maintain our focus on the particular patterns of attitudinal deviance exhibited by the individual countries, in order to produce an index of foreign policy deviance for attitudes analogous to that constructed for interactions.

DEVIANCE AND EVALUATED DEVIANCE

Deviance can be considered two different ways in summing up a country's reactions to international issues. On the one hand, pure or absolute deviance can be assessed to determine the degree to which each state differed from the rest, disregarding the direction in which it differed. On the other hand, the direction of this deviance, i.e. whether it is more positive or more negative in alliance terms, is an important parameter of deviance, as noted previously. Thus it will be useful to rank the states first in terms of the degree of their pure differences from the Pact, regardless of direction, and then to reorder these differences from the alliance perspective.

DEVIANCE (ABSOLUTE)

To produce a summary measure of absolute deviance four indicators will be combined for eight countries.[205] 1) The Kendall rank order correlation for all countries with the other pact countries. This indicator will be a rank ordering, lowest to highest, of the countries' correlations with variables WTOS through WTOR respectively, for the Pact members, and with WTO for Yugoslavia (see Table III.5). 2) The mean distance score (variable DISSOV through DISROM and DEVYUG) over all issues, rank ordered, highest to lowest. 3) Distance score rankings for twelve issue areas combined into an overall ranking, highest to lowest.[206] 4) Ranking of issue area z-scores (absolute value) combined into an overall ranking, highest to lowest.[207]

The two different assessments of distance rankings are included to insure the presence of two different important measures of the phenomena. The mean distance score gives an idea of the size of the average distance between one country's reactions and those of the others. Theoretically, though, this score may fail to account for breadth of deviance; that is, the number of issues on which distance was high or low. Thus if one country deviated infrequently but to a great degree its high score would pull the mean up. To balance this, the ranking of distance scores for each issue gives "credit," as it were, to distance scores that may have not been large but still farthest from the Pact, on that particular issue. This taps breadth of deviance.

The z-scores, in addition, standardize these deviations by comparing them to the standard deviation in each issue area.[208] A low z-score would result either when: the country has a high deviation (score minus group mean),[209] but the group is dispersed overall (high standard deviation); or when both the country and the group as a whole are close to the mean (low standard deviation). Conversely, the computation yields a high z-score for the country when its score differs greatly from both the group mean and group's standard deviation.

Finally, the Kendall rank order correlation is included as a measure of overall association of each country's reactions with the others. The four ranks are listed in Table III.21.

In order to determine if it is valid to combine these indicators into an overall index, that is, to be sure they are tapping the same substantive phenomena, it is necessary to determine if they themselves are strongly associated, which one would expect such indicators to be. Computing Spearman rank order correlation for the four indicators, we find them all to be significantly positively related.[210]

We see in Table III.20 that in terms of pure deviance, Romania is clearly the most deviant member of the Warsaw Pact in its international attitudes. It has the highest mean distance from the other Pact members and in fact its mean distance score (1.768) is closer to that of Yugoslavia than to that of any Pact member. Its distance from the other states was the highest of any state in five of twelve issue area groupings and its overall rank order correlation with the WTO states is the lowest in the Pact. Romania's deviance is spread over all issue areas and only in reacting to the United States, especially its actions in Vietnam, did the distance between Romania and the Pact close significantly.

Also as regards Pact countries, Czechoslovakia's high position is of interest. Its low correlation and high distances come substantially from deviation from its allies on the Czech issues and on issues relating to the United States. Czechoslovakia's reaction to the war in Vietnam were somewhat more moderate than those of her allies, as noted, and, in addition, Prague's response to activities by U.S. Presidents Johnson and Nixon were more moderate. Czechoslovakia's low z-score ranking reveals further details on Czech deviance. Prague's reactions differed greatly from the Pact on three issues on which the Pact itself was not well unified (the Czech events and related actions taken, and U.S. President Nixon), on another when the Pact was moderately unified (U.S. President Johnson), and on only one occasion when the Pact was well unified (U.S. actions in Vietnam). Thus only on two issue areas relating to one "target", the United States, did Czechoslovakia really differ from an otherwise unified alliance. As regards the other issues, Czech deviance was consistently present, as the earlier review pointed out and its ranks reveal, but was of a less substantial nature.

TABLE III.20
INDEX OF DEVIANCE (ABSOLUTE)
Indicators

A. Kendal tau (L-->H) [1]

Country	Score	Rank
YUG	(.1250)	1
ROM	(.2478)	2
CZE	(.4641)	3
SOV	(.6606)	4
BUL	(.6787)	5
POL	(.7418)	6
GDR	(.7438)	7
HUN	(.7545)	8

B. Mean Distance (H-->L) [2]

Country	Score	Rank
YUG	(1.954)	1
ROM	(1.768)	2
CZE	(1.170)	3
SOV	(0.926)	4
GDR	(0.828)	5
BUL	(0.683)	6
HUN	(0.592)	7
POL	(0.542)	8

C. Issue Area
Dist. Rnkings (H-->L)

Country	Rank
ROM	1
YUG	2
CZE	3
GDR	4
SOV	5
BUL	6
HUN	7
POL	8

D. Issue Area
Z-score Rnkings (H-->L)

Country	Rank
YUG	1
ROM	2
GDR	3
BUL	4
SOV	5
CZE	6
POL	7
HUN	8

COMPOSITE RANKING

Country	Rank
YUG	1
ROM	2
CZE	3
SOV	4
GDR	5
BUL	6
POL	7
HUN	8

1. Indicates ranking descends from Lowest to Highest Score.
2. Indicates ranking descends from Highest to Lowest Score.

The most consistently nondeviating Pact member in terms of their international attitudes, were Bulgaria, Poland and Hungary (ranks 6, 7, and 8 respectively). Only once did one of these allies score the largest distance from the others in its reactions. This was a somewhat more mild reaction by Budapest to the *Pueblo* affair—a single case and therefore of limited interpretative value. The only other instances of noticeable deviance by one of these three were Poland's less negative reaction to the Chinese party conferences (a mean of −1.4 compared with a Pact-without-Poland mean of −2.216), and Hungary's somewhat more mild reaction to the Czech events.[211]

The ranking of the Soviet Union illustrates the fact that often it is the "leading center" itself that is out of step with its comrades. On three disparate issue areas the Soviet Union's reaction score was the second largest distance from the mean of other Pact members. Moscow was more enthusiastic in urging a world communist parties conference than any of its allies (a Soviet mean of +2.64 vs. a WTO-without-Soviet mean of +1.28).[212] In its reaction to the border clashes with China, it was more supportive of itself than were any of its allies; and it reacted more positively to the Warsaw Pact Appeal and Tet offensive as well. It also reacted more positively to the April, 1968 bombing halt, and in doing so was for once abreast of the Romanians in its international attitude. These differences, which can no doubt be traced to specifically Soviet national—as opposed to Pact—perspectives, add some detail and support to existing suggestions concerning the qualitative differences in the foreign policies of larger states.[213]

For the whole group the clear outlier is Yugoslavia. Its rank on three of the four indicators is the highest (lowest in the case of the Kendall tau).[214] Its mean deviation is easily the largest. However, Yugoslavia did not deviate to such a great extent on all issues. On the Arab-Israeli War, as noted previously, Belgrade's reaction was virtually identical to that of the Pact (a deviation score of .054). This was true also with regard to the *Pueblo* incident (DEVYUG = .128). A less dramatic instance was Belgrade's overall reaction to President Johnson which, along with that of Romania, fell in between the Soviet and Czech reactions, to the more moderate side. In all other instances the Yugoslav reaction was substantially different than that of any Pact member save Romania and, on occasion, Czechoslovakia.

Albania's consistently extreme reactions to all issues insured that not even once were its reactions close to the average of the other states for any issue area.[215] On two occasions, though, (the *Pueblo* incident and the Tet offensive), Albania's reactions were similar to those of certain individual Pact members.

DEVIANCE (EVALUATED)

In order to further characterize the deviance that existed, it is important to specify the direction of that deviance. That is, deviation from the Pact could be in a direction more favorable from the alliance perspective or less favorable. As was done earlier, we will label the distance and z-scores to indicate *positive* and *negative* deviation on the part of the individual countries from their fellow allies.

For example, the Pact mean in the Arab-Israeli issue area was −2.270. Individual country issue area scores that were less negative than that, e.g. Romania, can be said to have a negative value from the alliance perspective. They do not support the alliance position. On the other hand, scores which were more negative, e.g. Bulgaria, are positively evaluated since they overfulfill the norm. Similarly, states which score high positive scores in their support of the Soviet position on the border clashes with China have their distances and z-scores positively evaluated; those who were less positive than average receive a negative evaluation on their scores.

In this way each distance measure and z-score for the twelve issue areas can be evaluated and assigned a positive or negative sign, as was done in Chapter II. The following breakdown occurs. For the issue areas of Arab-Israel (AIS), CCP conferences (CPRCO), Czech events (CZEEV), U.S. action in Vietnam (USVAC), the *Pueblo* incident (USPUEB), and U.S. Presidents Johnson and Nixon (USLBJ and USNIX), the Pact means were negative. Hence for these issue groups less negative scores or positive scores are given a negative evaluation; a more negative score receives a positive evaluation. (Note this is the opposite of the mathematical signs for the scores on these issues.) For the issue areas of Sino-Soviet border clashes (CPRBR), actions toward Czech developments (CZEAC), European peace and security (EUR), world communist party conference (WCMCF), and other Vietnam-related events—the Pact 1969 Appeal, Tet offensive and bombing halt (USVO) the Pact means are positive. Hence over-mean positive scores are positively evaluated for these issues, and under-mean, i.e. less positive or negative scores, are negatively evaluated. (These designations fit the mathematical signs for the z-scores on these issues.)

With the scores thus evaluated, the ranking of seven countries can be retaken on three indicators.[216] The resulting ranking is based not only on amount but direction of deviance—highest, most negatively deviant, to lowest, most positively deviant. Table III.21 shows the ranking of countries on the three new indicators: 1) mean evaluated distance over all issues; most negative to most positive; 2) evaluated distance scores on each issue area, ranked and combined into an overall ranking; most negative to most positive; and 3) evaluated z-scores ranked for each issue area and combined into an overall ranking; most negative to most positive.[217] When the indicators are combined a summary index of evaluated deviance is created.[218]

The changes produced in this evaluated index are not surprising. For Romania, Yugoslavia, and Czechoslovakia their positions stay the same. Romania's deviance, in addition to being consistently large was almost always in a negative direction form the alliance perspective. Nonsupport on China and Israel, nonsupport for a world communist conference or of Pact actions in Czechoslovakia, hesitancy on a European security conference and the NPT are key areas of this negative deviance. Romania's international attitudes on these issues paralleled those of Yugoslavia, with the exception of the Arab-Israeli issue. On this issue and on one other, the positive reactions to some events in Vietnam, Belgrade reacted in a positively evaluated manner. Romania also did so on these particular Vietnam issues and on the *Pueblo* incident.

At the other end we find East Germany consistently overfulfilling the norm, whether it was in a negative or positive direction. Not once did the GDR fail to be more negative or more positive than the Pact mean, whichever was appropriate. It was more harsh than average on the U.S. Presidents and the Chinese party, and was more supportive than the average towards a world communist conference and the actions taken regarding Czechoslovakia. The GDR was closely pursued for the distinction of most positively deviant ally by Bulgaria, who faltered only once (the U.S. bombing halt in Vietnam). The GDR, however, usually scored higher positive distances from the mean (average evaluated distance = +.8823).

TABLE III.22
INDEX OF DEVIANCE (EVALUATED)
Indicators

A. Mean Evaluate Distance

(Neg-->Pos) [1]

Country	Score	Rank
ROM	(−1.4942)	1
YUG	(−1.4119)	2
CZE	(−0.9798)	3
HUN	(+ 0.0615)	4
POL	(+ 0.1467)	5
BUL	(+ 0.5249)	6
GDR	(+ 0.8823)	7

B. Issue Area
Evaluated Distance Rankings
(Neg-->Pos)

Country	Rank
CZE	1
ROM	2
YUG	3
HUN	4
POL	5
BUL	6
GDR	7

C. Issue Area Evaluated
Z-scores Rankings
(Neg-->Pos)

Country	Rank
YUG	1
ROM	2
CZE	3
HUN	4
POL	5
BUL	6
GDR	7

COMPOSITE RANKING

Country	Rank
ROM	1
YUG	2
CZE	3
HUN	4
POL	5
BUL	6
GDR	7

1. Indicates ranking descends from most negative to most positive.

The ranking by direction of deviance also moved Hungary and Poland to the middle, as the states whose reactions generally fell closest to the mean of the others and who, when they did deviate, usually did so to the positive side. It is interesting to note that for both, negative deviations did occur in their reactions to both U.S. Presidents' activities and to the *Pueblo* incident.

Czechoslovakia's position is unchanged when evaluations are added to the rankings and this is not surprising as their substantial deviations in the areas mentioned were obviously in a negative direction from the Pact perspective. Added to these are Prague's minor negative deviation with regard to the world communist conference, the Arab-Israel and the Europe issue areas. But since Czechoslovakia's only instance of positive deviance is a slightly more negative than average reaction to the Chinese party conferences, her overall international attitudes rank high on the negative deviance index.

HYPOTHESES ON ATTITUDES

Armed with a summary index of deviance on international attitudes, it is possible to test some hypotheses as to the factors underlying the phenomena of attitudinal deviance in East Europe, in a manner similar to that executed for interactional deviance.

We should begin by testing those same factors that were tested against interactional deviance. It was shown in Chapter II that those states having a low level of economic development were the most negatively deviant in terms of interactions. This was explained in terms of desires for rapid industrialization combined with a relatively low vulnerablility to external economic dislocations. Will this same relationship hold true for the attitudes expressed during the same period? Table III.22 ranks the states according to deviance in attitudes (most negative to most positive). It also ranks them on GNP per capita. The statistical association between these two ranks (Spearman r) is not significant. Hence we can not say with any certainty that the two phenomena of level of economic development and attitudinal deviance are related.

Table III.22 also ranks the states on cumulative measures of trade/aid dependence and military position (the former adapted from Kintner and Klaiber).[219] By testing for the association of these items with negative attitude deviance we are repeating the procedure used with interactional

TABLE III.22
RANKINGS OF EAST EUROPEAN COUNTRIES

Country	Negative Deviance (Attitudes)	GNP Per Cap.[1]	Trade/Aid Dependence[2]	Military Position[3]
Romania	1	5	3.5	4
Yugoslavia	2	7	7	1
Czech.	3	2	5	3
Hungary	4	3	2	6
Poland	5	4	6	5
Bulgaria	6	6	1	2
GDR	7	1	3.5	7

1. Source: International Bank for Reconstruction and Development, *World Bank Atlas* (Washington: IBRD, 1970). Ranking based on 1968 figures.
2. Country ranked first is most dependent upon the USSR. For items making up this index see p. 235, note 60 and p. 251-52, note 19.
3. Source: Kintner and Klaiber, *op. cit.*, pp. 234-36. Country ranked first is in the most favorable military position vis-a-vis the USSR. Albania, which Kintner and Klaiber include with a ranking of 1, has not been included here and the other countries have been moved up one rank to maintain a 1-7 ordering. For items making up this index see p. 236 note 3.

deviance and hypothesizing for investigative purposes that either of these factors or both are significantly related to international attitudinal deviance. Neither is. Neither Spearman r correlation (deviance vs. trade/aid dependence, and deviance vs. military position) is statistically significant.

Thus something else has been learned about the deviance of these states. Unlike interactions, attitudes are evidently not stimulated in any direct way by needs related to a country's overall economic position. At least one cannot make that generalized statement about East Europe as a whole. Nor, surprisingly, is vulnerability of military position associated with deviance in the international attitudes of these states. This result does give further support to the proposition, noted earlier, that international attitudes and interactions differ qualitatively as well as quantitatively. In addition, of course, it renders untenable region-wide generalizations from the Yugoslav (and Albanian) cases with regard to geographic position and deviance.[220]

OTHER FACTORS

What of other factors? Perhaps owing to the difference in the nature of attitudes and interactions, separate causes need to be discerned for this aspect of international deviance in East Europe.

It has been noted in the literature on East Europe that deviant foreign policy has often been complemented at home not with "liberalization" or decentralization of party power, but with the reverse, a tighter policy at home. This has been offered chiefly in regard to Romania, though Albania would also fit the description.[221] In addition it would seem that the case of Czechoslovakia (1968) exemplifies the opposite situation.[222]

The rationale underlying this association is that deviants in foreign policy need to balance such deviance with internal consolidation. This is necessary, the reasoning goes, to: 1) placate Soviet fears of loss of party control of the domestic situation; 2) insure for the party leadership domestic control of the party and populace as a hedge against possible influence or faction-building attempts by Moscow; 3) create, insure or capitalize upon domestic support for a new foreign policy line. Furthermore, a reasonable inference drawn form the preliminary work of Salmore and Salmore[223] on the effect of regime constraints on foreign policy, would be that a regime engaging in a deviant foreign policy would strive to reduce internal "constraints" on that policy. For these reasons, one would expect, as in the Romanian case, that the party leadership engaging in potentially unsettling foreign policy behavior will need to insure greater rather than less party control over the economic, political and informational processes of the polity.

This can be tested for the group as a whole by using an index of "subsystem autonomy" developed by Triska and Johnson.[224] This index measures the degree of "liberalization" in the states by combining scores on various indicators of autonomy in the economic, political and educational realms of the country. The ranking of the seven states included in our study on this index is given in Table III.23.[225] Our hypothesis in this test is that there will be a negative relationship between degree of negative deviance on international attitudes and degree of subsystem autonomy or liberalization. This hypothesis is not borne out, however, as the Spearman r of .107 is not statistically significant. Generalization from the Romanian case is clearly not appropriate to the rest of the region.

However, there is still quite possibly a linkage between the states' foreign policies and their internal situations, even if it is not the one expected above. Instead of domestic measures, such as press and political restrictions being instituted, as it were, to "protect" a deviant foreign policy, perhaps it is the foreign policy which is pursued to protect the domestic position of the government.

It seems reasonable to assume, along with Salmore and Salmore, that "Leaders of a regime [will] conduct foreign policy so that it maximizes their likelihood of retaining power."[226] In other words, governments use foreign policy "outputs", like other policy outputs, to secure and increase public support.[227] The regimes of East Europe in particular are concerned with creating and increasing the authority, or legitimacy, of their rule, having already secured the monopoly of power which makes their rule possible.[228] They must in various ways convince the governed populace that their rule represents the genuine will of a significant majority (in numbers or world-historical terms) of that populace. Foreign policy can be an extremely useful "resource" in this campaign,[229] as part of a "national performance strategy,"[230] employed to improve their image as the legitimate holders of power.

TABLE III.23
RANKINGS OF EAST EUROPEAN COUNTRIES

Country	Negative Deviance (Attitudes)	Degree of Liberalization [1]	Ethno-linguistic Fractionalization [2]
Romania	1	7	3
Yugoslavia	2	1	1
Czech	3	3	2
Hungary	4	4	5
Poland	5	2	6
Bulgaria	6	6	4
GDR	7	5	7

1. Source: Triska and Johnson, *op. cit.*; for items making up this index see note 24 (below). Country ranked first has greatest degree of "liberalization."
2. Source: *World Handbook*, pp. 271-274.

In East Europe questions of legitimacy have generally involved the degree of autonomy the respective governments have had vis-a-vis the Soviet Union.[231] It is logical to assume that a government perceived as slavishly vassal to direction from Moscow would in certain situations have greater difficulty convincing the populace of their legitimacy than they would if a certain degree of independence were exhibited. For example, the Czech leadership in 1968 found itself in the classic bind between placating Soviet fears and satisfying its domestic constituency. Similarly, the outbrusts of violence, that took place in East Germany in 1953 and Hungary in 1956 were certainly as much anti-Soviet in character as they were anti-regime. A degree of foreign policy independence can be a vital way for the ruling party to respond to real or potential challenges to legitimacy by demonstrating the indigenous origins and results of their decision-making.[232]

This proposition can be tested with regard to deviance on international attitudes in the following manner. If we can assume that those countries having a greater degree of ethnolinguistic heterogeneity will be faced with greater real or potential challenges to the legitimacy of the ruling party, then amount of ethnolinguistic fractionalization operationalizes, albeit somewhat crudely, level of threat to the ruling group.[233]

In Table III.23 the countries are ranked by degree of ethnolinguistic fractionalization.[234] Testing for the correlation between this ranking and that on negative attitudinal deviance, we find a positive Spearman r of .786 (s<.05). This confirms our hypothesis that those states having a greater degree of population heterogenity are also those expressing more negatively deviant foreign policy attitudes.

Our understanding of foreign policy deviance is thus broadened through focusing on deviance in international attitudes. We should try to strengthen both of these conclusions- the one disconfirming a relationship, the other confirming one- by testing them with the earlier Index of Interactional Deviance.

When degree of subsystem autonomy (liberalization) is tested against the rankings of the states on interactional deviance, a similarly non-significant relationship is found. But when degree of fractionalization is compared to interactional deviance, such significance is found (r=.65, s=.06). Thus the relationship (or lack of same) of these two factors to international deviance seems consistent for both types of deviance.

CONCLUSIONS: FACTORS UNDERLYING DEVIANCE

What then are our conclusions about foreign policy deviance in East Europe as a region? Some are clear; others more problematical. Some are positive–showing a relationship; others are negative- disproving one. All offer interesting questions to guide research into individual cases or for further exploration of the region.

First, as must be clear from the lack of duplication of relationships, deviance in international attitudes and deviance in international interactions differ somewhat, both in terms of patterns for the region and for individual countries. This subject is explored further in the next chapter.

One factor, however, was seen to operate significantly in regard to both interactional and attitudinal deviance- that of ethnic fractionalization. Our strongest finding is that high levels of internal vulnerability of the various regimes is positively associated with pursuit of deviant foreign policies by those regimes. This was related in the discussion to the need for legitimacy and its consequent pursuit using a "nationalized" foreign policy vis-a-vis the dominating external power, the Soviet Union.

Second, international interactional deviance, but not attitudinal deviance, was seen to be related to level of economic development. The argument suggests that interactional deviance occurs because of the desire of the less developed states to develop broadly and rapidly and avoid a second class status within East Europe. Therefore one would expect international interactions, e.g. visits, trade agreements, to be deviant, but not necessarily international attitudes—except where such attitudes are related to the economic issue. There is no reason to expect, for example, that because Bulgaria was pursuing contacts with the West, it would stop criticizing United States policy in Vietnam. While as a whole international attitudinal deviance was seen not to be related to economic development, it may be that the attitudes of some countries (such as Romania) did deviate, and that such deviance was indeed related to economic development and the issues derivative therefrom. This is a question appropriate for investigation in a case study format (see Chapter V).

Third, the results show that while in general geographic position is directly related to interactional deviance, it is not related to attitudinal deviance. The inference drawn from this, i.e. that geographic position is only a weak factor operating against deviance, is further supported by the fact that the most actively deviant state in terms of both interaction and attitudes, Romania, lies in a very bad geographic position.

Fourth, we find that, overall, foreign policy deviance is not related to level of trade/aid dependence. This is a curious finding. We see, for example, that while Romania and the GDR were about equally dependent on the Soviet Union, their international attitudes differed diametrically. Similarly Poland, more independent of the Soviet Union in terms of trade than Romania, deviated in its interactions to a much smaller degree. It seems that the function of dependence as a factor limiting foreign policy deviance needs to be assessed in the individual countries with due cognizance taken of: 1) the existence and degree of the state's desire to change the structure of its economic and overall pattern of international ties—the GDR for example saw itself as profiting from its position in Comecon, while Romania did not; and 2) the level of economic development in the country; a higher level making the state more vulnerable to possible Soviet economic counter-measures, and thus acting as an energizer or multiplier to the limiting effects of economic dependency.

This finding on dependence can perhaps be refined somewhat by focusing briefly and specifically on energy dependence. We should be careful to recognize the difference between the situation in 1965 and that of the present in terms of energy price, supply and source.[235] Still, the crucial significance of dependable energy supplies for the progress and growth of the East European economies cannot be gainsaid, even if it received less publicity a decade ago. It does seem likely that foreign policy deviance would be facilitated, if not stimulated, by a natural resource position which allowed for a degree of energy self-sufficiency. A country more dependent upon the Soviet Union as its energy source would presumably be less likely to be deviant, fearing the Soviet displeasure could be translated into direct and serious dislocation of its economy. Conversely, the relatively good resource position of, say, Romania has been noted as important by students of that country's deviance.[236]

For a rudimentary test, we can place in matrix form the energy situations of the East European states as of 1965 (taken from Triska and Johnson)[237] compared with their deviance situation, positive or negative (Table III.24). The only states displaying negative attitudinal or interactional deviance are energy sufficient or surplus ones, while two states which are energy sufficient are average or positive deviants in attitudes (Poland, Bulgaria), and one is in interactions (Czechoslovakia). There are no energy deficient states who are negative deviants. Thus it seems that energy sufficiency is apparently a necessary factor for negative deviance, though not a sufficient one.

Finally, the states overall display no relationship between degree of internaal liberalization and foreign policy deviance. The situation of the number one and two deviants, Yugoslavia and Romania, having glaringly opposite domestic regimes typifies the lack of generalizability of this factor. Internally "tight" states are both deviant and non-deviant e.g. Romania and Bulgaria; and those with more "subsystem autonomy" are also. To wit, Yugoslavia and Poland. Thus this result, as for trade dependence, consists of a disproving of a relationship region-wide.

TABLE III.24
ENERGY DEPENDENCE AND DEVIANCE

	DEVIANCE			
	Attitudinal		Interactional	
ENERGY 1965	Negative	Average or Positive	Negative	Average or Positive
Sufficient or surplus	YUG, ROM CZE	POL, BUL	YUG,ROM BUL,POL	CZE
Deficient	ϕ	HUN, GDR	ϕ	HUN,GDR

IMPLICATIONS FOR THE CASE STUDY

In addition to the bearing of these results on the study of East Europe, the present conclusions also relate to the broader comparative study of foreign policy, and some thoughts on this will be offered in the concluding chapter. Presently, however, our task is somewhat more narrowly focused.

All of these results can be put to the test, in a way, through a case study. That is, by examining the case of one clear deviant with a view toward determining the underlying causes and development of that deviance, the usefulness (and validity) of the results of the region-wide study can be either supported or challenged. The results of one case either conforming to or differing from the system-wide trend would not, of course, indisputably confirm or refute the broader results. Rather, the conclusions adumbrated above provide us with certain expectations, or, more formally, hypotheses, with which to work in the examination of the causes and course of the individual deviant case, in this instance, Romania.

These expectations are:

1) that an important factor in either stimulating or allowing the Romanian deviance will be the regime's threat position vis-a-vis its own population. It is expected that a deviant foreign policy will be seen to have been used by the regime to improve its own legitimacy position domestically.

2) that an important factor will be the issue of economic development. Romania has a low level of economic development (ranking no.5 on GNP per capita of the seven East European states). We would expect to find that a desire to pursue broad, rapid economic development led, at the very least, to greater interactions with non-Soviet countries, and possibly, to the espousal of certain "deviant" foreign policy attitudes as well.

3) that for some reason poor geographic position did not operate against Romanian deviance.

4) that the possession of a sufficient energy or, more broadly, natural resource, base was important in allowing Romanian deviance to occur.

5) Finally, expectations regarding trade dependence and internal liberalization, if drawn strictly from the comparative research, must be open-ended, i.e. no clearly delineated relationship has been suggested by the region-wide comparison. Thus, previously noted expectations, such as the regime protecting its foreign policy deviance through tight internal control, may be retained. In these cases, though, we know in advance that our findings regarding this particular relationship between internal and international policy, and between that policy and dependence, while possibly useful in understanding the Romanian case, will, in the absence of broader proof, be limited to that case.

Of course we may also discover, to our horror, that other factors not included in the region-wide study are significant, even vital, to an understanding of an individual country's deviance. If this is the case, it is hoped that exploration of these factors, combined with a delineation of the effects of those factors already tested, will together produce a richer contextual frame and improve our understanding of the causes and course of foreign policy deviance in East Europe.

Before turning to the case study, however, the next chapter engages in a brief digression to discuss foreign policy deviance and its relation to issue areas for the region and the Pact, and explore the foreign policy consistency of the individual states.

CHAPTER FOUR

THE REGION AND THE PACT
CONSENSUS AND CONSISTENCY

CONSENSUS AND ISSUES

Notable first among the findings is the display of the wide range of foreign policy attitudes held by the states of East Europe. While this may conform to expectations about the whole region, i.e., with Albania and Yugoslavia included, the divergences found within the WTO itself are no less impressive. This is especially so when one considers the nature and seriousness of the issues upon which there was the least consensus.

In terms of individual countries, Czechoslovakia's deviance ranking is largely explained by its divergence on issues relating to the 1968 developments. As discussed, though, some spillover was shown to exist. On the other hand Romania's position as a nonconforming Pact ally is strikingly clear, with the scope and degree of deviance even outstripping non-Pact member Yugoslavia. At the other extreme the East German and Bulgarian regimes seem as fervent in the direction of positive "overfulfillment." And the largely centrist, though occasionally deviating, positions of Poland and Hungary are also evident.

By addressing the question from an issue-area perspective additional details can be discerned relating to the normative integration of the Pact and the region.[1]

As the scatterplots in the previous chapter indicate (Figures III.2-15) certain issues produced a high degree of consensus, others some uniformity with distinct deviants, and some a wide range of reactions. The fourteen issue areas can be roughly grouped according to the degree of consensus produced in the region and within the Pact itself (see Table IV.1). The numbers in parentheses indicate, for the region and Pact, the standared deviation (σ) of the countries' scores for each issue area. Owing to the diversity of communisms represented in the region it is not surprising that the nine countries are least unified in their reactions to issues which deal mostly with relations with other communist parties (EUR excepted). They are more unified, as is the Pact itself, when

TABLE IV.1
ISSUE GROUPS BY DEGREE OF CONSENSUS

REGION

High ($\sigma<.5$)	Medium ($.5<\sigma<1.0$)	Medium to low ($1.0<\sigma<1.5$)	Low ($\sigma>1.5$)
AIS (.369)	USVO (.549)	CPRCO (1.181)	CPRBR (1.654)
FRG (.421)	USLBJ (.748)	USNIX (1.155)	CZEEV (1.918)
USVAC (.367)			CZEAC (2.083)
USPUEB (.382)			EUR (1.771)
			WCMCF (1.931)
			WCMR (1.698)

PACT ONLY

High ($\sigma<.5$)	Medium ($.5<\sigma<1.0$)	Medium to low ($1.0<\sigma<1.5$)	Low ($\sigma>1.5$)
AIS (.446)	CPRBR (.920)	CPRCO (1.233)	CZEEV (2.032)
FRG (.287)	EUR (.581)	WCMCF (1.201)	CZEAC (1.684)
USVAC (.243)	USVO (.563)	USNIX (1.021)	WCMR (2.517)
USPUEB (.435)	USLBJ (.676)		

reacting to nations whose actions are seen as more clearly hostile or threatening (United States and West Germany). Similarly, for the Pact itself a relatively greater degree of consensus is displayed as well on the issue of Sino-Soviet border conflict. This finding conforms to the propositions suggested by Liska and others, and tested by Hopmann, to the effect that greater increases in international threat or conflict produces greater intra-group cohesion.[2]

WORD AND DEED

An issue-area focus also reveals some interesting dimensions of the relative consistency of the foreign policies of these individual states.

Two separate indices of deviance have been constructed: one for attitudes and one for interactions. Negative deviance for interactions, it will be recalled, refers to deviance from the "Average East European State" in a direction considered negative from the alliance perspective. Thus, below average interaction with the Soviet Union is negative deviance; and so is above average interaction with China. Similarly, as regards attitudes, less negative responses to China would be negative deviance, while negative responses to the call for a world communist party conference would also be negative deviance.[3] Rankings of the states on the two indices of negative deviance are presented in Table IV.2.

TABLE IV.2
DEVIANCE ON INTERACTIONS AND ATTITUDES

Interactions (Table II.9)			Attitudes (Table III.21)	
Country	Rank		Country	Rank
ROM	1		ROM	1
YUG	2	Neg	YUG	2
BUL	3	↓	CZE	3
POL	4	↓	HUN	4
CZE	5	↓	POL	5
HUN	6	Pos	BUL	6
GDR	7		GDR	7

As can be seen, the rankings are similar, though not identical ($r = .68$, $s < .05$). The strongest deviates, Romania and Yugoslavia to the negative, the GDR to the positive, are in the same positions on both scales; but the other countries are ranked differently. Especially noticeable is the difference in ranks for Bulgaria (third on interactions, sixth on attitudes). Clearly Bulgaria's international attitudes were not keeping pace with their actively non-Pact interaction patterns. Its behavior was, over all issues, inconsistent.

Looking at individual issue areas, we can compare a state's actions and attitudes across analogous issue categories in order to judge its behavioral consistency; that is, whether its deviance (or nondeviance) on interactions is matched on attitudes for the particular issue area. For each country, then a comparison of its performance will be made in the manner indicated in Table IV.3.

While this comparison is not always made across categories which are strictly equal, they are comparable. The comparison is not intended as a causal linking of certain Pact positions on certain issues with interaction levels toward relevant targets. The determination of this type of connection is the subject of studies other than this one. Rather, it assesses the degree of consistency of behavior of the East European states by asking if a state which supports the Pact on an issue will also support the Pact (by being average or positively deviant) in its interactions with certain related targets.[4] Table IV.3 lists six dimensions of each state's foreign policy behavior and, in the two columns to the right, the attitude and interaction areas to be compared in determining the degree of consistency of foreign policy behavior for each state.

BULGARIA

The most loyal of the Pact allies next to the GDR, Sofia supported the Pact position on eleven of the twelve attitude areas studied. In its interactions, however, it deviated to varying degrees in a negative way on fully half of the interaction targets. On the Middle East dimension Bulgaria's behavior was fully consistent; that is, its attitudes and actions matched. It supported the Pact hard line on Israel, had a low level of interactions (in fact none) with that state, and an above average level of interactions with the Arab states. On China, Bulgaria was also consistent in that it matched its support of Pact condemnation of the CCP and the border clashes with a properly low level of interactions with China. As regards its allies, however, Bulgaria's behavior was more complex. While it certainly supported Pact action on Czechoslovakia and was harshly critical of events there during the spring of 1968, its overall interaction levels with both the Soviet Union and East Europe were below average. In fact, for the latter target Bulgaria, along with Romania, had the lowest percentage of interaction of all the states. This lack of interaction with one's fraternal socialist comrades, combined with a negative deviance on the Other Communist category, also contrasts with Bulgaria's loyal urgings for the convening of a world conference of communist parties. On the Peace and Security dimension, while Bulgaria did see fit to support the Pact

TABLE IV.3
ASPECTS OF FOREIGN POLICY BEHAVIOR TO BE COMPARED FOR CONSISTENCY

Dimension of foreign policy behavior:	Deviance on *attitudes* toward: (issue area)	TO BE COMPARED WITH	Deviance on *interactions* with: (targets)
Middle East	AIS		Arab States Israel
Sino-Soviet	CPRCO CPRBR		CPR
Allies	CZEEV CZEAC		East Europe USSR
Peace & Security	EUR FRG[1]		West FRG
World Communist System	WCMCF		Other Communist East Europe; USSR
United States	USVAC USVO USLBJ USNIX		United States

1. If available.

method of promoting European unity (positive on EUR issues), Sofia nevertheless was busy pursuing such security for itself with a high level of bilateral interactions with the states of the West (negative deviance on the West target group).[5] Finally, Bulgaria was a better than average supporter of the Pact's responses to the United States (positive deviance in three of four issue areas), and matched this with a very low level of interactions with Washington.

Thus in three of six comparisons across comparable categories, Bulgaria's behavior was inconsistent. In these comparisons Bulgaria's fiercely loyal attitude was not matched with a loyal performance in terms of interactions. Each disparity of actions and attitudes contributes to the difference in the position of Bulgaria on the overall deviance rankings.[6]

CZECHOSLOVAKIA

Czechoslovakia's relative positions for attitudes and actions are somewhat the reverse case, i.e. overall positive interaction pattern but negative attitude pattern. This is due mostly to the significantly negative deviant position of the Czechs on all U.S.-related issues, and its definitely negative position vis a vis its allies on the Czech events and actions, both of which contrast with its average or positive performance with the associated interaction targets. The Czechs were, for the most part, consistent in all other areas. Their reaction to the Arab-Israel situation was average, as were their interaction patterns with the relevant states. It had a higher level of interactions with China matched by a somewhat less than average support of the Pact on the China issue. And its support of WTO positions on Europe and West Germany was matched by average or positive IP scores with the West and FRG targets.

THE GDR AND ROMANIA

These two states were the most consistent in their international behavior, though in dramatically opposite directions. The GDR verbally supported the Pact position on every single issue, usually reacting significantly more negatively or more positively than the other states. And its interaction pattern was similar, making it clearly the most positively deviant ally in the group. Romania, on the other hand, differed from the Pact on virtually every issue, and often held a position farther from the rest than did Yugoslavia. In its interactions, as seen earlier, it was a negative deviant toward all but two targets (one being Albania, toward whom it had, like two other allies, a zero IP). These strong opposites in the alliance saw their differences extend to virtually all areas of Pact concern during these years and in their relations with almost all other nations.

HUNGARY

The total of Hungary's evaluated z-scores is almost identical for both attitudes and interactions. In both categories the overwhelming impression is that in terms of foreign policy Hungary typifies the average ally. Its attitudes on all issues were nearly always very close to the Pact average, rarely being much more positive or negative. (Its reaction was negatively deviant only twice, toward both U.S. Presidents.) In its interactions its pattern was the same, notable only twice; more than average interactions with Yugoslavia and less than average with the Arab states. The actional and attitudinal components of Hungary's international behavior were consistent with each other regarding China, the Allies, and Peace and Security in Europe. They were only somewhat less consistent regarding the Middle East, the World Communist System and the United States.

POLAND

Though the difference in Poland's ranking on the interaction and attitude scales is only one rank, the position on the first is slightly negative (an evaluated z-score total of −1.530), while on the second it is closer to the positive deviance end.[7] This is due to frequently, but mildly inconsistent behavior by Poland in most areas. Poland's most significant areas of inconsistency lie in Europe. First, Warsaw had interaction levels with the Soviet Union and the rest of East Europe that were below average, but it supported Pact action in Czechoslovakia and was critical of events happening there. Second, though Poland supported Pact proposals on European security and disarmament and harshly rejected the FRG note, its interaction level with the West was highest in the Pact. Like Bulgaria, Poland was apparently pursuing detente through both unilateral and multilateral vehicles. Its behavior was only moderately consistent on the Middle East and China. Though it supported the Pact line in the 1967 War, its interaction with the Arab states was lower even than that of Romania.[8] And it failed to match its low interaction level with China with a condemnation of the CCP as harsh as those of its allies. In fact, Poland's only consistency in behavior was toward the United States, where its better than average IP was matched by its less harsh reaction to United States' activites. That Poland's ranks on the overall attitude and interaction scales do not differ greatly, despite its consistently inconsistent behavior, is thus somewhat misleading. This is due to the fact that though the direction of its deviance changes from issues to actions, the degree of deviance in both cases is usually small. Poland during these years was a rather lukewarm ally, giving the Pact lip-service on most (but not all) major issues, while still pursuing carefully many non-Pact and non-communist interactions.

YUGOSLAVIA

As noted previously, Yugoslavia's interaction pattern was on occasion as conforming as any of the Pact states. It had better than average interaction levels with East Europe and the Arab states, and low levels with China, Israel and the West. But in its attitudes it matched these positive deviances only with regard to the Middle East issue area. On virtually all other issues Yugoslavia's position was deviant from that of the Pact to the negative side. This would make its behavior inconsistent with regard to: China, where it failed to support the Pact hard line, but also failed to interact with Peking; the World Communist System, where it interacted with the other parties to a high degree (except the Soviet and Chinese), but opposed the convening of a world conference of parties; and the Allies dimension, where its high interaction level with the other East European states is not accompanied by support for their action in or attitudes on Czechoslovakia. Belgrade's behavior was consistent to the negative side with regard to the United States with whom it had the highest IP of the group, and toward whom it had the least negative reaction. Finally, its very negative reactions to the Pact proposals on European peace and security were nevertheless accompanied by low interaction levels with the West. Yugoslavia's IP with the West is in fact the lowest in the group, except for the GDR which had none.

In sum, not only is the Warsaw Pact (to say nothing of the region as a whole) made up of countries whose interactions with and reactions to the rest of the world differ from each other, but the actions of each member of the group frequently are not congruent with its own attitudes.

This latter conclusion reinforces previous research which has noted the difference in the action and attitude aspects of foreign policy.[9]

Moreover, for East Europe this within-nation inconsistency is perhaps as much to be expected as the among-nations diversity. It is not surprising that states which profess a motivating world outlook which is both an explanand of world history, a guide to present-day action, and a revelation of the future world order, should, in dealing on a day-to-day basis with problems of security, legitimacy and development, find themselves in violation of some of the tenets of their ideology. In such cases their professions of loyal adherence to the side of progressive forces may remain fervent, while their actions are based on other needs. Indeed, Tucker has noted, in discussing Marxism more generally, that it is precisely when ideological adherents need to accomodate themselves to realities which do not meet their expectations that reiteration of orthodox rhetoric becomes more fervent.[10] As interesting as this question and its relationship to East European foreign policies might be, it must remain for other studies.

CHAPTER FIVE

FOCUS ON ROMANIA

BACKGROUND

Exactly when Romania's deviance began is a matter of some dispute among Western analysts mainly because, until the early sixties, what conflict there was existed largely *sub rosa.* During the entire period of the nineteen fifties and early sixties Romanian foreign policy clearly, if not energetically, emulated Moscow. Party leader Gheorghe Gheorghiu-Dej consistently praised and supported all Soviet international initiatives and reiterated that,

> The friendship and fraternal cooperation with the Soviet Union, our liberator and staunch friend, was and is the fundamental principle of the foreign policy of our people and government.[1]

Bucharest applauded Soviet intervention in Hungary in 1956[2] and supported and reaffirmed the Moscow Declarations of 1957 and 1960 placing the CPSU at the center and head of the world communist movement.[3] While it may be true that Romanian criticism of Yugoslavia after 1956 was "relatively mild,"[4] that their acceptance of the Moscow declarations was "not unequivocal"[5] and that Romanian condemnation of Albania at the Twenty-second CPSU Congress in 1961 was "restrained,"[6] none of these divergences approached the level of those that were to emerge in the mid-sixties.

It was in economic policy that the basis for conflict was laid as early as 1955. In that year the Second Congress of the RWP set out to reverse the lull in development initiated in the post-Stalin "New Course" in 1953 and returned to plans for rapid, broad, heavy industrialization.[7] That this produced a condition of "latent conflict"[8] is clear from an examination of the Soviet position on East European development, explained by Khrushchev at the Twentieth CPSU Congress in February, 1956:

> each European country under the rule of the People's Democracy may specialize in the development of those particular branches of industry, in the production of those particular types of products, for which it disposes of the most favorable manual and economic conditions.[9]

The more developed members of a rather languid Comecon favored such specialization by which they would benefit from the manufacture and sale of high-profit manufactured goods, while securing their needed agricultural products and raw materials from their less developed allies, such as Romania and Bulgaria.[10] However, conflict, though it existed[11] stayed at a low level due to several factors: 1) Romania's ambitious goals had to be trimmed due to the unsettling events in Poland and Hungary in 1956; 2) for the time being such a development plan had to be based on improvement and mechanization of agriculture, Romania's key surplus-producing industry, which essentially meant that Romania would have to continue to specialize at least for a while, and moreover purchase equipment from the more industrialized Comecon members; and 3) by 1958 the Soviet position began to swing around to support broad industrialization for Comecon members. This shift was exemplified in an article by Ts. A. Stapanyan on the "simultaneous transition to communism of all socialist states," a doctrine explicitly included by Khrushchev in his speech to the Twenty-first CPSU Congress in 1959.[12] The Soviet position on development at the end of the decade was thus congruent with the ambitious plans laid out by the RWP at their November, 1958 plenum.[13]

Despite a nurmuring level of "discussion" between the Romanian and Soviet comrades on the extension of credits by the latter, and Moscow's failure to support the Galaţi steel mill project,[14] there still was no broader policy conflict between them until the Soviet position began in 1960-61 to take a harder line against the autarchic development of the other countries and in favor of "exploiting the possibilities of the international division of labor for the development of their socialist production."[15] This shift culminated in the joint declaration by CMEA on "The Basic Principles of the International Division of Labor," and an article by Khrushchev in *Kommunist,* both of which appeared in 1962, supporting specialization and increased economic coordination and planning on a supranational level in order to improve the performance of CMEA.[16]

The third Romanian Five Year Plan approved by the RWP at their congress in 1960 called for broad, rapid development of the Romanian economy, and the fifteen-year projections issued at the same time envisioned industrial production growing at 12 percent per year.[17] This could hardly be compatible with a supranational organization which assigned Romania the role of provider of agricultural and primary materials in an integrated regional economy. When by 1963 several inter-party visits had failed to iron out Romania-CMEA differences, the controversy emerged into the open. This was indicated by the failure of the Romanians to send a delegation to a conference of party leaders in Warsaw in June of that year,[18] by the success of the Romanian point of view on supranational planning achieved at the July CMEA meeting in Moscow,[19] and by the publication of the "Statement on the Stand of the Rumanian Workers' Party Concerning the Problems of the International Communist and Working Class Movement" in April, 1964."[20]

ANALYSIS

It is not the purpose of this discussion to investigate the various twists and turns of this controversy or even to pin down exactly its genesis in time. Other studies have ably covered that ground.[21] Rather, beginning from a concern with Romania's deviant foreign policy exhibited during the second half of the sixties, we are led to this controversy in a search for the roots of that policy. Therefore we must now examine the implications of these divergent views on development for Romanian foreign relations, especially keeping in mind the hypotheses generated in Chapters II and III.

In Chapter II the hypothesis was confirmed that, with regard to international interactions, those countries having a low level of economic development tended to be more deviant. The explanation offered was that this derived from a desire to develop more rapidly and from a fear of a permanent second-class status within the CMEA framework, combined with the relatively lower vulnerability of a less developed, less industrialized economy.

The Romanian case clearly illustrates the workings of the development factor. First, the Romanian drive for industrialization caused Bucharest to reorient its trade policies away from CMEA in order to: 1) secure from the West the technology and equipment necessary to achieve rapid industrialization; and 2) to secure alternative sources of supply of raw materials crucial to that industrialization drive. That trade was reoriented is clear. As early as 1958, according to Montias, the determination was made to seek credits and technology from the West.[22] In the period 1960-67 while overall Romanian commerce expanded 215% , its commerce with capitalist countries grew 385%.[23] The growth in Romania's trade with the developed capitalist countries can be seen in Table V.1. Overall, between 1960 and 1967 the capitalist countries' share of Romania's total trade turnover grew from 22.4% to 39.6%.[24] This was a period of striking growth for the Romanian economy—virtually the highest in Eastern Europe[25]—fueled by the massive importation of western technology. The west's share of Romania's importation of machines and equipment had been 25.8% in 1960; by 1967 its value increased over four times to account for more than 58% of that trade.[26] (See Table V.2).[27] In addition while in 1960 Romania had imported almost three times as many raw materials and semi-manufactures from CMEA as from the capitalist states, by 1971 their amounts were roughly equal.[28]

More specific still, West Germany, the leading western trading partner of Romania during this period, and its second leading partner overall after the Soviet Union, nearly doubled its trade during 1960-1965. And while in 1960 this trade had been almost evenly split between exports and imports, by 1966, of the total trade, more than two-thirds was Romanian importation of FRG goods, chiefly machines and equipment.[29]

Within the general exchange of views on specialization and economic development policy, the Romanians let it be known that they were securing from the West what they could not get in the East. Writing on "The tendency of equalization of the economic development of the Romanian People's Republic at the level of the more developed socialist states." I. Rachmuth emphasized the importance of obtaining, "the most advanced technologies on the world plane" in an era of scientific and technological revolution.[30] That this search for the best technology was to continue into the period of our study, despite a levelling off in the share of non-CMEA trade,[31] is evident from an increasing number of statements on the need for taking advantage of "the contemporary technological-scientific revolution."[32]

TABLE V.1
ROMANIAN TRADE WITH DEVELOPED CAPITALIST COUNTRIES (DC) 1960-69
(in millions of current dollars)

Year	Imports From DC	% of Total Imports	Exports To DC	% of Total Exports
1960	152.5	23.5	152.1	21.2
1961	228.3	28.0	192.9	24.3
1962	271.0	28.8	193.1	23.6
1963	275.6	27.0	219.7	24.0
1964	326.4	28.0	233.1	23.3
1965	330.8	30.7	255.2	23.2
1966	411.2	33.9	332.4	28.0
1967	731.4	47.3	443.3	31.8
1968	701.0	43.6	448.2	30.5
1969	758.6	43.6	520.4	31.9

Source: Paul Marer, *Soviet and East European Foreign Trade, 1946-1969, Statistical Compendium and Guide* (Bloomington, IND.: Indiana University Press, 1972), pp. 30, 40.

TABLE V.2
ROMANIAN IMPORTATION OF MACHINES AND EQUIPMENT 1960-69
(in millions of devisa lei)

Year	From CMEA	% of Total M&E Imports	From Developed Capitalist Countries	% of Total M&E Imports
1960	924.4	73.2	325.3	25.8
1961	1238.8	62.6	687.9	34.8
1962	1534.4	62.6	874.1	35.7
1963	1726.4	65.1	813.5	30.7
1964	1933.9	70.3	798.4	29.0
1965	1568.2	62.3	920.0	36.6
1966	1595.4	53.5	1351.8	45.3
1967	1826.1	40.3	2662.2	58.8
1968	2040.6	45.2	NA	--------
1969	2331.1	50.4	2222.6	48.1

Source: From tables in John M. Montias, "Romania's Foreign Trade: An Overview," in Joint Economic Committee, *East European Economies Post-Helsinki: A Compendium of Papers*, 95th Congress, 1st Session (Washington: U.S. Govt. Printing Office, 1977), pp. 884, 885.

In addition, the adjustment of trade was designed to limit Romanian dependence upon its CMEA (chiefly Soviet) suppliers of raw materials and hence its vulnerability to disruptive delays or pressures directed at its development program. Though as late as 1967 Romania was still a net exporter of energy supplies—chiefly petroleum products and natural gas[33]—its dependence on the Soviet Union for industrial raw materials, e.g. rolled steel, iron ore, coke, cotton, was still substantial.[34] But by 1965 Brazil and India were supplying over 38% of Romania's iron ore imports; in 1970 these two suppliers, plus Algeria and Yugoslavia supplied 32% of these imports. In cotton imports, the Soviet share fell somewhat from the 1962 level (44%) to 42.5% in 1965, with Egypt and Syria together supplying 44%. By 1970 Romania's disparate group of non-Soviet cotton suppliers included China, Egypt and the United States who together supplied 38%. In addition, in a crucial sphere, Romania reduced the almost total control of its external supply of metallurgic coke by East European states—Czechoslovakia, Poland and the Soviet Union supplied 85% of Romania's imports in 1965—by adding significant supplies from the United States, China, Italy, and the FRG. In 1970 the three East European countries' share of Romanian coke imports was down to 67.8%. In rolled ferrous metals, the Soviet share went down from 74% to 45.3%, but this is misleading since other East European suppliers more than doubled their share (11.8% to 24.6%) in this time, although France, Britain and Japan also made significant gains.[35]

In 1967 Ceausescu stated the Romanian needs openly,

> In the domain of the import of raw and auxiliary materials—a highly important component of our foreign trade activity—steps must be taken for ensuring future supplies over a longer period of time. This is required by the sustained rate of development of the national economy.[36]

In August, 1968 Ion Olteanu writing in *Lupta de Clasa,* explained Romanian policy. "A prime problem," he wrote, "which occupies the organs of foreign trade is diversification to the maximum of the markets of provisions. This constitutes a stringent necessity."[37] Calling for a judicious choice of foreign partners based on level of resources, quality, and payment possibilities, Olteanu also noted the importance of developing contacts, "which present assurances of delivery on time of prime materials."[38]

Thus Romanian dependence on the Soviet Union, already lower than that of the more advanced CMEA members, was further reduced. Moreover, Levesque makes the cogent point that the lower level of development in Romania meant that less of the diversification, stratification, and economic/political "forces sociales d'opposition" had developed within the country, giving the leadership greater freedom of action, and making the nation less *internally* vulnerable.[39]

Finally, the reorientation of Romanian *exports* which was somewhat less successful (See Table V.3) was designed to further limit economic vulnerability to fickle or slow-paying buyers by developing a wider range of markets and at the same time earn much needed hard currency.[40]

It seems clear that a significant part of Romania's international deviance was stimulated by the development drive and its technical and economic needs. Further, this drive was specifically directed at rectifying or preventing the continuation of a second-class position within the group of CMEA states vis a vis the more developed members. Writing in a volume on *The Economic Development of Romania 1944-1964*, M. Horovitz, I. Burstein and I. Lemnij pointed out,

> socialism is not compatible with the existence of an international division of labor which perpetuates the maintenance of an unfavorable structure of the national economy, which characterizes the countries poorly developed from the point of view of economics, under conditions of capitalism.[41]

Writing in the same volume, which may be considered a seminal statement of the Romanian economic position, I. Rachmuth wrote of the necessity for each socialist country to develop the means of production, i.e. heavy industry, and to overcome the differences of economic development that result from "historical conditions."

> Socialism, through its very essence, generates in an objective manner the necessity of liquidating these differences formed as a result of historical conditions, [and] the necessity of equalizing the economic levels of the socialist countries.[42]

TABLE V.3
ROMANIAN EXPORTS 1960-69
(in millions of current dollars)

Year	All Communist Countries*	% of Total Exports (TE)	Non-Communist Countries Dev.	%TE	Less Dev.	% TE
1960	523.2	73.0	152.1	21.2	41.7	5.8
1961	556.2	70.2	192.9	24.3	43.3	5.5
1962	559.6	68.4	193.1	23.6	65.2	8.0
1963	639.4	69.9	219.7	24.0	56.0	6.1
1964	687.8	68.8	233.1	23.3	79.2	8.0
1965	755.9	68.6	255.2	23.2	90.5	8.2
1966	735.4	62.0	332.4	28.0	118.4	10.0
1967	786.0	56.3	443.3	31.8	166.0	11.9
1968	869.4	59.2	448.2	30.5	151.0	10.3
1969	970.4	59.4	520.4	31.9	142.3	8.7

Source: Marer, *Soviet and East European Foreign Trade,* p. 40.
*Includes Asian Communist Countries.

After flatly stating that this equalization represents a "law of economic development," Rachmuth states pointedly,

> To the removal of inequalities of levels of economic development of the socialist countries can not contribute any division of labor between these countries, any specialization among them. . . . It cannot be doubted that equalization of the levels of economic development of the socialist countries has to be considered as an essential economic and social-political task, in which not only the less developed countries are interested, but the whole socialist camp. It is clear that in this important problem the requirements of the law of value will have to yield, as is the case, in view of necessities resulting from this law of equalizing the level of economic development of the socialist countries.[43]

The Romanian reorientation of international contacts, based on the need for involving the country's economy more fully in the world economy as a way of securing rapid and broad industrial development, necessitated a repudiation of those who might have preferred to keep the Romanian economy specialized in agriculture and in a state of enclosure and dependence upon CMEA industrial suppliers.[44] In the "Statement" of April 1964, the Romanian party declared,

> . . . the socialist international division of labor cannot mean isolation of the socialist countries from the general framework of world economic relations. Standing consistently for normal, mutually advantageous economic relations, without political strings and without restrictions or discriminations, the Rumanian People's Republic, like the other socialist states, develops its economic links with all states irrespective of the social system.[45]

Writing on the role of foreign trade in the country's economic development, G. Surpat and N. Ionel noted the importance of "recognizing the right of each people to control their resources and the fruits of their labor," and stated, straight to the point,

> In this sense, Romania has rejected, on different occasions, any kind of "theories" or "plans," "theses," or "suggestions," which come into contradiction with the principles of international collaboration and offend the major interests of one people or another.[46]

ECONOMICS AND SOVEREIGNTY

Assertion of Romanian economic independence was inextricably tied in with assertion of overall national independence. Economic sovereignty could not be separated from political sovereignty. In an editorial in *Probleme Economice* commemorating the twentieth anniversary of the liberation of the country, the anonymous author wrote that first and foremost the liberation, "meant on the external plane the total and definitive liquidation of economic and political dependence of the country and establishment of real independence and sovereignty of the people and the state."[47] In a volume elucidating *The Principles of Relations*

Between States, published in 1966, Alexandru Aureliu explicitly linked economic and political sovereignty,

> Directing the economy in a planned manner, continuing unflinchingly the course of industrialization, and developing intensively as well a complex agriculture, the Socialist Republic of Romania supports its political independence through economic independence.[48]

The point, in sum, is that not only did Bucharest assert its independence by resisting supranational control over its economic processes, but it did not hesitate—indeed it sought and advanced—the international political consequences of such a policy.

Concurrent with the Romanian drive for economic sovereignty, stimulated by a desire for industrial development and economic equality, was a growth in overall Romanian nationalism. This was evident in two forms: first, the virtues of general national sovereignty were expressed, as a goal to be pursued in international politics, having at least as much value as, if not more than, that of proletarian internationalism; and second, a Romanization of and concomitant de-Russification of the national culture and politics.

Early (i.e. pre 1964) Romanian foreign policy statements had, of course, listed "respect for national sovereignty" as one of the cardinal principles of socialist international relations, along with equality of rights, nonaggression, peaceful settlements of all disputes, peaceful coexistence, noninterference in states' internal affairs, mutual assistance and proletarian internationalism.[49] But in the mid-sixties, the notion of the sovereign state as the chief arbiter of the national destiny, based on its position as the cornerstone of international politics, supported by historical tradition, international law and even Marxism-Leninism itself, began to be adumbrated more clearly and fully.

The Romanians had seen their role in international politics in the 1950's as that of staunch member of the Soviet camp in a divided world. To wit, this statement by Gheorghiu-Dej:

> The Romanian state of people's democracy is part of the great camp of democracy and socialism and is striving by its entire policy

> to contribute to the strengthening of this camp and to the success of its efforts for safeguarding peace and the security of the peoples.[50]

Then, during the late fifties and early sixties, when Khrushchev's de-Stalinized foreign policy was being pursued by the Kremlin, "peaceful coexistence" became the Romanian watch word.[51] This policy, combined with "the indestructible cohesion of the socialist camp" was considered "the essential prerequisite for the safeguarding of independence and the successful building of a new system in each socialist country."[52]

By the time of the 1964 "Statement," however, the Romanian party saw fit to underline that

> The strict observance of the basic principles of the new-type relations among the socialist countries is the primary prerequisite of the unity and cohesion of these countries and of the world socialist system performing its decisive role in the development of mankind.[53]

According to the Romanian view, beginning and continuing socialist construction in conditions of economic diversity requires the recognition of native divergencies and differences. No less an authority than Lenin had recognized this.[54] In order to implement a specific approach to socialism which recognizes the distinct problems of the country, a national plan, based on those distinct needs is required and, further, the national perogatives of a sovereign state must in no way be infringed. The rejection of the supranationalization of Comecon is based upon, and indeed tied up inextricably with, the assertion of state sovereignty.

> Our party has very clearly expressed its point of view, declaring that, since the essence of the projected measures lies in shifting some functions of economic management from the competence of the respective state to that of superstate bodies or organisms, these measures are not in keeping with the principles that underlie the relations among the socialist countries.
>
> The idea of a single planning body for all CMEA countries has the most serious economic and political implications. The planned management of the national economy is one of the fundamental,

> essential, and inalienable attributes of the sovereignty of the socialist state—the state plan being the chief means through which the socialist state achieves its political and socio-economic objectives, establishes the directions and rates of development of the national economy, its fundamental proportions, the accumulations, the measures for raising the people's living standard and cultural level. The sovereignty of the socialist state requires that it effectively and fully avail itself of the means for the practical implementation of these functions, holding in its hands all the levers of managing economic and social life. Transmitting such levers to the competence of superstate or extrastate bodies would turn sovereignty into a meaningless notion.[55]

Beginning with the Ceauşescu regime the Romanian party began to more fully and explicitly express a commitment to the nation-state as the central phenomenon in international politics—as opposed to the international class struggle—and to the safeguarding of the prerogatives of that entity. This shift can be seen in a cross-time comparison of Romanian writers on foreign policy.

Edwin Glaser, one of the chief scholar-spokesmen of Romanian foreign policy, writing in 1960 stressed the central place of peaceful coexistence in Romanian policy. After tracing the historical and Leninist traditions of peaceful coexistence, Glaser quotes Gheorghiu-Dej,

> The foreign policy of the People's Republic of Romania is a policy of active participation in the struggle for peace in the whole world. For the peaceful resolution of all litiguous international problems, for peaceful coexistence among all countries, indifferent to social system, for the continual strengthening of the unflinching unity of the community of socialist countries with the Soviet Union in front.[56]

However in 1965 Glaser's efforts are directed toward tracing and elaborating a different principle, that of sovereign equality. While not denying the importance of peaceful coexistence, Glaser at this time stresses the importance of world recognition of the sovereign equality of all states. Calling it the "primary prerequisite" for unity of the socialist states, he notes,

> States may differ as to their territorial extent, population, economic level, ancienty[sic], military power, influence in international relations, but they are equal in their right to sovereignty, to independence, to all rights inseparable from state sovereignty . . . .[57]

The purpose of this and other articles appearing at this time was to broaden and reiterate the party positions expounded in the April, 1964, "Statement."[58] In doing so Glaser, in another article, illustrates how the conception of Romanian foreign policy has shifted:

> The firm permanent foundation of the foreign policy of our country is formed by the principles of sovereignty and the independence of peoples, equality of rights, noninterference in internal affairs, mutual advantage.[59]

A comparison of this statement with that of Gheorghiu-Dej in 1955 (see p. 248 above) makes it evident that in the intervening years, concern with national sovereignty, stemming from concern with economic sovereignty, had replaced broader, international socialist concerns as the "foundation" of Romanian foreign policy.[60]

CEAUŞESCU ON SOVEREIGNTY

Before 1968 Nicolae Ceauşescu's expressions in support of national sovereignty took on three forms. First, he stressed national differences in building socialism and the right of each party to develop its own policy:

> It is known that the diversity of economic, political, social and national conditions, the historical characteristic features and the different stages of development of each socialist country generate differences in the forms of building up socialism, in the methods of organizing [the] economy and social life. Such differences are but natural and inevitable in such a complex process like the building of socialism. Life and experience demonstrate that the advancement of the socialist countries imperiously requests the application of the general truths of Marxism-Leninism to the concrete conditions of each people, both in the interest of each and every socialist country, as well as for the strengthening of the world socialist system.[61]

Second, he blamed military blocs for continuing international discord and the lack of world peace, and called for their abolition.

> One of the barriers in the path of cooperation among peoples are the military blocs and the existence of military bases and troops of some states on the territories of other states. The existence of these blocs, as well as the dispatching of troops to other countries, represent an anachronism which is incompatible with the national independence and sovereignty of peoples, with normal inter-state relations. Ever widening circles of public opinion and a growing number of states manifest inclinations which are gaining more and more ground lately, for the abolition of the military blocs, the dismantling of foreign bases and the withdrawal of troops from the territory of other countries. The fulfillment of these burning desires of peoples would be of special importance and would give a powerful impulse to the development of trust among peoples, to the relaxation of the international situation, and the consolidation of world peace.[62]

In addition to supporting nations as opposed to supranational groups in world politics, this position is certainly different from laying the blame for international tension at the doorstep of world imperialism.

Third, Ceauşescu began to stress the role and responsibility of small and medium-size states for producing and protecting world peace.

> One of the features of our era is greater and greater participation of the small and middle-sized countries in the settlement of international issues. . . . Reality demonstrates that the settlement of international disputes can no longer be decided only by the big powers.[63]

All of these positions boost the idea of national—as opposed to supranational—decision-making in various realms, and in doing so, praise the inviolability and primacy of national, state sovereignty.[64]

In 1968 the increasing pressure upon the Czech regime and the ultimate invasion by WTO troops occasioned a more explicit, forceful assertion by Romania of these principles, especially that of national sovereignty for all states, fraternal allies included, and the rejection, after the invasion, of any form of "limited sovereignty."[65]

At the tenth Congress of the RCP in August 1969, Ceaușescu stated,

> We consider that in the conditions of today, when the socialist countries are developing impetuously, registering again and again successes on the road of material, political and spiritual progress, the peoples who have won their liberty, who have taken their destiny into their own hands, who have created a dignified, prosperous life, can not be diverted by anyone or anything from the road upon which they have embarked—the road of socialism and communism. The development and strengthening of the new social order, the defense of the revolutionary victories of socialism, represent the sacred right and duty of each people, of each communist party in the socialist countries. Of course, in the spirit of proletarian internationalism, in case of an imperialist attack the peoples of the socialist countries have to assist each other, fighting shoulder to shoulder for the defeat of the aggressor. The establishment of the forms and manners of assistance in such cases has to be the result of agreement between the party leaders, between the leading constitutional organs of each country. The solidarity and mutual assistance of the socialist countries presupposes relations of equality among all socialist nations, [it] must not lead to interference in the internal affairs of any people since this would profoundly prejudice the cause of socialism.[66]

Finally, the reaction to events in Czechoslovakia brought the Romanian exegesis full circle. A volume published in 1968 by Mircea Malița, Costin Murgescu and Gheorghe Surpat expressed in the clearest and most direct manner the evolution of the Romanian position from concern with economic sovereignty—in the form of demands for total control over the national economic plan, ability to engage in the world scientific-technical revolution, recognition of national differences and a unity of world socialism only on that basis—to the overall assertion of national sovereignty.

This volume proceeds "backwards" with the Introduction and Chapter I focusing on the basic problem of national sovereignty and drawing support from both contemporary and past Romanian diplomats.[67] Costin Murgescu, the author of Chapter I states,

> The right of self determination implies the right of peoples to constitute independent states themselves, to choose freely their

> economic, political and social system, to dispose freely of their national resources, to determine for themselves in a free manner, the entire course of their lives. . . . it is an axiomatic thesis that through interference in the internal affairs of others in any form, the essential premise of sovereignty and equality of states is destroyed.[68]

In Chapter 2 Nicolae Belli explicitly connects both socialism and national development to "active participation in the world circulation of the material and spiritual values of society."[69] Further, the theme of national assertion is expressed in another way.

> . . . socialism creates the conditions for all countries, including small and medium sized countries, to assert themselves in the world arena with an independent policy, with their own initiatives, destined to contribute to the closeness of peoples, to the development of economic collaboration between them, serving, in this way, the cause of peace and socialism in the world.[70]

Chapter 3 reiterates and elaborates the Romanian position on the need for full participation in the world economy and furthermore makes it clear that the pressures for economic development and the manner of its achievement spring from essential national, unique situations.

> Modernization of the economy, superior use of material and human resources depends in the first place on the efforts of the people themselves. The modern structure of the economy—as the experience of our country has shown, developing from objective requirements, has to be the result of concrete demands, specific to the countries; it can not be copied.[71]

Chapter 4 becomes even more specific, when in discussing Romanian relations with other socialist countries, the authors continually emphasize the necessity of receiving in their commerce high quality technological material "at a world competitive level." This chapter is a none-too-subtle criticism of CMEA for failing to provide the less developed members, e.g. Romania, with high quality industrial goods.

> With all the results obtained up until the present in the activity of international specialization in production, one can appreciate that these are inferior to the possibilities and necessities of the socialist countries. . . .
>
> At present, and even more in the future, taking into consideration the tendencies of world technical development, the realization of products which keep pace with the world technical level becomes a necessity for any producer who presents their products for export.[72]

Chapter 5 then goes on to state the obvious conclusion: that the entry of Romania into the world market, and most specifically that with the world capitalist countries, is based on "objective necessity" deriving from socialist development needs.

> The enlarging of the economic exchanges of Romania with the capitalist countries expresses in the first place the sustained industrial and technical progress of our national economy; it follows in good measure from the necessity of providing through imports our national economy, and in particular, industry, with machines and equipment on the highest level of world technology, with a view toward the continued growth of the economic potential of the country.[73]

These elaborations on the notion of sovereignty, its place in international politics and in Romanian foreign policy serve to further strengthen the evidence that the roots of Romanian international activities lie in its drive for economic development.

NATIONALISM

This general concern with state sovereignty was paralleled by a resurgence of Romanian nationalism. Given Romania's history, and especially its post-war period, Romanian nationalism means anti-Russian nationalism. The Romanians' feeling of being "Latins in a sea of Slavs," the *terra irredenta* of Bessarabia and Northern Bukovina, and the disembowelling of the Romanian economy by the Soviets after World War II, are all aspects of this hostility.[74] Its expression since the early sixties has taken a variety of forms. On one level, it meant a de-Russification of the national culture. In 1963, Paul Lendvai writes,

> Through a series of seemingly small but politically extremely important gestures, the Rumanian Communist leadership began to feed the strong anti-Russian feelings of a population that since 1945 had been ordered to look to Russia as liberator, guide, and model. Almost overnight, the lavish demonstrations of cultural solidarity with Russia were replaced by a deliberate policy of "Rumanianization," a new emphasis on national traditions with anti-Soviet overtones. Streets, movie houses, theaters, and cultural centers were renamed. The A. Popov movie theater became the Dacia; the Maxim Gorky, the Union. A Russian language institute in the capital was closed and a large Russian book store demolished; the Rumanian edition of the Soviet propaganda monthly *New Times* was discontinued; schools dropped compulsory Russian-language studies; Rumanian orthography was "re-Latinized," eradicating the previous "Slavification" of the alphabet and of culture in general.[75]

In addition the Romanians began to reclaim their history.Explanations of Romanian foreign policy appearing in the fifties had stressed the dramatic break with the pre-war diplomatic situation.[76] In the sixties Romanian diplomatic historians began to stress the intimate connection of the current independent policy line with traditional, independent, active world-involvements of past regimes.[77] Further, in 1962, for the first time since the establishment of the communist regime, a generally favorable article appeared on Nicolae Titulescu, a reknowned Romanian foreign minister and diplomat of the interwar years.[78] In the succeeding years until the end of the decade, increasing, and increasingly favorable, attention began to be focused on pre-regime Romanian diplomats.[79] Also, an interest in prewar Romanian Balkan diplomacy was evidenced, with several volumes being published on the Little and Balkan Ententes.[80] The message sent north by these works was that Balkan cooperation had a long and favorable history in Romanian diplomacy and was therefore always a potential region for exploitation of support by Bucharest.[81]

Finally, in a key reinterpretation of history abounding with contemporary implications, the newer investigations of Romania's actions in

World War II made it plain that Romania's liberation was indigenous, effective, almost total in the country, and most importantly, pre-dated the entry of Russian troops into the heart of the country.[82]

This growth of nationalism relates to the heterogeneity-legitimacy link found in Chapter III. The Romanian party, it is generally recognized, has "the weakest indigenous roots," of the East European communist parties.[83] The country also includes a large minority of Hungarians, living mostly in once Hungary-controlled Transylvania, ethnically and culturally separate and potentially hostile to the regime in Bucharest.[84] The regime's legitimacy and power position, especially after 1956, was enhanced by being able to exploit a long tradition of anti-Russian national feeling.[85] By skillfully doing so, i.e. with certain strict limits on anti-Russian expression, and by casting this nationalist expression within an overall self-assertion internationally, the regime's position vis a vis its population was improved greatly. Francois Fejtö writes of Ceauşescu's predecessor, Gheorghe Gheorghiu-Dej, "The more independently Dej acted, . . . the more his country approved of him."[86] Both regimes used latent anti-Russian nationalism to solidify their own positions.[87] The independent foreign policy was and is popular; hence the regime's position has been improved to the degree that it can practice its deviant foreign policy line without bringing in Russian troops. Furthermore, the presence of potential and actual anti-Russian feeling among both the Romanian and minority Hungarian population enabled the Gheorghiu-Dej and Ceauşescu regimes to pursue an independent policy, secure that they would not decrease—indeed that they would stimulate—popular support with such policies.[88] Ionescu sums up this policy:

> . . . . unlike the Yugoslav Party, the Rumanian Party did not have in its background a myth of past identity with 'the people.' Their coexistence had been a sullen affair. The new 'theme' of national pride and self-assertion might prove to be popular, but past memories of indifference towards the nation's feelings and of political external allegiance did not warrant a sudden widespread campaign.
>
> Slowly and indirectly more and more people would come to realise what was going on. The party reckoned that once the expanding circle of people with vested interests had got the idea it would spread to wider and wider circles. They also reckoned that if, even so late in the day, they could present the party as the

> champion of Rumania's rights, and implicitly as the instrument of some measure of withdrawal from the Russian embrace and of reassociation with the West, it could not go wrong with the Rumanian people.[89]

To sum to this point, the present research shows and other observers agree that the overriding stimulus to the Romanian deviation in foreign policy was economic. The desire to develop broadly and rapidly, pursued unsteadily since the early fifties came into conflict with renascent Soviet plans of the early sixties for increased supranational planning and integration through Comecon. A desire to both avoid a second-class, primarily agricultural, status within Comecon, plus a need for western technology and resources led to increasingly non-pact oriented foreign relations. It also led the Romanians to reject as invalid any positions or actions tending to assert that any one approach to socialism or socialist construction, e.g. the Soviet one, was paramount and correct. Hence Romanian opposition to pressure and actions directed against both the Czech spring and the People's Republic of China.[90]

A second factor seen to have had both a stimulating and enabling effect on Romanian deviance in foreign policy is that of party legitimacy and its enhancement through the vehicle of Romanian nationalism.

OTHER FACTORS

We must consider certain other aspects of Romania's internal and international milieu, before being satisfied that we have explored fully the underlying parameters of Romanian deviance. Therefore we will examine briefly certain other enabling factors.

First, there is geography. The analysis in Chapter II showed a strong positive association between geographical position and a deviant interaction pattern. Yet the Romanian case stands as a notable exception to this overall relationship; the most deviant case being in a very bad geographical position. An eight-hundred mile border with the Soviet Union, with few natural barriers separating the Ukraine from Moldavia and Northern Transylvania—especially after the reabsorption by the Soviet Union of Bessarabia and northern Bukovina—makes Romania especially vulnerable to threatened or actual military pressure of the type Moscow has used before. Moreover, Romania is bordered on all sides by socialist states, of which two (Bulgaria and Hungary since 1956) have not only pursued rather cautious foreign policies, but which agreed with and participated in the actions relating to Czechoslovakia in 1968. Furthermore, the small border with Yugoslavia can provide little comfort, especially with the post-Tito future unclear.[91]

On the other hand, several observers have pointed out that it is precisely this relatively isolated geographic position that enables Romania to act in a deviant way on foreign policy issues. Having no border with the western nations is a strength of sorts. According to Peter Bender,

> NATO cannot break into the Eastern alliance there, either militarily or politically, and Bucharest cannot break out to the West. . . . Because they are less important to their great ally, because it is "sure" of them, they can allow themselves more independence of action.[92]

While it seems reasonable to conclude that Romania has made a virtue out of this particular necessity—compare its strategic significance with that of Czechoslovakia, for example—it should also be recognized that the Romanian geographical position does pose clear limits on its policy positions. Note that Bucharest, for example, espouses not withdrawal from the Warsaw Pact but abolition of all blocs. With Bulgaria to the south bordering on two—presently conflicting—NATO states, it seems unlikely that Moscow would tolerate a non-Pact state between itself and its Slavic fraternal ally.

A second physical factor to be recalled is the relatively good position of Romania in terms of three aspects of its internal natural resources. First, important minerals such as coal, manganese, and bauxite as well as abundant supplies of crucial natural gas and of course oil[93] allowed it to be much less dependent upon foreign sources for these important development-related items. Second, the nature of Romania's major export products—corn, meat, timber, as well as primary petroleum products—was such that a steady supply of foreign, i.e. western, exchange was assured.[94] And third, the Romanians had the ability to pay for western goods by mining their own gold and silver—a critical ability to a country with a nonconvertible currency.[95]

An additional significant and unique aspect of the Romanian case must be stressed. While for the overall group of states studied, domestic orthodoxy was shown not to be associated with foreign policy deviance,[96] it cannot be doubted that Romanian domestic policy during the period of our study was strictly orthodox. It was an exceedingly tightly controlled regime in terms of civil liberties (ranking last among CMEA states on Triska and Johnson's scale as least "liberalized.") The state of the domestic system thus presented no threat to the Soviets through "infection." As Larabee puts it, ". . . Ceausescu's brand of national communism has avoided the political pluralism that threatened the Leninist party state system in Czechoslovakia and with it the Soviet view of 'legitimacy.' "[97]

Further, the Romanians offered their plan for a "multilaterally developed society" based on a quite orthodox view of an industrialized socialist economy and society, and buttressed their arguments with the holy words of the masters themselves.[98]

It has been the Romanian Communist Party which has directed and guided the development of Romanian internal and international policy, retaining at all times policy initiatives, cadre leadership and political control. The face shown to the people of Romania has been that of Ceauşescu, not that of "humanism."

A related factor noted by Jowitt, Fischer-Galati and others is that the leadership of the RCP has been for the most part well united and under the control of the party boss, be it Gheorghiu-Dej or Ceauşescu. Gheorghiu-Dej had gained predominance and total control of the Romanian party by the end of the fifties, after purging, among others, a rival group of "Moscovite" party leaders.[99]

Upon his ascension, Ceauşescu began to both replace people who owed their position to Gheorghiu-Dej with his own patronees, and to assert himself as undisputed party leader.[100] In December, 1967 he was elected President of the Council of State and had thereby, in only two years, duplicated Gheorghiu-Dej's feat of holding the top posts in both the government and the party. The acquisition by Ceausescu of more and more power continued through the end of the decade.[101] The effect of this has been to give the Romanian Party, in Jowitt's words, "A fairly united elite, a coherent organization."[102]

Low potentiality for the developing of a rival faction within the RCP, whether due to the pure power dominance of the party leader or based on a broad consensus on the party's goals and their pursuit, is of undeniable significance as an enabling factor explaining the continuation of deviant policy. The well-known cases of the use of internal factions by the Kremlin to limit deviant policies in Hungary and Poland in the fifties gives evidence to this point.[103]

An additional factor related to party leadership deals with the CPSU. During the "covert" years of the Comecon dispute, Khrushchev was in a defensive position internationally as a result of several policy failures: first, the inability to prevent a widening Sino-Soviet rift, taking in its wake the small but noisily defiant Albania; second, failure of the Yugoslav rapprochement policy–pursued on and off since 1956–to bring about a significant reabsorption of that party or state into the Moscow-dominated sector of the communist movement; and third, the dramatic diplomatic defeat on the Cuban missile issue. Further, Khrushchev's domestic leadership role was weakened by serious internal policy and power struggles.[104] These defections and pressures made his position less than dominant vis a vis the Romanians. Jowitt states,

> Not only was this set of interacting problems, issues, and challenges a threat to Khrushchev but also it constrained and limited the ways in which he could attempt to secure his position, while simultaneously placing decisive action at a premium.[105]

Similarly, during Ceauşescu's assertion and restatement of the deviant Romanian line, the new Soviet collegial leadership was itself fresh to its tasks. Neither Brezhnev nor Kosygin could dramatically invoke a hard line policy toward the Romanians without a firm supportive power base at home. Moreover, they inherited the same problems and risks, at least as regards the international communist movement, as had their predecessor.[106]

Chief among these problems and clearly one of the most significant factors, was China. There are several aspects of this factor to be considered. First, by its defection, combined with a vigorous assertion of an altervative method of construction of communism, the Chinese provided an example to the Romanians. They offered an example of a defiant communist party, declaring its own primacy and prerogatives in its internal affairs, against the controlling center of Moscow. They thus weakened Moscow's claim to be the sole papal center of ideological orthodoxy. If the Chinese could not only dispute this claim, but even suggest that the CPSU had deviated from true Marxism-Leninism, the

Romanian Party's demand that it be allowed to apply orthodox Marxist doctrine, i.e. broad industrial development, in its country, was strengthened.[107] Though the Romanians held no particular brief for the specifics of the Chinese ideological position, they did, for their own reasons, support the contention of the CCP that Moscow could not be the all-powerful controlling center of international communism.[108] The Romanians took careful note of the Chinese "twenty-five points" letter, issued in June, 1963, which outlined the points of disagreement between the CCP and the CPSU. Bucharest, alone among CMEA states, published both the Soviet position *and* long sections of the Chinese letter.[109]

More specifically, China provided an example of a country committed to rapid, broad industrial development, based as much as possible on its own resources, dependent as little as possible on outside (Soviet) ones. That the Chinese party was doing this while retaining full control over the economic, political and social processes of the country, to say the least, may have provided an example even more compelling to Bucharest than the more familiar, but unwieldly Yugoslav approach.[110] In addition, the Chinese did provide explicit moral and ideological support to Bucharest, in a clear attempt to woo yet another Balkan ally away from the Soviet Union.[111] Romanian-Chinese trade, though remaining a tiny proportion of total Romanian trade turnover, did return to and surpass 1960 levels by the end of the decade, as seen in Table V.4.

TABLE V.4
ROMANIAN-CHINESE TRADE 1960-69
(in million lei valuta)

Year	Exports to China	Imports from China	Total Trade	% of Total Trade
1960	200.0	141.6	341.6	4.1
1961	55.7	118.4	174.6	1.8
1962	13.0	63.2	76.3	.7
1963	82.9	84.7	167.6	1.4
1964	95.6	107.8	203.4	1.6
1965	159.8	131.2	291.0	2.2
1966	204.6	190.4	395.0	2.7
1967	236.6	198.9	435.5	2.5
1968	245.6	260.8	506.4	2.7
1969	231.0	254.1	485.1	2.4

Source: *Anuarul Statistic*, 1971, pp. 608-09.

At the very least, the Sino-Soviet split provided the RCP with greater "room for manoeuver."[112] Needing support against the aggressive Chinese and fearful of further defections—especially in light of Albania's adherence to the Chinese line—the Kremlin was loathe to react in too stern a manner to Romania's relatively innocuous deviations. Fischer-Galati notes,

> Though Khrushchev's hand was stayed by fear of totally alienating Rumania at a time of increased strain in Sino-Soviet relations, he remained fundamentally opposed to Gheorghiu-Dej's independent course. Certainly the Russians would have exerted pressure on Rumania to bring her back in line had it not been for the Chinese issue.[113]

Thus on the one hand Soviet need for conciliation, coupled with the apparent availability of another "option" for Romania, made a certain degree of compromise—in this case acceding to Romanian objections to CMEA supranational integration—necessary for Moscow.[114] On the other hand, in light of previous defections, Khrushchev was in the contradictory position of desiring greater control over the states of East Europe. This was apparently one reason for his push for greater CMEA integration in the early sixties, precipitating the public escalation of the dispute and hardening respective positions for some time to come.[115] Thus, in a number of ways, the Sino-Soviet split can be seen as having been a factor in Romania's continued opposition to Moscow.[116]

THE ROMANIAN CASE: STIMULATING AND ENABLING FACTORS

The deviance of Romanian foreign policy highlighted in this study, was a product, a result. The phenomena apparent in the second half of the sixties was due to factors having origins going back into the previous decade, and in some cases beyond that. Though its genesis may be found there, its continuance must be considered in terms of the specifics of the internal and international situation during its development.

Two factors can be judged as stimulating the Romanian foreign policy deviance. First, the desire on the part of the party leadership to pursue a rapid course of broad industrialization, to end Romania's position as breadbasket and gas station of Europe; in short, to take seriously the communist theoretical commitment to revolutionary economic, state and societal development. This position, brought the Romanian party into conflict with the CPSU and other East European communist states, both on a theoretical and policy level. Specifically it made Bucharest oppose Moscow's push for supranational planning through CMEA and implicity its hegemony in the East European and world communist movement. Further, it meant that Romania should and would participate as fully as possible in the international movement of goods, especially scientific and technical goods. This would mean an end to both the economic dependence on CMEA and the political isolation with it.[117]

Intertwined with the pursuit of this goal was the regime's desire and ability to use Romania's independent policy to improve its political position at home. This was possible through the utilization of latent Romanian nationalism, especially its traditionally strong anti-Russian aspects. In addition, it is not unlikely that such aspects influenced, in a personal way, the party leadership itself.[118] While it would be begging the question merely to label the Romanian self-Assertion as "[d] ressing up an essentially nationalist practice in orthodox ideology and representing real national interests while verbally upholding 'proletarian internationalism,' "[119] it does seem justified to conclude that the regime's desire to improve its domestic standing, its legitimacy as a Romanian party, contributed to its cumulative decision to stake out an independent line. Certainly the popularity of their position contributed to both its desire and ability to pursue the consequences of that position.[120]

If nationalism can thus be seen as both a stimulating and enabling factor underlying the Romanian deviance, the other factors noted all fall clearly in the enabling category, with one exception. The presence of China must be considered a chief, perhaps the chief, factor enabling Romanian foreign policy to deviate as it did. But it also had some stimulating effect as an example to the Romanian party leadership of a form of international political deviance and internal political and economic development.

The unity of the Romanian party leadership, concomitant with the weakness of the Soviet leadership, especially at key times, is a second significant aspect of the political situation during this period that allowed, if not provoked, Romanian policy deviance.

Finally, Romania's vulnerable and strategically less significant geographic position, served by very virtue of its lack of high strategic value, to give the Romanians greater breadth of action in pursuing a foreign policy at variance with that of Moscow. And the fact that this deviance was within the realm of foreign policy, with no serious destabilizing domestic political implications inimical to the Soviet Union, also provided the Romanian leadership with a latitude of action which has not been tolerated by Moscow in the realm of domestic political activities by its East European allies.

Every social phenomenon that occurs in man's history is, of course, unique in some aspects, international political phenomena no less than others. However, there are in the case of such phenomena aspects of comparability and aspects of uniqueness in explaining their occurrence. In the case of foreign policy deviance in East Europe in 1965-1969 certain factors were present in all cases. For instance, the example of China was there for all the East European states to see. And the insecurity of the Soviet leadership was also a "constant" for all these states. Yet such factors, while qualifying as sufficient to allow foreign policy deviance, were not integral to the genesis of such a policy. On the other hand, other states, such as East Germany and Czechoslovakia, were similarly desirous of pursuing broad, industrial socialist development and may have thus been stimulated to pursue international actions at variance from their allies, but they were enjoined from doing so by certain internal or international political, or geopolitical, factors–i.e. "disabling" factors.

It was thus in the Romanian case that the precise combination of necessary and sufficient factors adumbrated above came together to both stimulate and allow the continuation of significant international political deviance.

CHAPTER SIX

CONCLUSION

The time for assessment is upon us. What has been proven or disproven, suggested or dismissed by this investigation? We will consider the question in terms of: the phenomenon of foreign policy deviance in East Europe, as revealed by the comparative and case study findings; the international relations and level of integration of the of the region; and the relationship of the results to the comparative study of foreign policy. Some additional thoughts will also be offered on the individual countries, and, finally, the limitations of the work will be recalled as an eye is cast toward future tasks.

DEVIANCE IN EAST EUROPE AND ROMANIA

The Romanian case illustrates quite forcefully the operation of some of the factors which the broader study led us to expect would be important in understanding international political deviance.

That economic development was significant in stimulating interactional and attitudinal deviance on the part of Romania is clear. Its international interactions were designed to secure for the Romanian economy: 1) a high level of quality and quantity of technical goods, available from non-Soviet suppliers; 2) other sources of important raw materials vital to Romanian development; 3) other, reliable markets for Romanian exported goods, especially those where hard currency would be earned. At the same time, deviance in attitudes and interactions was in directions which supported the party's internal and external autonomy, and the national sovereignty of the state; and which might secure acceptance, if not support for, alternative methods of building socialism. The implications of this last were that the RCP thus found it in its interest to support the position of the Czech and Chinese parties. It these parties and their policies could be branded as non-Marxist, so could any other party including the Romanian.

The internal position of the Romanian government was significant in two ways, one expected, one not. The regional results indicated a relationship between level of threat based on the regime's need for legitimacy, and deviant foreign policy. In the Romanian case, the desire and ability of the party leadership to stimulate and capitalize on popular approval through an anti-Soviet policy was clearly an important factor. The de-Russification and general nationalistic nature of the campaign is proof of the importance of the regime's desire to achieve such support. After all, hypothetically, the international deviance need not have been accompanied by such activity. By the same token, the results did not suggest—and thus we would have no reason to expect- that in the Romanian case the regime would accompany international deviance with relaxation of internal subsystem controls. On the contrary, both within the RCP and in the nation at large, orthodoxy and elite control were strictly enforced. The Romanian leaders saw this as an important hedge against Soviet interference or threats against their policies, and little relaxation of either intra- or extra- party control occurred.

The comparative results suggested that overall there was no relation between trade/aid dependence and international deviance. Thus in investigating an individual case, we would expect that manipulation of such dependence by the Soviet Union was either 1) not utilized, or at least not utilized to a degree sufficient to be effective; or 2) circumvented in some way. In the case of Romania we found that both were true to some extent. We must be careful to recognize that since no cross-sectional association was found for the region, the precise working of this factor, indeed its significance, may vary from case to case, as with internal liberalization.

Two other expectations were that Romania's relatively strong natural resource position would be important in allowing deviance, and—a somewhat more equivocal expectation—that its weak geographic position would act as a braking factor on some aspects of deviance. The first was found to be true. As regards the second, geographic position did not act as a braking factor, due evidently to the very weakness and therefore insularity and "security" (from the Soviet standpoint) of the Romanian position. This would suggest, parenthetically, that, graphically represented, the relationship between geographic position and deviance is curvilinear.

In addition, investigation of the Romanian case demonstrated the importance of two other factors, not delineated in the comparative framework. One was the effect of the Chinese deviation: as an example to other potential deviants, as a divisive and weakening force within the world communist movement, and as a source of direct political and ideological support. The other was the internal weakness of the Soviet regime. Both of these are categorized as enabling factors for explaining deviance.

In sum, those results of the broader study that suggested the presence of a relationship were, for the most part, supported by the investigation of the individual case. And the non-generalizability that the comparative study suggested regarding certain other factors seems confirmed by the uniqueness of their character in the Romanian situation. Finally, the investigation of the case produced two additional factors not tested for the region, but found to be important in understanding this instance of international deviance.

THE INTERNATIONAL RELATIONS OF EAST EUROPE

The few studies there are on the international relations of the East European region have produced essentially two kinds of empirical results. On the one hand, some have shown that the region is, by measures of trade, UN voting, and in some cases interstate visits, relatively more integrated, quite cohesive and, moreover, quite separate from the rest of the world.[1] At the same time other studies have documented the degree of diversity in the foreign policy of these states.[2]

The results of our interaction study support the first of these conclusions to some degree. In Table VI.1 we see that of the six states of the WTO, two (GDR and Hungary) have a majority of their relations-as measured by Interaction Percentage (IP), 1965-1969—with the other states in East Europe and the Soviet Union; one (Czechoslovakia) has almost half; and two (Poland, Bulgaria) have around thirty percent. All but Romania spend a substantial portion of their time interacting with each other.

However, when these portions are compared with the shares of the states' *trade* with each other, a new perspective is given to the conclusions of Modelski[3] and Miles and Gillooly[4] regarding East Europe. Using these authors' method for 1967,[5] we find that five of the six states of East Europe carried on more than sixty percent of their trade with each other or the Soviet Union. But for every single state, with the exception of the GDR, the percentage of trade conducted with their allies was substantially larger than their interaction percentage with them.[6]

TABLE VI.1
EAST EUROPEAN TRADE PERCENTAGES AND INTERACTION PERCENTAGES

Percentage of Trade—1967[1]			Interaction Percentages—1965-69[2]		
BULGARIA			GDR		
	with EE & SOV	71.3		with EE & SOV	63.7
	with WCS[3]	74.5		with WCS[4]	76.7
GDR			HUNGARY		
	with EE & SOV	69.0		with EE & SOV	52.0
	with WCS	72.8		with WCS	67.4
CZECH			CZECH		
	with EE & SOV	66.5		with EE & SOV	47.8
	with WCS	70.3		with WCS	62.6
HUNGARY			POLAND		
	with EE & SOV	64.2		with EE & SOV	30.9
	with WCS	66.7		with WCS	41.5
POLAND			BULGARIA		
	with EE & SOV	60.7		with EE & SOV	28.1
	with WCS	64.4		with WCS	38.2
ROMANIA			ROMANIA		
	with EE & SOV	47.2		with EE & SOV	22.9
	with WCS	52.5		with WCS	37.7

1. Source: *Yearbook of International Trade Statistics*, 1969 (New York: United Nations, 1971).
2. From Table II.3.
3. Includes, in addition to East Europe and the USSR: Albania, China, Yugoslavia, North Korea and North Vietnam.
4. Includes, in addition to East Europe and the USSR: Albania, China, Yugoslavia and other Communist.

This means, it seems, several things. First, East Europe as a region is less integrated within itself and less separate from the rest of the world than Modelski and Miles and Gillooly suggest. The states therein may be heavily involved with each other in terms of trade, but they are much less so in terms of overall interactions. Moreover, one should not fail to note that Romania, Poland, and Bulgaria had as high or higher interaction percentages with the West as they did with a combination of the Soviet Union and East Europe.[7] The states of East Europe may be comparatively more integrated and more separate from the rest of the world than are other regions; but to confirm this, similar interaction studies would have to be done with other regions, such as Western Europe, Southeast Asia, or Africa.

Further, this points up some of the "slippage" involved in using trade as an indicator of political cohesion, as Hopmann and Hughes suggest.[8] The rankings of states by degree of involvement according to trade and according to interaction percentage are similar, though not identical (Spearman r = .607; s = .074). It might be safe to say, as do Clark and Farlow, that trade is suggestive of broad political involvement.[9] But a more comprehensive and sensitive indicator, used perhaps in combination with pure trade, would more nearly tap the totality of phenomena involved in the relations between these states.[10]

What about the conclusions of both Modelski and Miles and Gillooly that the entire world communist system is tightly integrated and separate from the rest of the world? As measured by overall Interaction Percentages (Table VI.1), three of the East European states (GDR, Hungary, Czechoslovakia) were involved mostly with other communist states, while three others (Poland, Bulgaria, Romania) were less so.[11] In terms of trade with other communist countries, and all except Romania conducted more than 60% of their trade with these states. Overall, however, the values of these two indicators are closer regarding the state's relations with the world communist system than for purely East European relations. (WCS trade mean = 66.9; WCS IP mean = 54.0). Thus the conclusions of Modelski and of Miles and Gillooly for the entire communist system are supported more strongly.

In addition to interactional evidence, the evidence drawn from the international attitudes of the East European states seriously calls into question the degree of Pact integration, and thus supports broadly similar findings offered by Hughes and Volgy,[12] Harle[13] and Kintner and Klaiber.[14] A distinct lack of consensus was revealed on a variety of issues. In fact, as Table IV.1 shows only four issues (three of them rather distant) produced high consensus within the Pact; four others, of which two were similarly distant, evoked somewhat less consensus; and on six others, of which four can be considered crucial Pact issues, the consensus was medium to low. The three lowest were all key Pact issues. While no data was taken for the whole communist system, there were data for the Eastern European region, i.e. including Albania and Yugoslavia, and the degree of consensus was even lower.

In this connection our results also support previous work which has indicated that greater inter-alliance conflict produces greater intra-group cohesion.[15] Looking back at Table IV.1, we see that for East Europe, of the issues producing high or medium cohesion, five of six involved situations of inter-camp hostility and conflict and the sixth involved open though indirect inter-alliance conflict (Arab-Israeli war). There is lower cohesion on issues relating almost completely (six of eight) to issues lying within the purview of the world communist movement. Further, if we look at the WTO only, the flareup of the Sino-Soviet border clash is added to the group of issues which produce relatively higher intra-group cohesion.

This result differs from previous tests of the conflict-cohesion hypothesis,however, by emerging from a comparison across issues rather than over time. In doing so a refinement is suggested in this relationship. Greater conflict or threat does indeed produce greater cohesion, but only in terms of that threat or conflict. There still may be, as indeed there was in East Europe, other areas where the group remained less cohesive. Thus, for example, the alliance was quite united in responding to US action in Vietnam, but quite varied in its response to developments in Czechoslovakia.

Examining the integration of the region in the particular manner that we have, i.e. by looking for conformity and deviance in the individual states' behavior, more depth and texture has been added to assesments of the degree of the region's integration. Not only do the levels of conformity vary across issues and forms of behavior (interaction *vs* attitudes) but these levels also differ from the degree of integration present or indicated by the institutions of the area. The approach offered in this work may thus bring us a step closer to tapping the informal, hegemonic political processes which are, as Clark put it, "where the action is" in East European integration.[16] An illustrative example is the case of China. The Soviet Union clearly sought a "well integrated" alliance to wield against its heretical ex-comrades in Peking. Institutionally neither the WTO nor Comecon were the appropriate vehicle; and investigating the locus of decision-making therein would be unlikely to reveal the success or failure of Soviet-led integration efforts. Looking at meetings of the world communist parties provides a convenient focus for such an investigation, but it is the dialogue and positions taken on these meetings, the movement, and on China itself which reveal the level of integration among Moscow's junior allies.

EAST EUROPE AND FOREIGN POLICY STUDIES

How do the present results relate to work accomplished in the comparative study of foreign policy? As indicated by the earlier review, there is little direct complementarity, there being few comparative studies of the foreign policies of East Europe. The present work supports and broadens the findings of those who have demonstrated the diversity of East European foreign policies. In addition, the specific hypothesis relating level of conflict to level of cohesion has been confirmed and embellished.

But the relationship to the findings of those who work with global aggregates of nations can be approached only inferentially, indirectly, though it is hoped, heuristically.

For example, Moore found that for "closed" nations regime accountability and level of Christian culture were important factors related to foreign policy alignment.[17] East Europe certainly qualifies as a set of "closed" nations, but the two independent variables, "accountability", and "Christian culture," and the dependent variable, "alignment" are not particularly sensitive or powerful when applied to East Europe. The distinctions designed for global categorization are too gross for use in this particular region.

Still, we did find that level of potential threat to a regime, resident in the ethnic heterogenity of its population, was an important factor underlying international deviance; both for interactions and attitudes; for the region and in the Romanian case. This would seem to indirectly support Moore's finding, if the leap from accountability to legitimacy and threat is not to great to make. On the other hand, the degree of "internal subsystem autonomy" also taps aspects of accountability, and the disproving of a relationship between this and foreign policy runs counter to the relationship posited by Moore. This poses, it would seem, the necessity for students of comparative foreign policy to revise and improve variables such as "accountability", and dichotmous designations such as "open-closed" to render these more sensitive and powerful as variables and classifications.

In a similarly broad study, McGowan and Kean determined that for all states, size and moderization, through indirect effect on a states needs and resources, were related to three of four dimensions of foreign policy: involvement in international organizations: degree of focus of foreign policy, i.e. the breadth or narrowness (in terms of targets) of foreign policy actions: and overall international involvement.[18] However these two attributes were found not to be related to level of regional involvement. These authors suggest that,

> Perhaps national attributes make a difference when foreign policy participation relates to the entire system, but not with respect to regional matters where needs, resources, and level of modernization tend to be similar because of historical and geographical diffusion.[19]

We have examined closely a region which, viewed from outside, seems to contain roughly similar nations but which upon closer scrutiny reveals important differences which are related directly to the states' patterns of regional involvement. The reduced regional involvement of Romania, for example, *was* related to its level of economic development, and to the peculiarities of its political authority and culture.

Further, McGowan and Kean themselves recognize that their results for the entire universe of nations do not stand up as well for the group of small states as for other clusters. For these, size and modernization explain a much smaller proportion of the variance in foreign policy focus and international involvement.[20] Studying East Europe has shown that these states' modernization level was related to their pattern of international involvement, but negatively, and only for one type of involvement, interactions. In addition, due to the nature of the subsystem, the "focus" of the relations of the less developed states was affected as well. Thus it would seem that the increasing attention being paid to the peculiarities of small states' foreign policies, mentioned earlier, seems justified.

Along these lines, McGowan and Gottwald, in a study of African states found that small, less modern states tend to follow either "promotive" or "acquiescent" foreign policies. That is, they will either try to deal with their internal and external environments by changing both of them to their satisfaction (promotive), or by changing the domestic environment to the satisfaction of external forces (acquiescent).[21] These authors found that, holding size constant, it was the more modernized of these states which were more likely to follow acquiescent policies, and moreover this was especially true of those states whose leadership was "other directed," i.e. attentive to external demands and forces, rather than "inner directed," attentive to domestic demands.

Our results seem to be at least analogous to those of McGowan and Gottwald, in the following way. The more modernized states of East Europe were those whose foreign policies did not deviate, in terms of interactions. They were not following the promotive foreign policies; of the less developed states such as Romania, who sought to change both their domestic structures—through broad and rapid development- and their external environment.

Had the latter been more modernized and hence burdened by an increased level of "stress sensitivity",[22] they would have been more likely to have been obliged to follow policies aimed at "modifying domestic structures to agree with external demands and changes."[23] As for the leadership, Ceauşescu seems clearly to qualify as more "inner directed" i.e. attuned to his domestic needs than "other directed" attuned to external- Soviet- demands. For East Europe then, the results tend to support those of McGowan and Gottwald for Africa, and to give empirical grounding to the expectations voiced by Morse on the relationship of modernization to foreign policy.[24]

In a similar vein, the present work also confirms a proposition stated by East relating modernization and societal stress to foreign policy. Focusing upon a state's "capacity to act," East divides this independent variable into two parts, size and social organization. The latter is itself divided into level of modernization and level of societal stress. Lower levels of these two result in lower levels of social organization and hence a lower capacity to act in foreign relations. Under these conditions, East suggests

> . . . it can be reasonably argued that the foreign policy behavior of such nations will reflect great concern for an interest in issues that might have an effect on their already low capacity to act. Issues and problems relating to economic growth and development will be particularly salient. Nations will seek out relationships with those other international actors that are most likely to increase their capacity to act.[25]

In East Europe it was precisely those states with a lower capacity to act—both due to lower levels of economic development (modernization) and higher levels of ethnic heterogenity (societal stress)—which deviated in their foreign policies. Furthermore, as was seen in the Romanian case, economic development was a primary concern of their foreign policy. Here East's specific investigative proposition seems to have been supported.:

> The lower a nations capacity to act, the higher its proportion of foreign policy behaviors related to those substantive problem areas involving economic matters.[26]

In addition, East suggests that there will be a higher level of committment and greater negative affect in the foreign policies of states with a lower capacity to act.[27] These cases of Romania and Yugoslavia, and to a lesser extent, Czechoslovakia—deviants one, two and three in international attitudes— do support this expectation. However, the stronger negative attitudes demonstrated by them did not occur for the reasons suggested by East; that is, that a lower capacity to act allows less monitoring of international relations and thus produces delayed and necessarily more forceful reactions.[28]

Finally, another of East's propositions, that "the lower a nation's capacity to act, the narrower the scope of action (both substantively and geographically) of its foreign policy behavior," is clearly disproven in the case of East Europe.[29]

Results such as these, based on smaller groups of states, partly supporting, partly undermining findings generated through universal data, should give a strong fillip to those who argue the merits of focusing on subsets. The results are indicative of ways in which broad-based findings may have to be qualified when students get down to specific cases or groups of cases. Such qualification may entail improvements in an independent or dependent variable, or may be as "simple" as explaining why some results are right for the wrong reasons!

INTERESTING COUNTRIES

As for the individual states we have studied, Romania was shown to be consistently and strongly deviant, on both attitudes and actions. This is not surprising. The fact that it deviated on both indices more than did Yugoslavia is. What does this mean? First and most simply, both indices revealed that Yugoslavia is somewhat more conforming than its formal position would suggest.[30] In terms of interactions its behavior was norm-adhering toward several targets (Israel, the Arab states, the FRG, the West, East Europe, China), while its attitudes were occasionally, though much less frequently, non-deviant (the Arab-Israeli war and two U.S.–related issues). That this was so and that Romania was generally more deviant than was a nonaligned state highlights the degree of Romanian foreign policy deviance.

More than that, a behavioral, outcome-oriented approach to integration has drawn attention to the fact that some nations which are formally outside a regional alliance structure frequently behave in a more allied manner than some of the allies, a lack of institutional rewards and constraints notwithstanding. Judgments of alignment, and of regional integration, then, should be based on an accurate observation of international behavior as well as on formal policy pronouncements.

The phenomenon of positive deviance was also highlighted in this analysis. It was interesting to see, for example, that on international attitudes the Soviet Union was out in front of its allies in hostility to China, hostility toward Czech reforms, and encouragement for a world communist conference. Pact norms on international attitudes did not necessarily equal the Soviet position. Alternatively, the presence of allies who are "holier than the Pope," was clearly evident. Bulgaria and the

GDR for example, were more hostile toward the United States and China than was the Soviet Union. It seems quite likely that this is due to a certain moderating effect operating on the Soviet Union in its position as world superpower. This conclusion, then, would tend to support those studies noted earlier which stress the difference in the international relations of the large states vs small ones. The Pact, it is clear, contains the whole spectrum of communist states' perspectives on the world. The analysis showed nothing if not that.

In addition to the alliance itself displaying a variety of reactions to the world, the individual countries were often inconsistent in their own international relations, at times supporting Pact positions while deviating in their interactions, and at times doing the reverse. Bulgaria was least consistent, and ironically the two opposites, East Germany and Romania, were most consistent in their external relations. If there was any "typical ally," and a consistently typical one, it was Hungary. There is irony in that too perhaps.

MILES TO GO . . .

The weaknesses of this analysis, though concealed as cleverly as possible should be recalled in order to make clear the limitations of the above conclusions, and illustrate possible improvements by which the work might benefit.

Methodologically, the scaling system in the interaction analysis needs improvement. It may reflect too great a degree of observer judgment and therefore might be too much "softer" than a less obtrusive measure might be. But there is no getting around the fact that not all international interactions are of the same worth. Students of East Europe, indeed students of international relations in general, make such judgments as a matter of course. They must, in order to determine, quite literally, what in the world to study. The strength of the system used herein lies in making the assumptions underlying such judgments explicit. Further, as seen in the use of the trade data, a low level of obtrusion is no guarantee that the results will not be misleading. Indicators based on well-considered judgments of international phenomena in fact, can correct such errors.

There are, as well, other, probably less serious, methodological disputes that may be offered. The grouping of targets in the interaction study according to formal alliances seems to be contrary to the very purpose and indeed conclusions of this study. Yet, if one may be so bold, until other similar studies are prepared for the rest of the world, how else could one proceed to group the 150-odd nations with whom the East European states interacted during our time period? The periodization, as well as the unweighted adding of evaluated z-scores in Chapter II may, as was recognized, have added a certain degree of distortion to the analysis. The arguments offered supporting these determinations, upon rereading, still seem sound.

As for the attitudinal analysis, it is hoped that, as in all content analysis, possible weaknesses of the observer measurement technique were mitigated by, paradoxically, utilizing more observers. The most serious problem in this part of the study, however, is that of data insufficiency, discussed earlier. The goal of this research, as in all research, was to state conclusions justified by a sufficient number and spread of reporting cases and to avoid those conclusions not so justified. It is to be remembered that what was studied, attitudes, were reactions to international issues or events, used to operationalize that maddeningly fuzzy term, foreign policy. In doing so, a certain amount of generality has been lost. But by doing so in a rigorous, structured manner, the method of comparative analysis is rendered more powerful and the results therefrom, more robust.

This leads to a final point. This was a comparative study of foreign policy. That it was comparative cannot be disputed. That it was foreign policy that was studied may be. Aside from issues of events data or document analysis, it may be argued that foreign policy is somehow more than the sum of its parts; that by stringing together issue responses and interactions over time and geography one gets the trees of foreign policy but loses the forest.

Getting at the depth of a country's foreign policy, ferreting out whatever texture and form lies apart from the desiderata studied herein is of undeniable significance. It is the proper work of case study. To fully and completely comprehend the cause and course of one nation's foreign policy, one can focus on but one country at a time, in East Europe no less than elsewhere. This work, it is hoped, gave support to that approach as well by offering an examination of the Romanian case that both fit into and embellished upon the comparative study of the region.

It should be noted, further, that the approach offered here is confined in its structure neither to East Europe nor to the phenomena studied therein. There are a variety of mutations, and, it is to be expected, improvements, which would add depth and breadth to the determination of the patterns of and factors related to the international relations of states. Other forms of interaction, e.g. multilateral meetings, attitudes on other issues, other forms of comparison, other factors, other statistical investigations, and of course other groups of countries, regional, ideological or petrochemical, could be utilized fruitfully. The injunction of this writer, then, would echo that of Maurice Sendak's character, Max: "And now," cried Max, "Let the wild rumpus start!"[31]

NOTES

NOTES TO CHAPTER ONE

1. (New York: The Free Press, 1971).

2. (Chicago: Rand McNally, 1969).

3. See the April, 1971 (IV, 2); Spring/Summer, 1973 (VI, 1&2); and Autumn, 1975 (VIII, 3); issues of *Studies in Comparative Communism*; and the *Newsletter on Comparative Studies of Communism*, published from February, 1968 through August, 1973.

4. (Indianapolis: Bobbs-Merril, 1969).

5. (Baltimore: Johns Hopkins University Press). The eight individual works are: Arthur Hanhardt, Jr., *The German Democratic Republic*, 1968; James Morrison, *The Polish People's Republic*, 1968; M. Georgе Zaninovich, *The Development of Socialist Yugoslavia*, 1968; Nicholas Pano, *The People's Republic of Albania*, 1969; Stephen Fischer-Galati, *The Socialist Republic of Rumania*, 1969; Bennet Kovrig, *The Hungarian People's Republic*, 1970; and Nissan Oren, *Revolution Administered: Agrarianism and Communism in Bulgaria*, 1973.

6. (New York: Praeger, 1977).

7. (Stanford: Stanford University Press, 1970).

8. (New York: David McKay Co., 1973).

9. (Pittsburgh: University Center for International Studies, University of Pittsburgh, 1975).

10. Ghita Ionescu, *The Politics of the European Communist States* (New York: Praeger, 1967); Richard C. Gripp, *The Political System of Communism* (New York: Dodd, Mead, 1973); Jan F. Triska and Paul M. Johnson, *Political Development and Political Change in Eastern Europe* (Denver: University of Denver, 1975).

11. On the development of this field see James N. Rosenau, *The Scientific Study of Foreign Policy* (New York: The Free Press, 1971); Rosenau, "Comparing Foreign Policies: Why, What, How" in Rosenau (ed.) *Comparing Foreign Policies* (Beverly Hills, Calif.: Sage Publications, 1974), pp. 3-22; and Rosenau, "Comparative Foreign Policy: One-Time Fad, Realized Fantasy, and Normal Field," in Charles W. Kegley, Jr., Gregory A. Raymond, Robert M. Rood, and Richard A. Skinner, *International Events and the Comparative Analysis of Foreign Policy*

(Columbia, S.C.: University of South Carolina Press, 1975), pp. 3-38.

12. *Studies in Comparative Communism*, VIII, 1&2 (Spring/Summer, 1975). The conference was held in Marina del Rey, California, May 14, 1974.

13. (New York: Praeger, 1976).

14. (New York: Praeger, 1976).

15. (New York: Praeger, 1974).

16. McGowan and Shapiro's review of the field of comparative foreign policy as of 1973 notes only two studies which include East Europe as a substantial portion of their investigative domain: P. Terrence Hopmann, "International Conflict and Cohesion in the Communist System," *International Studies Quarterly*, XI, 3 (September, 1967), pp. 212-36; and Johan Galtung, "East-West Interaction Patterns," *Journal of Peace Research*, III, 2 (1966), pp. 146-77; and none at all focusing on East Europe alone. See Patrick J. McGowan and Howard B. Shapiro, *The Comparative Study of Foreign Policy* (Beverly Hills, Calif.: Sage Publications, 1973). Studies to date which do focus on East Europe are: Barry Hughes and Thomas Volgy, "Distance in Foreign Policy Behavior: A Comparative Study of Eastern Europe," *Midwest Journal of Political Science*, XIV, 3 (August, 1970), pp. 459-92; William R. Kintner and Wolfgang Klaiber, *Eastern Europe and European Security* (New York: Dunellen, 1971); and Harvey J. Tucker, "Measuring Cohesion in the International Communist Movement," *Political Methodology*, II, 1 (1975), pp. 83-112. Some comparative content-analysis can also be found in Robin A. Remington, *The Warsaw Pact* (Cambridge: M.I.T. Press, 1971), Chapter 3. In addition, P. Terrence Hopmann and Barry B. Hughes, "The Use of Events Data for the Measurement of Cohesion in International Political Coalitions: A Validity Study," in Edward Azar and Joseph Ben-Dak, (eds.) *Theory and Practice of Events Research* (New York: Gordon and Breach, 1975), pp. 81-94; Vilho Harle, "Actional Distance Between the Socialist Countries in the 1960s" *Cooperation and Conflict*, 14, 3&4 (1971), pp. 201-22; and the chapters in Jan F. Triska, Jan F. (ed.) *Communist Party-States* (Indianapolis: Bobbs-Merril, 1969), by Finley, Miles and Gillooly, Cary, and Simon, include East Europe in their studies of various aspects of the international relations of all communist states.

17. Henry Krisch states the case as follows: "Only a comparative approach can move us beyond the level of understanding reflected in the plethora of descriptive single-country studies. One thinks of attempts to conceptualize Soviet foreign policy under such varied headings as 'Muscovite', 'Slavic', 'Eurasian','continental', 'great power', 'international

communist', and so forth." "Some Undone Jobs," in *Studies in Comparative Communism* VIII, 1&2 (Spring/Summer, 1975), p. 32. For a review of recent work done on Soviet foreign policy, see William Welch and Jan F. Triska, "Soviet Foreign Policy Studies and Foreign Policy Models," *World Politics*, XXIII, 4 (July, 1971), pp. 704-33.

18. Charles Gati, "Area Studies and International Relations: Introductory Remarks," *Studies in Comparative Communism*, VIII, 1&2 (Spring/Summer, 1975), pp. 9-10.

19. Roger Kanet, "Is Comparison Useful or Possible?" *Studies in Comparative Communism*, VIII, 1&2 (Spring/Summer, 1975), p. 23.

20. *Ibid.*, p. 26. Cf. Alvin Z. Rubinstein, "Comparison or Confusion?" *Studies in Comparative Communism*, VIII, 1&2 (Spring/Summer, 1975), p. 45.

21. David Finley, among others, makes this point in relation to the study of communist foreign policies. See his discussion of the "mutual dependence" of idiographic and nomothetic work in "What Should We Compare, Why and How?" *Studies in Comparative Communism*, VIII, 1&2 (Spring/Summer, 1975), pp. 12-19; and the support for this position voiced by nearly all of the contributors to the "Symposium on the Comparative Study of Communist Foreign Policies" *Idem*. pp. 5-65.

22. See Chapter 5.

23. On this point, in addition to the "Symposium" (see footnote 21), see Zvi Gitelman, "Toward a Comparative Foreign Policy of Eastern Europe," in Peter Potichnyj and Jane Shapiro, (eds.) *From the Cold War to Detente* (New York: Praeger, 1976), pp. 144-65.

24. See e.g. David W. Moore, "National Attributes and Nation Typologies: A Look at the Rosenau Genotypes" in Rosenau (ed.) *Comparing*, pp. 251-67. Moore states in concluding that "The empriical results demonstrate that the relationships between foreign policy and national attributes in one genotype tend to be very different from the relationships in another genotype, and that these differences are obscured in an analysis across all nations. Thus, it would appear that sub-models of foreign policy behavior for each genotype may need to be developed."(p. 264). Cf. David W. Moore, "Governmental and Societal Influences on Foreign Policy in Open and Closed Nations," *Ibid.* pp. 171-199; Maurie A. East and Charles F. Hermann, "Do Nation-Types Account for Foreign Policy Behavior?" in *Ibid.*, pp. 269-303; and Barbara Salmore and Stephen Salmore "Political Regimes and Foreign Policy" in Maurice A. East, Stephen Salmore and Charles F. Hermann, *Why Nations Act* (Beverly Hills, Calif.: Sage Publications, 1978), pp. 103-22. Rosenau's original delineation of the genotypes can be found in "Pre-Theories and Theories of Foreign Policy," in R. Barry Farrell (ed.) *Approaches to Comparative and International*

Politics (Evanston: Northwestern University Press, 1966), pp. 27-92.

25. James E. Harf, David G. Hoovler, and Thomas E. James, Jr., "Systemic and External Attributes in Foreign Policy Analysis," in Rosenau (ed.) *Comparing*, pp. 246 (emphasis in original). Similarly, Kean and McGowan suggest "that although most theorizing in international relations and in the comparative study of foreign policy purports to be general, and therefore applicable to all states, it is implicitly modeled on the behavior of "great" powers. The particular behavioral patterns of smaller states are only now beginning to get the attention they deserve from theorists in foreign policy studies." James G. Kean and Patrick J. McGowan, "National Attributes and Foreign Policy Participation: A Path Analysis" in Patrick J. McGowan (ed.) *SAGE International Yearbook of Foreign Policy Studies*, Vol. I (Beverly Hills, Calif.: Sage Publications, 1973), p. 246.

26. See James N. Rosenau, "Comparison as a State of Mind," *Studies in Comparative Communism*, VIII, 1&2 (Spring/Summer, 1975), pp. 57-61.

27. Roger E. Kanet, "Integration Theory and the Study of Eastern Europe," *International Studies Quarterly*, XVIII, 18 (September, 1974), pp. 368-92.

28. Walter C. Clemens, Jr., describes the Warsaw Pact as a case of "enforced integration." "The historical record suggests that WTO is unique- not for being a voluntary alliance of equals dedicated to enhancing their common aims, but for providing the legal and military framework that for decades helps a hegemonical power to impose its will upon weaker neighbors who, given a free choice, might well opt for non-alignment or even participation in the security operations of the opposing camp." "The European Alliance Systems: Exploitation or Mutual Aid?" in Charles Gati (ed.), *The International Politics of Eastern Europe* (New York: Praeger, 1976), p. 235.

29. Cal Clark, "The Study of East European Integration: A Political Perspective," *East Central Europe*, II, 2 (1975), pp. 133-51. As an example see David D. Finley, "Integration Among the Communist Party-States: Comparative Case Studies," in Jan F. Triska, (ed.) *Communist Party-States*, pp. 57-81. This is not to gainsay the value of such studies for what they reveal about the institutions themselves or Soviet foreign policy. On COMECON, see Henry W. Schaefer, *Comecon and the Politics of Intergration* (New York: Praeger, 1972); and Andrzej Korbonski, "COMECON: The Evolution of COMECON," in *International Political Communities* (Garden City, N.Y.: Doubleday, 1966), pp. 351-405. The latter admits in a later study: "More and more, Comecon seemed to act

primarily as a clearing house for ideas and suggestions, providing an institutional umbrella for bilateral and occasionally multilateral agreements; serving as a forum for the exchange of economic and technical information; conducting research, and acting as an arbiter in cases of disagreement or nonfulfillment of contracts. However important and useful, all these functions could be seen as a terminal rather than an initial stage in the East European integratition process." Andrzej Korbonski, "Theory and Practice of Regional Integration: The Case of Comecon," in Leon H. Lindberg and Stuart A. Scheingold, *Regional Integration Theory and Research* (Cambridge: Harvard University Press, 1971) pp. 338-74 (quote is p. 364). Clark agrees with this characterization for both COMECON and the WTO and concludes that integration viewed in this manner appears to have stopped at a "halfway house," *Idem.*, p. 149. Two more recent works on CMEA are Paul Marer, "Prospects for Integration in the Council for Mutual Economic Assistance (CMEA)" *International Organization*, XXX, 4 (Autumn, 1976), pp. 631-48; and Dina Spechler, "The Council for Mutual Economic Assistance: An Analysis of Soviet Motives in the Making of Foreign Policy," *Slavic and Soviet Series* (Tel Aviv), II, 3 (Fall, 1977), pp. 3-39. On the Warsaw Pact see Remington, *op. cit.* and A. Ross Johnson, "Has Eastern Europe Become a Liability to the Soviet Union? (II) The Military Aspect," in Gati, *op. cit.* pp. 37-58; and Lawrence T. Caldwell, "The Warsaw Pact: Directions of Change," *Problems of Communism* XXIV (September/October, 1975), pp. 1-19.

30. Karl W. Deutsch, "A Comparison of French and German Elites in the European Political Environment," in Deutsch, et. al., *France, Germany, and the Western Alliance* (New York: Scribner's, 1967), pp. 213-314.

31. See Nye's dissection and review of the study of the "confusing concept of integration;" Joseph S. Nye, *Peace in Parts* (Boston: Little, Brown, 1971). pp. 21-54. Cf. Leon Lindberg, "Political Integration as a Multidimensional Pehnomenon Requiring Multivariate Measurement," in Lindberg and Scheingold, *op. cit.*, pp. 45-127. For an empirical assessment of Nye's disaggregation, see Charles W. Kegley, Jr. and Llewellyn D. Howell, Jr., "The Dimensionality of Regional Integration: Construct Validation in the Southeast Asian Context," *International Organization*, XXIX, 4 (Autumn, 1975), pp. 997-1020. For a brief argument that integration is in fact unidimensional, see Robert A. Bernstein, "International Integration: Multidimensional or Unidimensional?" *Journal of Conflict Resolution*, XVI, 3 (September, 1972), pp. 403-08.

32. Deutsch, *op. cit.*, pp. 218-40; Donald J. Puchala, "International

Transactions and Regional Integration," in Lindberg and Scheingold, *op. cit.*, pp. 128-59.

33. Ernst B. Haas and Philippe C. Schmitter, "Economics and Differential Patterns of Political Integration," in *International Political Communities* pp. 259-301; Nye, *op. cit.*, pp. 30-32 and 48ff; Lindberg, *op. cit.*,; and Lindberg and Stuart Scheingold, *Europe's Would-Be Polity* (Englewood Cliffs, N.J.: Prentice-Hall, 1970).

34. Karl W. Deutsch, et. al., "Political Community and the North Atlantic Area," in *International Political Communities*, pp. 1-92; Haas and Schmitter, *op. cit.*; Deutsch, "A Comparison,"; Donald J. Puchala, "Integration and Disintegration in Franco-German Relations, 1954-65," *International Organization*, XXIV, 2 (Spring, 1970), pp. 183-209. Llewwllyn D. Howell, Jr., "Attitudinal Distance in Southeast Asia," *Southeast Asia*, 3, 1 (Winter, 1974), pp. 577-605. In addition, students who view international law sociologically have studied this type of integration. See Stanley Hoffmann, "International Systems and International Law," in Klauss Knorr and Sidney Verba (eds.), *The International System* (Princeton: Princeton University Press, 1961) pp. 205-39.

35. See Warner S. Landecker, "Types of Integration and Their Measurement," in Paul F. Lazarsfeld and Morris Rosenberg (eds.), *The Language of Social Research* (New York: The Free Press, 1955), pp. 22-3.

36. Ole R. Holsti, P. Terrence Hopmann, and John D. Sullivan, *Unity and Disintegration in International Alliances: Comparative Studies* (New York: John Wiley, 1973), p. 176. Zimmerman, in discussing the "boundaries" of hierarchical regional systems, agrees: "boundaries may similarly be established by emphasizing behavioral criteria, i.e., by identifying norms especially pertaining to conflict management and resolution which are specific to a group of states," William Zimmerman, "Hierarchical Regional Systems and the Politics of System Boundaries," *International Organization*, XXVI, 1 (Winter, 1972), p. 18.

37. Julius Gould and W.L. Kolb (eds.), *A Dictionary of The Social Sciences* (New York: The Free Press of Glencoe, 1964), p. 472.

38. On this problem see Myres S. McDougal, "Some Basic Theoretical Concepts About International Law; A Policy Oriented Framework of Inquiry," and Morton Kaplan and Nicholas deB. Katzenbach, "Law in the International Community," both in Richard A. Falk and Saul H. Mendlovitz (eds.), *The Strategy of World Order*, Volume II, *International Law* (New York: World Law Fund, 1966), pp. 116-34 and 19-45, respectively. Cf. K. Goldman, *International Norms and War Between States* (Stockholm: Laromedels Forlagen, 1971), who attempts to solve the problem by using states' justificatory statements as evidence for the

existence of international norms.

39. Goldmann, *op. cit.*, pp. 13, 34-5. On the sources of international norms of this type see Louis Henkin, *How Nations Behave* (New York: Praeger, 1968), pp. 13-23; Gerhard von Glahn, *Law Among Nations*, 2nd ed. (New York: Macmillan, 1970), pp. 10-21; Kaplan and Katzenbach, *op. cit.*; Hoffmann, *op. cit.*; Hans Kelsen, *Principles of International Law*, 2nd ed. rev. and ed. by Robert W. Tucker, (New York: Holt, Rinehart & Winston, 1966), pp. 437-508.

40. See e.g. von Glahn, *op. cit.*, pp. 23ff.; Henkin, *op. cit.*, pp. 31ff.; Kelsen, *op. cit.*, passim; Edward McWhinney, "Soviet and Western International Law and the Cold War in the Era of Bipolarity," and Louis Henkin, "Force, Intervention and Neutrality in Contemporary International Law," both in Falk and Mendlovitz, *op. cit.*, pp. 189-231 and 335-52, respectively; Myres S. McDougal and associates, *Studies in World Public Order* (New Haven: Yale University Press, 1960), *passim*.

41. See Anthony A. D'Amato, *The Concept of Custom in International Law* (Ithaca: Cornell University Press, 1971), pp. 128-38; cf. Wesley L. Gould and Michael Barkun, *International Law and the Social Sciences* (Princeton: Princeton University Press, 1970), pp. 179-87, 205-209; and Kelsen, *op. cit.*, pp. 440-54. Sociological interpretations of international law especially emphasize this aspect of norm creation; see McDougal, "Some Basic Theoretical Concepts," *op. cit.*.

42. Goldmann, *op. cit.*, p. 37 (emphasis in original).

43. Steven Box, *Deviance, Reality and Society* (London: Holt, Rinehart, and Winston, 1971), p. 9. Albert K. Cohen agrees, "Behavior is deviant, then, only if the actor is subject to the jurisdiction of the rules that the behavior contravenes," *Deviance and Control* (Englewood Cliffs, N.J.: Prentice-Hall, 1966), p. 13. See also Howard S. Becker, *Outsiders* (New York: The Free Press, 1963), p. 9; John I Kitsuse, "Societal Reaction to Deviant Behavior," in Earl Rubinton and Martin Weinberg, *Deviance: The Inter-Actionist Perspective* (New York: Macmillan, 1968), p. 30; Marshall Clinard, *Sociology of Deviant Behavior* (New York: Rinehart and Company, 1957), pp. 13-14; and Rubington and Weinberg, *op. cit.*, p. v.

44. See Bruce Andrews, "Social Rules and the State as a Social Actor," *World Politics* XXVII, 4 (July, 1975), pp. 521-40; Goldmann, *op. cit.*, pp. 17, 22-8; Henkin, *op. cit.*, pp. 65-83; von Glahn, *op. cit.*, pp.5-6, 18-20; W. Michael Reisman, "Sanctions and Enforcement," in Cyril E. Black and Richard A. Falk (eds.), *The Future of the International Legal Order*, Vol. III: *Conflict Management* (Princeton: Princeton University Press, 1971), pp. 273-335. See also Wallensteen's discussion of the

"expressive" character of sanctions; Peter Wallensteen, "Characteristics of Economic Sanctions," in William D. Coplin and Charles W. Kegley, Jr. (eds.), *A Multimethod Introduction to International Politics* (Chicago: Markham Publications, 1971), pp. 128-54; and on positive sanctions, David A. Baldwin, "The Power of Positive Sanctions," *World Politics* XXIV, 1 (October, 1971), pp. 19-39.

45. Goldmann, *op. cit.*, p. 43.

46. Even if, as is frequently the case, the explicit deviance-sanction process were not traceable, or if, as is rarely the case, it could be shown to be of trivial importance in the group, the normative integration approach to the study of foreign policy would still be significant due to its sketching of the context of the state's international behavior. "State actions," Andrews notes, "resemble language . . . . The significance (or 'motivation') of words, for example, is not intrinsic. It derives from their place within a framework of governing rules and expectations." Actions of states thus have "referential dimensions" which are the "patterns (outside the regularities made by the actions themselves) which these actions both point to and implicate . . . . " They have significance and can be understood only in certain contexts: domestic, regional, world. "Without these rule-guided elements state behavior would be socially unintelligible, even where it might be causually predictable." *op. cit.*, p. 525.

47. Gould and Barkun point out that deviance on a specific issue can be, on the contrary, norm-creating or reinforcing, since it calls forth responses, e.g. sanctions, which thereby define the norm. (*op. cit.*, pp. 197-209). This seems logical and, assuming one could extract the specific evidence required to sustain this expectation, not necessarily mutually exclusive of the position stated above. For it seems equally logical that a deviant behavior pattern on the part of a group member over a broad time and issue area domain will have a harmful effect upon the group's norms. Indeed this seems even more likely since, as Gould and Barkun themselves indicate, such behavior tends to create a new and conflicting norm which competes with the old, further eroding the level of normative integration of the group. *Idem.* On norm conflict see also Goldman, *op. cit.*, p. 32.

48. Nye, *op. cit.*, pp. 43-57.

49. Kegley and Howell, *op. cit.*, pp. 1016-17.

50. *Ibid.*, p. 1015. Such testing also would presumably utilize a somewhat broader empirical domain than that employed by Kegley and Howell to test policy identity (one general assembly session of the UN). These authors are themselves aware of the limited nature of their findings; see pp. 1017-18.

51. Roger D. Hansen, "European Integration: Forward March, Parade Rest, or Dismissed?" *International Organization*, XXVII, 3 (Spring, 1973), pp. 225-54. (Emphasis added).

52. Gitelman, *op. cit.*, George Modelski, *The Communist International System* (Princeton: Center for International Studies, 1960); Zimmerman, *op. cit.*

53. Lindberg suggests, "A more direct measure (of integration) might be derived from analyses of the content of governmental foreign and domestic policies. Are they becoming more similar in areas in which there has been collective action? I know of no systematic effort to compare public policies and to develop measures of similarity although some have sought to measure the similarities between in terms of increasing homogeneity of voting behavior in the United Nations. Nevertheless, it should not be an insuperable task if we turn our minds to it." *op. cit.*, p. 107.

NOTES TO CHAPTER TWO

1. Phillip E. Uren, "Patterns of Economic Relations," in Adam Bromke and Teresa Rakowska-Harmstone, *The Communist States in Disarray, 1965-1971* (Minneapolis: University of Minnesota Press, 1972), p. 311. Bogdan Denitch notes similar pressures stemming largely from the needs of the more developed states to satisfy the increasing demands- material, cultural, and educational- of their new technocratic elites and their new working classes. See "The Domestic Roots of Foreign Policy in Eastern Europe," in Gati, *The International.*

2. For a discussion of these aspects of modernization and development as they apply to East Europe, see Charles Gati (ed.) *The Politics of Modernization in Eastern Europe* (New York: Praeger, 1974).

3. Edward L. Miles and John S. Gillooly, "Processes of Interaction Among The Fourteen Communist Party-States: An Explanatory Essay," in Triska, *op.cit.*, p. 127.

4. Uren, *op. cit.*, p. 315.

5. *Idem.*

6. Edward L. Morse, "The Transformation of Foreign Policies: Modernization, Interdependence, and Externalization," *World Politics*, XXII, 3 (April 1970), pp. 371-92.

7. Economic sanctions and rewards are examples of "selective incentives" employed to improve group performance. See Mancur Olson, *The Logic of Collective Action* (New York: Schocken Books, 1968), pp. 60-65.

8. For a discussion of the dependence of East Europe on the Soviet Union, see Paul Marer, "Has East Europe Become a Liability to the Soviet Union? (III) The Economic Aspect," in Gati, *The International* pp. 31-59; Roger Kanet, "East-West Trade and the Limits of Western Influence" in *Ibid.*, esp. pp. 201-207; Christopher Joyner, "The Energy Situation in Eastern Europe: Problems and Prospects" *East European Quarterly* X, 4 (Winter, 1976), pp. 495-516. On Soviet willingness to use such economic levers see Robert O. Freedman, *Economic Warfare in the Communist Bloc* (New York: Praeger, 1970).

9. Miles and Gillooly, *op. cit.*, p. 110.

10. As Teune and Synnestvedt put it, "Economic needs may condition a nation's decision-makers to commit themselves to one or the other of the major powers on which their economic future depends. In short, perceived economic dependence may issue into political sympathies." Henry Teune and Sig Synnestvedt, "Measuring International Alignment," in Julian R. Friedman, Christopher Bladen and Steven Rosen, *Alliance in International Politics* (Boston: Allyn and Bacon, 1970), p. 331.

11. John M. Montias, "The Structure of Comecon Trade and the Prospects for East-West Exchanges," in Joint Economic Committee, *Reorientation and Commercial Relations of the Economies of Eastern Europe, A Compendium of Papers*, 93rd Congress, 2nd Sess., (Washington: US Government Printing Office, 1974), pp. 662-81; Kanet, "East-West Trade". Marer sums up the situation as follows: "Within the bloc, the Soviet Union is the only net supplier of primary products and the principal net importer of machinery and other manufactures. Thus the commodity composition of trade should reinforce the strong potential bargaining power inherent in its superpower status.""Has Eastern Europe," p. 61.

12. Kintner and Klaiber *op.cit.*, found such a relationship between trade dependence and their "Index of Conformity." See pp. 226-34 and discussion below.

13. See, e.g. George Liska, *Nations in Alliance* (Baltimore: The Johns Hopkins University Press, 1962), pp. 12-13. For an inventory of such propositions, see Holsti, Hopmann and Sullivan, *op. cit.*, pp. 17-18.

14. Clemens, *op. cit.*,; Zimmerman, *op. cit.*

15. In analytical terms, Scott points out that credibility of threat is dependent upon the capability, intent, and past behavior of the threatener. Andrew Scott, *The Functioning of the International Political System* (New York: Macmillan, 1967), pp. 184, 186.

16. Among those who have noted the importance of this variable are: Jan F. Triska, "The World Communist System," in Triska, *op. cit.*, p. 34; Vernon V. Aspaturian, "The Soviet Union and Eastern Europe: The

Aftermath of the Czechoslovaki Invasion," in I. William Zartman (ed.) *Czechoslovakia: Intervention and Impact* (New York: New York University Press, 1970), p. 36; and Peter Bender, *East Europe in Search of Security* (Baltimore: The Johns Hopkins University Press, 1972), *passim.*

17. This proposition can also be derived from that of Holsti, Hopmann, and Sullivan: "The more dispersed the members of an alliance, the less effective it will be." See *op. cit.*, p. 57, and also their inconclusive findings pp. 68-72. Kintner and Klaiber, *op. cit.* test a similar proposition; see pp. 234-35 and the discussion below.

18. Charles W. Kegley, Stephen A. Salmore and David Rosen, "Convergences in the Measurement of Interstate Behavior," in Patrick J. McGowan (ed.) *SAGE International Yearbook of Foreign Policy Studies*, Vol. II (Beverly Hills: Sage Publications, 1974), p. 336n1. Edward L. Morse has illustrated that the term "foreign policy" has been conceived of in three broad categories—as substance, i.e. goals and purposes, as processes, and as effects—leading to no less than ten different sub-categories of definition. "Defining Foreign Policy for Comparative Analysis: A Research Note," unpublished mimeo, (Princeton, June 1971).

19. For discussions of this problem see Edward L. Morse, *A Comparative Approach to the Study of Foreign Policy: Notes on Theorizing* (Princeton: Center of International Studies, 1971); and James N. Rosenau, "Moral Fervor, Systematic Analysis and Scientific Consciousness," in Rosenau, *The Scientific Study of Foreign Policy* (New York: The Free Press, 1971), pp. 23-65.

20. Rosenau, "Moral Fervor," p. 24. Cf. Kegley, et. al., *op. cit.*

21. It would be possible to utilize as an aggregable, though unilateral, foreign policy act Rosenau's concept of an undertaking, i.e. "a course of action that the duly constituted officials of a national society pursue in order to preserve or alter a situation in the international system in such a way that it is consistent with a goal or goals decided upon by them or their predecessors," if one removed from it the phrases that imbue purpose, e.g. " in such a way that it is consistent with a goal . . . ." "Moral Fervor," p. 50. On Rosenau, see also Morse, "Defining Foreign Policy," pp. 35-39.

22. J. David Singer, "The Level of Analysis Problem in International Relations," in Knorr and Verba, *op. cit.*, pp. 77-93.

23. John C. McKinney, *Constructive Typology and Social Theory* (New York: Appleton-Century Crofts, 1966), pp. 6, 12. "In contrast to the concept," McKinney points out, "a constructed type is determined to a greater degree by the selective and creative activity of the scientist. The primary distinction, however, is that its value as a component of

knowledge is not to be measured by the accuracy of its correspondence to perceptual experiences (although some degree of correspondence is essential), but *in terms of its capacity to explain*, however tentative and preliminary that explanation may be. A scientific function of the constructed type is to order the concrete data so that they may be described in terms that make them *comparable*, so that the experience had in one, despite its uniqueness, *may be made to reveal with some degree of probability what may be expected in others*." (pp. 11-12, emphasis in original.) McKinney furhter notes that "[i]mplicit reference here is to a statistical probability," (p. 12 n.3).

24. Albania, though formally a Pact member until 1968, in fact stopped or was excluded from participating in WTO meetings in 1961 and thus will be considered in this analysis as a non-member. See Robin Remington, *The Warsaw Pact* (Cambridge: M.I.T. Press, 1971), pp. 48-55, 167-69.

25. Galtung, *op. cit.*, found, for example, that the big powers accounted for most of the fifteen different types of interactions measured between the NATO alliance and the WTO. Hermann found, in studying the period 1959-1968, that 11 large nations initiated 6532 foreign policy events (average = 593.8), while 24 small nations accounted for 5085 (average = 211.9). Charles F. Hermann, "Comparing the Foreign Policy Events of Nations," in Charles W. Kegley, Jr., Gregory A. Raymond, Robert M. Rood, and Richard A. Skinner, *International Events and the Comparative Analysis of Foreign Policy* (Columbia, S.C.: University of South Carolina Press, 1975) pp. 115-58. For similar results, see Maurice A. East, "Size and Foreign Policy Behavior: A test of Two Models," in *Ibid.*, pp. 159-178. See also Mancur Olson, and Richard Zeckhouser, "An Economic Theory of Alliances," *Review of Economics and Statistics*, XLVII, 3 (August, 1966), pp. 266-79; Charles A. McClelland and Gary D. Hoggard, "Conflict Patterns in the Interactions Among Nations," in James N. Rosenau (ed.) *International Politics and Foreign Policy* rev. ed. (New York: The Free Press, 1969), pp. 711-24; Steven J. Brams, "The Structure of Influence Relationships in the International System," in *Ibid.*, pp. 583-99. In addition a small but growing body of literature has focused upon the distinctiveness of the role and performance of small and medium-size states. See Robert Rothstein, *Alliances and Small Powers* (New York: Columbia University Press, 1968); David Vital, *The Inequality of States* (Oxford: Oxford University Press, 1967); and Edward Azar, *Probe for Peace* (Minneapolis: Burgess Publishing Company, 1973).

26. See Ole R. Holsti, Richard A. Brody, and Robert C. North, "Measuring Affect and Action in International Reaction Models: Empirical Materials From the 1962 Cuban Crisis," in Rosenau *International*

Politics and Foreign Policy, rev. ed. (New York: The Free Press, 1969), (hereafter cited as *IPFP*), pp. 679-96; Ole R. Holsti, Robert C. North, and Richard A. Brody, "Perception and Action in the 1914 Crisis," in J. David Singer, (ed.) *Quantitative International Politics* (New York: The Free Press, 1968), pp. 123-59; McClelland and Hoggard, *op. cit.*; Phillip M. Burgess and Raymond W. Lawton, *Indicators of International Behavior: An Assessment of Events Data Research* (Beverly Hills, Sage Publications, 1972).

27. An example of a study using high-level visits only is Brams, *op. cit.* A study of East Europe using "top-level" visits as one indicator is Hughes and Volgy, *op. cit.* For a criticism of this study see Kenneth S. Hempel, "Comparative Research on Eastern Europe: A Critique of Hughes and Volgy's 'Distance in Foreign Policy Behavior'" *American Journal of Political Science*, XVII, 2 (May, 1973), pp. 367-93.

28. Note that it is trade *agreements* (and protocols) as opposed to trade itself, which are measured by this scale. This is intended to more closely tap the political and intentional nature of the states' foreign policy behavior, by focusing on a government's intended, formal actions rather than on the mere exchange of commodities which may fluctuate for a variety of reasons, political and nonpolitical. On this point, see Charles Ransom, *The European Community and Eastern Europe* (Totowa, N.J.: Rowman and Littlefield, 1973), pp. 53; cf. note 66 and pp. 206-08 below.

29. Charles W. Kegley, Jr., "A Circumplex Model of International Interactions" in Edward Azar and Joseph Ben-Dak (eds.) *Theory and Practice of Events Research* (NY: Gordon and Breach, 1975), pp. 217-218. Some event interaction studies, e.g. those designed to delineate global patterns such as the World Event Interaction Survey (WEIS) Project, do not scale their data, though they do categorize it by event type. See McClelland and Hoggard *op. cit.* Others, such as Edward E. Azar, "Analysis of International Events," *Peace Research Reviews*, IV, 1 (November, 1970) scale their events along conflict-cooperation continua. For a discussion of scaling and a review of events data research in general, see Burgess and Lawton, *op. cit.* For a more detailed discussion of scaling and weighting see Theodore J. Rubin and Gary A. Hill, *Experiments in the Scaling and Weighting of International Event Data* (Brentwood, Calif: Consolidated Analysis Centers, Inc., 1973); Herbert L. Calhoun, "Exploratory Applications to Scaled Event Data," (Dallas: paper prepared for delivery to the 1972 ISA convention, March, 1972); and the selections by Sloan; Goodman, Hart and Rosecrance; Havener and Petersen; and Lanphier in Azar and Ben-Dak, *op. cit.* For an unscaled (all weighted equally) treaty-count study of communist states for the period 1953-62,

see Miles and Gillooly, *op. cit.*, esp. pp. 117-25. A more differentiated discussion of Soviet treaty-making behavior for that period is found in Charles D. Cary, "Patterns of Soviet Treaty-Making Behavior with Other Communist Party-States," in Triska, *op. cit.* pp. 135-60. Items scaled for cooperation and conflict are also included in Hopmann and Hughes, *op. cit.* pp. 81-95.

30. Azar argues for coding events only according to the "Primary components"—actor, target, date, issue and action- and leaving the context of the interaction for analysis. See "Ten Issues in Events Research," in Azar and Ben-Dak, *op. cit.*, pp. 1-17.

31. Azar "Analysis of International Events."

32. Lincoln Moses, Richard Brody, Ole Holsti, Joseph Kadane and Jeffrey Milstein, "Scaling Data on Inter-Nation Action," *Science*, May 26, 1967, pp. 1054-59; see also the discussion in Thomas Sloan, "The Development of Cooperation and Conflict Interaction Scales: An Advance in the Measurement and Analysis of Events Data," Azar and Ben-Dak, *op. cit.*, pp. 29-39.

33. Singer and Small utilize a similar reliance upon "scholarly consensus" to help determine international system membership and status in their Correlates of War project. See J. David Singer and Melvin Small, *The Wages of War 1816-1965: A Statistical Handbook* (New York: John Wiley, 1972), pp. 19-30. Goodman, Hart, and Rosecrance also make the case for a broad use of scholarly sources: "Even if it were possible to study the entire output of diplomatic documents for a particular period, a mere listing would not distinguish the important from the unimportant events. The total diplomatic record would provide in addition to the significant acts,masses of irrelevant detail . . . . Thus the question crops up again. How do we screen the important from the relatively unimportant events? Who should be our guide to such a task? Once more, the diplomatic historian serves as an indispensible filter." See their "Testing International Theory: Methods and Data in a Situational Analysis of International Politics," in Azar and Ben-Dak, *op. cit.*, p. 44.

34. See the discussion in Thomas Havener and Alan Peterson, "Measuring Conflict/Cooperation in International Relations," in Azar and Ben-Dak, *op. cit.*, pp. 57-61.

35. See McClelland and Hoggard, *op. cit.*

36. See R. J. Rummel, "Dimensions of Conflict Within and Between Nations," in Jonathan Wilkenfeld (ed.) *Conflict Behavior and Linkage Politics* (New York: David McKay, 1973), pp. 59-106.

37. Azar, "Analysis of International Events."

38. McClelland and Hoggard, *op. cit.*, p. 714ff.

39. Rummel, *op. cit.*, p. 69ff.

40. Azar, "Analysis of International Events" p. 21ff.

41. Hopmann and Hughes, *op. cit.*

42. *Ibid.*

43. Havener and Peterson, *op. cit.*; Kegley, "Circumplex Model."

44. Hermann, *op. cit.*, pp. 153-157.

45. The scale employed by Azar has 13 categories, expanded to22 by Sloan, *op. cit.*; The WEIS formulation, 63 types of events, reworked by several others, including Hermann, *op. cit.* and Kegley "Circumplex Model;" and the DON scheme includes 94 categories, also reworked by several others.

46. Kegley ("Circumplex Model", p. 228) rates as "dubious" the "plausibility of all-purpose single dimension scales."

47. Havener and Peterson, *op. cit.* report the not surprising result that a categorization system with fewer categories (that of Azar) has higher coder reliability. Lanpheir found that a 600-pt scale created by Walter Corson was largely unusable beyond the first 10 points. See Vernard A. Lanpheir, "Foreign Relations Indicator Project (FRIP)" in Azar and Ben-Dak, *op. cit.*, pp. 161-74.

48. Needless to say, suggestions for improvements and modifications in this respect are welcomed by the author.

49. Traditional sources for events data such as the *New York Times* or *The Times* of London systematically underreport East European events. See Gary Hill "Dyad Coverage in *New York Times*," (Los Angeles: WEIS Project Memorandum, September 10, 1973). In discussing source problems, Azar suggests using both a regional and standard global events data source. See "Ten Issues," pp. 3-5. Hoggard agrees, especially for "in depth analyses of regional interaction"; see Gary Hoggard, "An Analysis of the 'Real' Data: Reflections on the Uses and Validity of International Interaction Data," in Azar and Ben-Dak, *op. cit.*, pp. 19-27; and Hoggard, "Differential Source Coverage in Foreign Policy Analysis," in Rosenau (ed.) *Comparing*, pp. 353-81. On the problems of source coverage see also Burgess and Lawton, *op. cit.*, pp. 63-7; Goodman, Hart and Rosecrance, *op. cit.*; Gary A. Hill and Peter H. Fenn, "Comparing Event Flows- *The New York Times* and *The Times* of London: Conceptual Issues and Case Studies," *International Interactions*, 1 (1974), pp. 163-86; and Robert Burrowes, "Mirror, Mirror on the Wall . . . A Comparison of Events Data Sources," in Rosenau (ed.) *Comparing*, pp. 383-406.

50. Actual judgement reliability was probably higher since much of the error in the test sample was mechanical, e.g. incomplete event description. The total number of interactions for each country and their

total weighted value by year and partner are given in the Appendix I, Table A.1.

51. While McKinney, *op. cit.*, pp. 12-15, indicates that Max Weber did not envisage the ideal type- "a special case of the generic constructed type"- as an average, its use here seems valid based in part on McKinney's own support for the quantification and measurement of degree of deviation. Further support for the use of an average is drawn from Durkheim: "The state of health, as defined by science, cannot fit exactly any individual subject, since it can be established only with relation to average circumstances, from which every one deviates more or less; nevertheless, it may serve as a valuable point of reference for regulating our conduct." Emile Durkheim, *The Rules of Sociological Method* (New York: The Free Press, 1964) (orig. pub. 1938), p. 49. Azar devises a similar method, known as the "Normal Relations Range (NRR)" for studying dyadic interaction; see "Ten Issues," pp. 10-14. For foreign policy deviance in East Europe, Hempel has demonstrated the necessity of clearly specifying the norm for measurement and the differing results that can occur depending on the norm that is employed. See Kenneth S. Hempel, *op. cit.*, pp. 370-71. The present method has the strength of comparing the small East European states only to themselves (as opposed to the Soviet Union) for reasons stated earlier (see p. 13), and in addition providing a comparative assessment based on empirical findings rather than one observer's designation of one state as "loyal" or "deviant." For an attempt to measure distance between these states on various internal variables, using a "focal state" as a norm, see William A. Welsh, "Toward An Empirical Typology of Socialist Systems," in Carl Beck and Carmelo Mesa-Lago, *Comparative Socialist Systems* (Pittsburgh: University Center for International Studies, University of Pittsburgh, 1975), pp. 52-91.

52. The formula for standardizing the deviations is

$$z = \frac{x-\bar{x}}{\sigma} \text{ or in words } z = \frac{\text{IP} - \text{ave. IP}}{\text{Standard dev.}}$$

53. Because of the low level of interaction with Albania and China, a z-score dispersion for these targets would present a distorted picture; see note 58 below. For inclusion of the z-score values for these targets in a summary measure, see pp. 46-49.

54. See pp. 19 above.

55. It should be recalled that the US and the West are separate potential partners. Thus a high score on one does not necessarily produce a high score on the other, as Bulgaria's scores show.

56. This type of chart is taken, *mutatis mutandis* from Bruce M. Russett, *International Regions and the International System* (Chicago:

Rand McNally, 1967). Z-scores for the China and Albania targets are not included. For their inclusion in a summary measure of deviance, see p. 46-49.

57. Note that for this latter group of potential partners this labelling is the opposite of its mathematical sign.

58. The following are the proportions of the total WTO interaction level directed toward each target:

TABLE II.7

PROPORTION OF WTO INTERACTION WITH EACH TARGET

Total Interaction Score, all WTO with all partners = 3937.
(1308 Interactions)

All WTO with:	ALB	CPR	OTH	USSR	EE	YUG	NON
Inter. Score:	10	36	175	568	877	295	506
Proportion of total:	.003	.009	.044	.144	.223	.075	.129

All WTO with:	AR/IS	WEST	US	FRG
Inter. Score:	187/44	1078	91	82
Proportion of total:	.047/.011	.274	.023	.021

59. Nissan Oren, for example, sees Bulgarian diplomacy as characterized by "docility," "untainted discipline as a Russian satellite," and "immobilism". *Revolution Administered: Agrarianism and Communism in Bulgaria* (Baltimore: The Johns Hopkins University Press, 1973), p. 182. Clark and Farlow state that "Over the past two decades Bulgaria has been one of the most docile and devoted of the Soviet satellites." Cal Clark and Robert L. Farlow, *Comparative Patterns of Foreign Policy and Trade: The Communist Balkans in International Politics* (Bloomington, Ind.: International Development Research Center, Indiana University, 1976), p. 15. On the basis of a model of "subalignment" these authors predict that "involvement in foreign affairs (will be) relatively low as a result of subalignment's one country focus." (p. 16) The present findings indicate that this is not the case; and indeed Clark and Farlow themselves found that "the difference between Romania and Bulgaria [was] not as great as expected," (p. 97). Other studies noting Bulgaria's behavior are

Hughes and Volgy, *op. cit.*, pp. 477-81; and Michael Costello, "Bulgaria," in Bromke and Rakowska Harmstone, *op. cit.*, pp. 135-158. Cf., F. Stephen Larrabee, "Bulgaria's Politics of Conformity." *Problems of Communism*, XXII (July/August, 1972), pp. 42-52.

60. This index is a sum of the rankings on the following four indicators: 1) mean trade value with the USSR as a percentage of total trade value, 1956-67; 2) mean trade value with the USSR as a percentage of Net Material product, 1956-67; 3) mean trade value with East Europe as a percentage of total trade value, 1956-67; 4) mean trade value with East Europe as a percentage of Net Material product, 1956-67; Kintner and Klaiber, *op. cit.*, pp. 226-34. It will be noted that while the present Index of Deviance includes actions taken during the years 1965-69, the Kintner and Klaiber Trade Dependency Index is a mean compiled from values for the years 1956-67. Thus trade dependence for the year 1968 and part of 1969 are not included. While this time-lag is in fact necessary for our hypothesis,–for dependence to have the effect expected it would have to precede deviance–a correlation was run with 1968 trade figures included in the dependency ranking. The correlation with deviance was driven even closer to zero (.14).

61. Kintner and Klaiber found a Kendall rank correlation of .52, significant at the .068 level, between the two Indices' ranks. *Ibid.*, p. 234. The Kintner and Klaiber Index of Conformity is based on: 1) membership in the Soviet bloc; 2) participation in COMECON; and 3) the Warsaw Pact; 4) whether or not it has been subject to the use of force against it by the USSR; 5) instances of critisism of the USSR by the country; 6) instances of criticism of the country by the USSR; 7) observers' judgement of its policy conformity in several key issue areas. The scores for each indicator for each country are summed to produce a ranking of states, from least conforming to most (pp. 219-26). The Kintner and Klaiber index has the advantage of a greater time-span, 1956-68, and takes notice of some types of behavior not included in the present Index of Deviance. However, it can be criticized on the following grounds: 1) it is overly institutional; of the seven indicators, two are purely trichotomous designations of participation in East European organizations (COMECON and the WTO), and a third, the one weighted most heavily, is membership in the bloc itself; 2) this last category is not only redundant but its heavy weighting almost predetermines the outcome; moreover it is tautologous, i.e. conformity is indicated by, among other things, membership in the Soviet bloc; 3) indicators No.2 and especially No.4 are rather blunt instruments for measuring conformity, this last by a dichotomous scoring of a drastic Soviet response to other factors, as well as policy conformity and

which in any case has only occurred twice in the twelve years studied; 4) the indicator "policy conformity" (No.7), would in many instances be a repeat of No.1 (criticism of the USSR) and would thus be scored twice (each time, with a different weight). 5) Further, relying upon the investigator's own judgements of policy conformity in individual cases substantially predetermines the outcome of the overall index and seriously undermines its reliabiltiy; 6) the weights of the several indicators vary greatly in the construction of the overall index, with little or no justification of how the various weights are determined; 7) finally, as the authors admit, no measure of direction of noncomformity is included, the result being that "the 'Stalinist' Albania ranks next to 'liberal' Yugoslavia." (p. 226). Thus it seems clear that while the Kintner and Klaiber index does alert one to some key types of policy behavior, a less subjective, less redundant, more precise index (both behaviorally and in terms of measurement) is necessary.

62. This conforms to Gehlen's finding that such dependence and his "level of integrative activity" are not associated. Michael P. Gehlen, "The Integrative Process in East Europe: A Theoretical Framework," *Journal of Politics*, XXX, 1 (February, 1968), pp. 111-12. For a criticism of this study see Kanet, "Integration Theory," pp. 379-82. Hopmann and Hughes, *op. cit.* argue that the problem may be with the validity of the trade indicator, see note 66 below.

63. This index is based on: 1) the number of Soviet divisions deployed in the country; 2) the geographic distance separating the state from the USSR; and 3) the number of intervening states which would have to be crossed by the Soviets to deploy troops into the country.*op.cit.* pp.234-36.

64. The time difference noted previously (see note 60 above) would be significant in this case since Czechoslovakia was invaded in 1968, a year and fact not included in Kintner and Klaiber's military position index. However, as can be seen this would serve to lower Czechoslovakia's ranking on the index and thereby *increase* the positive correlation between this ranking and that of the Index of Deviance.

65. Robert Freedman notes, for example, the following tactics the Soviet Union has employed: 1) delay in trade negotiations upsetting the smaller government's planning program (used against Albania in 1960); 2) a refusal by the USSR to purchase key outputs of the target nation (used against Yugoslavia in 1958); 3) a refusal to sell to it as much of a key commodity as it wants (used against Cuba in 1968); 4) refusal to sell a certain good at all to the target nation (used against China since the mid-sixties); 5) delay or refusal to sign the final trade agreement (used against Romania in early 1960); 6) delay in the delivery of crucial goods

used against Romania in 1967); and 7) a general embargo (used against Yugoslavia, 1949-53 and Albania 1962-70). See pp. 7-9.

66. See *Ibid., passim.* Cf. Peter Wiles *Communist International Economics* (Oxford: Basil Blackwell, 1968), pp. 499-503, 507-14. Hopmann and Hughes, *op.cit.*, argue, though, that trade data are not necessarily a good indicator of cohesion in an international coalition because, "they may respond heavily to economic considerations, so that trade may increase substantially even in the presence of some political conflicts. Second, they tend to be highly trended and to change slowly, thereby reducing the variability which is often required to get sensitive tests of our hypotheses," (p. 93). A similar argument is found in Hempel, *op. cit.*, pp. 374-5. For an alternative view see Cal Clark, "Foreign Trade as an Indicator of Political Integration in the Soviet Bloc," *International Studies Quarterly*, XV, 3 (September, 1971), pp. 259-95. Clark and Farlow, *op. cit.* found that for the Balkans trade was a better "result" indicator than causal indicator, and furthermore trade was a better indicator of the foreign relations of Bulgaria and Albania than for Romania or Yugoslavia.

67. On the measures taken against Yugoslavia see Freedman, *op. cit.*, pp. 18-58. On the nature and level of Romanian economic dependence on the Soviet Union, see John M. Montias, *Economic Development in Communist Rumania* (Cambridge: M.I.T. Press, 1967), pp. 135-86. On the relatively mild tactics used against Romania, see *Ibid.*, pp. 204-5 and David Floyd, *Rumania: Russia's Dissident Ally* (New York: Praeger, 1965), pp. 60-1, 67.

68. For a further discussion of resource position, see pp165-66below.

NOTES TO CHAPTER THREE

1. For a review of foreign policy definitions see Edward L. Morse, "Defining Foreign Policy." McClelland and Hoggard found, for example, that verbal behavior made up 65.5% of all international activity during the time period they studied. Charles A. McClelland and Gary D. Hoggard, "Conflict Patterns in the Interactions Among Nations," in Rosenau *International Politics and Foreign Policy*, p. 711-725. An exception to this approach would be Rudolph J. Rummel, "Some Dimensions in the Foreign Behavior of Nations," *Journal of Peace Research*, 3 (1966), pp. 201-224, which virtually ignores verbal behavior.

2. Charles W. Kegley, Jr., "A Circumplex Model of International Interactions," in Azar and Ben-Dak, *op. cit.*, pp. 217-232.

3. Katarina Brodin, "Belief Systems, Doctrine, and Foreign Policy,"

Cooperation and Conflict. VII, 2 (1972), p. 104.

4. Two such studies are Hopmann, *op. cit.* and Jeanne K. Laux, "Intra-Alliance Politics and European Detente: The Case of Poland and Romania," *Studies in Comparative Communism*, VIII, 1/2 (Spring/Summer, 1975), pp. 98-122. Less systematic content analysis can be found in Robin Remington, *The Warsaw Pact* (Cambridge: M.I.T. Press, 1971), pp. 41-46. For narrative discussions comparing the views of these states on particular issues, see Henry W. Schaefer, *Comecon and the Politics of Integration* (New York: Praeger, 1972); Robert R. King and Robert W. Dean, *East European Perspectives on European Security and Cooperation* (New York: Praeger, 1974); and Jeffrey Simon, *Ruling Communist Parties and Detente: A Documentary History* (Washington: American Enterprise Institute, 1975), pp. 3-65. For a validity study comparing *inter alia* event-interaction data with content analysis data for the region, see Hopmann and Hughes, *op. cit.*

5. Ole R. Holsti, Richard A. Brody, and Robert C. North, "Measuring Affect and Action in International Reaction Models: Empirical Materials from the 1962 Cuban Crisis," in Rosenau, *IPFP*, pp. 683-84.

6. See R. C. Snyder, H.W. Bruck and B. Sapin, *Foreign Policy Decision-Making: An Approach to the Study of International Politics* (New York: The Free Press, 1962). Cf. Robert Jervis, *Perception and Misperception in International Politics* (Cambridge: Harvard University Press, 1977). Examples are Richard Cottam, *Foreign Policy Motivation* (Pittsburgh: University of Pittsburgh Press, 1977); Ole R. Holsti, "The Belief System and National Images: A Case Study," in Rosenau, *IPFP*, pp. 543-51; Alexander L. George, "The Operational Code: A Neglected Approach to the Study of Political Leaders and Decision-Making," in Erik P. Hoffmann and Frederic J. Fleron, Jr., *The Conduct of Soviet Foreign Policy* (Chicago: Aldine, 1971), pp. 165-91; Ole R. Holsti, "Individual Differences in the Definition of the Situation," *Journal of Conflict Resolution*, XIV, 3 (September, 1970), pp. 303-10.

7. See Graham T. Allison, *Essence of Decision* (Boston: Little, Brown and Co., 1971), pp. 67ff. Cf. Glenn D. Paige, "The Korean Decision," in Rosenau, *IPFP*, pp. 461-73; and Paige, "Comparative Case Analyses of Crisis Decisions: Korea and Cuba," in Charles F. Herman, (ed.) *International Crises: Insights from Behavioral Research* (New York: The Free Press, 1972), pp. 41-55; Chihiro Hoyosa, "Characteristics of the Foreign Policy Decision-Making System in Japan," *World Politics*, XXVI, 3 (April, 1974), pp. 353-69.

8. Holsti, Brody and North, *op. cit.*, p. 683.

9. Ole R. Holsti, *Content Analysis for the Social Sciences and*

Humanities (Reading, Mass: Addison-Wesley, 1969), p. 14.

10. Content analysis has been used in the study of: elite perceptions, Robert C. Angell, "Content Analysis of Elite Media," *Journal of Conflict Resolution*, VIII, 4 (December, 1964), pp. 330-86 and J. David Singer, "Content Analysis of Elite Articulations," *Journal of Conflict Resolution*, VIII, 4 (December, 1964), pp. 424-85; positive and negative affect, Dina A. Zinnes, "The Expression and Perception of Hostility in Prewar Crisis: 1914," in J. David Singer, *Quantitative International Politics* (New York: The Free Press, 1968), pp. 85-123; Richard A. Brody, "Some Systemic Effects of the Spread of Nuclear Weapons Technology," *Journal of Conflict Resolution*, VII, 4 (December, 1963), whole issue, and Holsti, Brody and North, *op. cit.*; cooperation and conflict, Ole R. Holsti, Robert C. North and Richard A. Brody, "Perception and Action in the 1914 Crisis," in Singer, *op. cit.*, pp. 123-59, and John H. Sigler, "Cooperation and Conflict in United States-Soviet-Chinese Relations, 1966-71," Peace Research Society, *Papers*, XIX (1971), pp. 107-128; and interaction frequency, Richard L. Merritt, "Distance and Interaction Among Political Communities," *General Systems Yearbook*, IX (1964), pp. 255-63.

11. Use of computer technology has improved these methods however. See Ole R. Holsti, "An Adaptation of the 'General Inquirer' for the Systematic Analysis of Political Documents," *Behavioral Science*, IX, 4 (October, 1964), pp. 382-88, and Chapter 7 of Holsti, *op. cit.* Examples of studies using the General Inquirer system are Holsti, Brody and North, *op. cit.*; and Hopmann and Hughes, *op. cit.*

12. Holsti, North and Brody, *op. cit.*, p. 138; Holsti, *Content Analysis*, p. 33.

13. Robert Jervis, "The Costs of the Quantitative Study of International Relations," in Klauss Knorr and James N. Rosenau (eds.) *Contending Approaches to International Politics* (Princeton: Princeton University Press, 1969), pp. 190-95.

14. Rosenau notes, for example, ". . . like any individual, the decision-maker is not likely to be aware of all the stimuli to which he is in fact responding, and thus what he reports perceiving may not be the entire basis of his actions." James N. Rosenau, "Comparing Foreign Policies: Why, What, How," in James N. Rosenau (ed.), *Comparing Foreign Policies* (Beverly Hills: Sage Publications (John Wiley Distributor), 1974), p. 19. Cf. Robert Jervis, "Hypotheses on Misperception," in Rosenau (ed.) *IPFP*, pp. 239-54.

15. Robert E. Mitchell, "Content Analysis for Explanatory Studies," *Public Opinion Quarterly*, XXXI, 2 (Summer, 1967), p. 235. An example of this type of study would be Merritt, *op. cit.* For a similar criticism

see Jervis, "The Costs," pp. 185-188.

16. Charles F. Osgood, George J. Suci, Percy H. Tannenbaum, *The Measurement of Meaning* (Urbana: University of Illinois Press, 1957). Holsti, Brody and North, *op. cit.*, and Hopmann and Huges, *op. cit.* are studies using this technique.

17. Holsti, Brody and North, *op. cit.*, p. 685 and ff.; Holsti, Brody and North, *op. cit.*, pp. 138-145; Hopmann and Hughes, *op. cit.*

18. Thus, for example, the problem of distortion of interpretation caused by the decision-maker's intentional misrepresentation of his intent or attitude is avoided. It may be that such intentional misrepresentation occurred in the states during the period of this study. If the attitudes expressed are not the "true" ones- and such a determination is the proper function of case study, assuming that data sufficient to prove a disparity are both extant and available- then the independent factors utilized in the study underlie both the misrepresentation of attitudes and/or the attitudes themselves. In terms ofthe immediate concern of this study, it is highly unlikely that decision-makers in East Europe would intentionally misrepresent their attitudes in a deviant direction. They might, on the contrary, misrepresent them in a nondivant direction, especially if their actions were at the same time deviant. This makes the deviance in attitudes that is evident all the more significant, and in addition shows the value of using indicators of both action and attitude. For a study relating expressed perceived "maps" to foreign policy motivation,see Cottam,*op.cit.*

19. Mitchell, *op. cit.*, p. 234.

20. See, e.g. Holsti, "The Belief System," *op. cit.*

21. See John E. Mueller, "The Use of Content Analysis in International Relations," in George Gerbner, Ole R. Holsti, Klaus Krippendorff, William J. Paisley and Phillip J. Stone, *The Analysis of Communication Content* (New York: John Wiley, 1969), pp. 187-97. He notes, "As an analytic convenience, the communication scrutinized by the content analyst can be viewed as a *response* by a certain type of person in written *form* to a *stimulus* in a specific *context*." (Emphasis in original) p. 187. Hopmann and Hughes, *op. cit.* use a similar approach for somewhat different purposes, see pp. 315-23.

22. See pp. 12.

23. W. Andrew Axline, "Common Markets, Free Trade Areas, and the Comparative Study of Foreign Policy," *Journal of Common Market Studies*, X, 2 (December, 1971), p. 168 (emphasis in original). For Teune and Przeworski's discussion of a "most similar systems" reseach design, see Henry Teune and Adam Przeworski, *The Logic of Comparative Social Inquiry* (New York: John Wiley, 1970) pp. 32-34.

24. Theodore M. Newcomb, "An Approach to the Study of Communicative Acts," *Psychological Review*, LX, 6 (November, 1953), pp.393-404.

25. Hopmann, *op. cit.*, and Hopmann and Hughes, *op. cit.* Harle, *op. cit.* employs a variant of this model to study "actional distance" between the socialist states.

26. F. Heider, "Attitudes and Cognitive Organization," *Journal of Psychology*, XXI (January, 1946), pp. 107-12.

27. Leon Festinger, "Informal Social Communication," in L. Festinter, K. Back, S. Schacter, H. Kelley, and J. Thibaut, *Theory and Experiment in Social Communication* (Ann Arbor: Institute for Social Reseach, University of Michigan, 1950), For a discussion of this approach and related studies, see, George D. Wright, "Inter-Group Communication and Attraction in Inter-Nation Simulation" (St. Louis: unpub. Ph.D. dissertation, Washington University, 1963).

28. Newcomb, *op. cit.*, pp. 393-394.

29. *Ibid.*, pp. 394-95.

30. Theodore M. Newcomb, "Individual Systems of Orientations," in Sigmund Koch (ed.), *Psychology: A Study of a Science*, Vol. III, *Formulation of the Person and the Social Context* (New York: McGraw Hill, 1959, pp. 393-94.

31. Newcomb found the following to be related to strain: 1) degree of perceived discrepancy, 2) sign and degree of attraction, 3) importance of the object of communication, 4) certainty ("committedness") of own orientation, and 5) object relevance." *Ibid.* p. 394 and ff.

32. One case with four responses was also included, the only other exception to this inclusion rule were issue areas in which substantial documentation already was present. Thus, for example, for United States' actions in Vietnam certain events, considered not likely to add notably to the information field, were eliminated.

33. An additional point to note concerning the difference between the foreign policy behavior of large and small states is East's finding that for large states, the ratio of verbal to nonverbal behavior is larger than that ratio for smaller states. ("Size and Foreign Policy Behavior," pp. 166-68). Since the present study involves a comparison of theme and not frequency, this difference is unlikely to affect the results.

34. *Notes and Etudes* is a documentation and research serial of *Direction de la Documentation*, Secretariat General du Gouvernement, Paris. The *Yearbook*, edited most recently by Richard F. Staar, is published by the Hoover Institution, Stanford, California.

35. See Appendix II for a list of all sources of statements for all countries.

36. How would one know whether *FBIS*, *Deadline Data* or *Notes and Etudes* simply failed to monitor and record the country's reaction that day? The author apologizes for failing to master nine different East European languages in pursuit of total data; but in the absence of such skill or indeed, access to such sources, the above caution is mandatory.

37. For a discussion of various statistical approaches to missing data, see Raisa B. Deber, "Missing Data in Events Analysis: The Problem and Suggested Solutions," in Azar and Ben-Dak, *op. cit.*, pp. 63-79.

38. For a discussion of the "Language of Conflict" and examples of this type of analysis, see Sidney I. Ploss, *The Soviet Political Process* (Waltham: Ginn and Co., 1971). Cf. William E. Griffith "On Esoteric Communications," *Studies in Comparative Communism*, III, 1 (January, 1970) pp. 47-54. The most complete discussion of differing views on international relations in the Soviet Union is William Zimmerman, *Soviet Perspectives on International Relations* (Princeton: Princeton University Press, 1969).

39. See Morse, "Defining Foreign Policy," and *A Comparative Approach to the Study of Foreign Policy* (Princeton: Center for International Studies, 1971), by the same author.

40. A frequency distribution listing the number of cases reported for each country is found in Appendix VII.

41. "Expert" refers to the fact that all coders were knowledgeable about East Europe and familiar with governmental statements and jargon, but none were foreign policy specialists. This was preferred in order to avoid a coder whose previous knowledge or familiarity with East European foreign policy would bias his coding judgements.

42. Examples of several coding descriptions can be found in Appendix III.

43. For a complete description of the intercoder reliability tests, see Appendix IV.

44. TASS News Agency (hereafter, TASS), Moscow, June 6, 1967.

45. Prague Domestic Service (hereafter Prague D. S.), June 8, 1967.

46. Bucharest D.S., June 5, 1967.

47. *Ibid.*

48. TASS, Moscow, June 9, 1967.

49. *Borba*, September 7, 1967. For a discussion of this Statement see Paul Yankovitch, "La Conference des pays communistes sur l'aide aux etats arabes s'est abstenue de condamner 'l'agression israelienne,'" *Le Monde*, September 8, 1977, p. 4.

50. TASS, Moscow, December 22, 1967. It is interesting to note, however, that though this meeting was held at the Foreign Minister level,

the Romanians were represented by a Deputy Foreign Minister, M. Mihai, and the Czechs, the only other state whose earlier position could be considered somewhat more mild, sent a First Deputy Foreign Minister, T. Pudlak. Moreover, in a coincidence that "could not have been accidental" the Romanian Minister of Foreign Trade, G. Cioara, arrived in Israel to sign a three-year trade agreement with Tel Aviv, comprising large increases in trade levels, on the very day that the Foreign Ministers' meeting began in Warsaw.

51. *Ibid.*

52. For a discussion of this plenum see William F. Dorrill, "Power, Policy and Ideology in the Making of the Chinese Cultural Revolution," in Thomas W. Robinson (ed.) *The Cultural Revolution in China* (Berkeley: University of California Press, 1971), pp. 105-12; cf. G. Dutt and V. P. Dutt, *China's Cultural Revolution* (New York: Asia Publishing House, 1970), pp. 42-46.

53. New China News Agency (NCNA), International Service, August 13, 1966.

54. *Neues Deutschland*, September 4, 1966.

55. *Trybuna Ludu*, September 7, 1966.

56. On Romanian mediation efforts see the "Statement on the Stand of the Rumanian Workers' Party Concerning the Problems of the International Communist and Working-Class Movement," issued in April, 1964; text in William E. Griffith, *Sino-Soviet Relations, 1964-65* (Cambridge: M.I.T. Press, 1967), (hereafter "Statement"), pp. 269-96; cf. the discussion in Floyd, *op. cit.*, pp. 110-12.

57. *Nepszabadsag* (Hungarian Party daily), March 16, 1969.

58. *Rude Pravo*, March 4, 1969.

59. *Scinteia*, March 4, 1969.

60. Tanyung International Service (hereafter Tanyung), Belgrade, March 7, 1969.

61. Tanyung, Belgrade, April 16, 1969.

62. *Scinteia,* April 3, 1969; April 26, 1969; Bucharest D.S. April 14, 1969.

63. The most complete study of this period is H. Gordon Skilling, *Czechoslovakia's Interrupted Revolution* (Princeton: Princeton University Press, 1976).

64. TASS, Moscow, March 23, 1969. The full text of this communique can also be found in Robin Remington, *Winter in Prague* (Cambridge: M.I.T. Press, 1969), pp. 55-57.

65. *Scinteia*, March 29, 1968.

66. *Ibid.*

67. *Review of International Affairs*, April 5, 1968 , p. 6.

68. The "Two Thousand Words" document criticized past leaders' "mistaken policies;" praised the "regenerative process" which "revives ideas and topics, many of which are older than the errors of our socialism;" urged a wider and further pressing of the party's "Action Program" for removing inequities; called for further political and economic democratization; and pledged to support the government "if it will do what we give it a mandate to do." The statement was signed by seventy party and non-party Czech intellectuals and reprinted widely in the country, though it was disavowed by the government and party. For a text of the statement see *Times Literary Supplement*, July 18, 1968 (also in Remington, *op. cit.*, pp. 196-203). For examples of critical reactions directed toward the "Two Thousand Words," see S. Lonyai, "A Few Remarks on 'Two Thousand Words,'" *Kisalforld*, July 7, 1968 (RFE Hungarian Press Survey, 16 July 1968), and I. Aleksandrov, "Attack on the Socialist Foundations of Czechoslovakia," *Pravda*, July 11, 1968.

69. *Pravda*, July 11, 1968.

70. *Ibid.*

71. *Trybuna Ludu*, July 15, 1968. *Neues Deutschland*, July 13, 1968.

72. Speech to Parliament, July 13, 1968.

73. CTK International Service (hereafter CTK), Prague, July 12, 1968.

74. MTI Domestic Service (hereafter MTI D.S.), Budapest, July 17, 1968. For the Czech reply see Diplomatic Information Service, Prague, July 18, 1968. Both the "Letter" and the reply ("Standpoint of the Czechoslovak Communist Party Central Committee Presidium . . . on the letter of the five communist and workers' parties") (hereafter "Standpoint") are also in Remington, *op. cit.*, pp. 225-31 and 234-43, respectively.

75. *Scinteia*, July 15, 1968.

76. See Bucharest D.S., July 18, 1968.

77. *Politika*, July 14, 1968; Tanyung, July 18, 1968.

78. *Neues Deutschland*, July 23, 1968.

79. *Pravda*, July 28, 1968.

80. *Trybuna Ludu*, July 25, 1968; *Nepszabadsag*, July 25, 1968.

81. *Nepszabadsag*, July 25, 1968.

82. See CTK, Prague, July 16, 1968; and "Standpoint," Diplomatic Information Service, Prague, July 18, 1968.

83. *Scinteia*, July 31, 1968.

84. *Borba*, the Yugoslav Communist League organ, did so by way of condemning and refuting the views of the East Germans (July 26, 1968) and the Bulgarians (July 28, 1968). The Romanians, in addition to two

front page supportive articles in *Scinteia*, also indulged their taste for tweaking the Kremlin's tail by reprinting under its own title the declaration of the communist party of Great Britain stating that "Only the Czechoslovak People and the Communist Party Can Decide How to Approach Their Internal Problems." See *Scinteia*, July 27, 28, and 31, 1968.

85. *Magyar Hirlap*, August 5, 1968. The Bratislava statement can be found in Remington, *op. cit.*, pp. 256-61.

86. Prague D.S., August 4, 1968.

87. *Kommunist* (Belgrade), August 4, 1968.

88. *Scinteia*, August 7, 1968.

89. MTI, Budapest, August 5, 1968.

90. Sofia D.S., August 21, 1968.

91. In protest the Albanians formally withdrew from the WTO in September. See *New York Times*, September 13, 1968, p. 8 and September 14, 1968, p. 12.

92. Bucharest D.S., August 21, 1968.

93. *Ibid.*

94. Indeed rumors abounded, in that rumor-prone capital, that the Red Army was poised in the southern Ukraine for a drive further south. See Peter Bender, *East Europe in Search of Security* (Baltimore: The Johns Hopkins University Press, 1972), p. 117. Cf. M.K. Dziewanowski, "The Pattern of Rumanian Independence," *East Europe*, XVIII, 6 (June, 1969), p. 10; George Cioranescu, "Rumania After Czechoslovakia: Ceauşescu Walks the Tightrope," *East Europe*, XVIII, 6 (June, 1969), p. 3. On August 25, after a meeting between Romanian Party leader Ceauşescu and Soviet Ambassador to Romania, A.V. Basov, the Soviet Union announced on Moscow television, responding to "provocations" and "imperialist propaganda" that it did not intend to invade Romania. (*Keesing's*, October 26-November 2, 1968, p. 22994.).

95. Bucharest D. S., August 21, 1968.

96. These events are described in Skilling, *op. cit.*, pp. 759-810; cf. *Czechoslovakia Since World War II* (New York: The Viking Press, 1971), pp. 377-443; and in Pavel Tigrid, *Why Dubcek Fell* (London: MacDonald and Company, 1971), pp. 99-123.

97. *Pravda*, August 28, 1968.

98. Warsaw D. S., August 27, 1968.

99. Tirana D. S., August 27, 1968.

100. For Czechoslovakia, see speech by Premier Oldrich Cernik, Radio Czechoslovakia, August 28, 1968; for Romania see 'Statement of the Executive Committee of the Romanian Communist Party Central

Committee," Agerpress International Service, Bucharest, August 29, 1968.

101. The Political Consultative Committee (PCC) consisted until 1969 of the respective parties' First or General Secretary, and the governments' Prime Minister, Foreign Minister and Defense Minister.

102. For a text of the Bucharest Declaration see TASS, July 11, 1966; *Pravda*, July 9, 1966; Remington, *The Warsaw Pact*, pp. 209-21.

103. *Zolnierz Wolnosci*, July 9, 1966.

104. *Scinteia*, July 25, 1966, p. 6.

105. See Jeanne K. Laux, "Intra-Alliance Politics and European detente: The Case of Poland and Romania," *Studies in Comparative Communism*, VIII, 1-2 (Spring/Summer, 1975), pp. 98-122; and Robert Association, New Orleans, Louisiana, September 4-8, 1973; and Robert R. King, "Rumania: The Difficulty of Maintaining An Autonomous Foreign Policy," in Robert R. King and Robert W. Dean (eds.) *East European Perspectives on European Security and Cooperation* (New York: Praeger, 1974), pp. 168-91.

106. *Review of International Affairs*, June 20, 1967, p. 28.

107. For more on the Yugoslav view see Slobodan Stankovic, "Yugoslavia: Ideological Conformity and Political-Military Nonalignment," in King and Dean, *op. cit.*, pp. 191-210.

108. Bucharest D.S., April 10, 1969.

109. *Borba*, March 19, 1969.

110. *Pravda*, May 14, 1966.

111. *Lumea*, Nov. 21, 1966. In addition, less than a week before the WTO anniversary, Ceauşescu, in a speech marking the 45th anniversary of the Romanian Communist Party, declared,"One of the barriers in the path of cooperation among peoples are the military blocs and the existence of military bases and troops of some states on the territories of other states. The existence of these blocs, as well as the dispatching of troops to other countries, represent an anachronism which is incompatible with the national independence and sovereignty of peoples with normal inter-state relations." *Scinteia*, May 8, 1966, p. 6.

112. Belgrade D.S., June 19, 1966.

113. Budapest D.S., March 18, 1969.

114. Interview with *Rude Pravo*, March 19, 1969.

115. See the report of a Ceauşescu speech to a joint session of the State Council and the Council of Ministers, and of the decisions of these bodies approving the Budapest Appeal, Bucharest D.S., April 10, 1969. In his speech, Ceauşescu particularly noted that "in accordance with the constitution and the laws of the country, our army cannot be engaged in any action except by the constitutional organs." This is a reference to

attempts by the Soviet Union to have the renewed twenty-year mutual aid treaties include clauses obliging the allies to aid the Soviet Union in a military clash with China. See Remington, *The Warsaw Pact*, pp. 141-45; and M.K. Dziewanowski, "The Rumanian-Soviet Treaty," *East Europe*, Vol. 20 (January, 1974), pp. 19-22.

116. *Borba*, March 19, 1969.

117. For a text of the German note see *Keesing's* for week of July 9-16, 1966, pp. 21498-99.

118. For the Polish reply, see *Ibid.*, pp. 21500-01.

119. *Ibid.*, p. 21501.

120. *Borba*, July 28, 1966.

121. For a text of the Statement, see *Pravda*, March 9, 1968, p. 1. For a discussion of the draft treaty see Walter C. Clemens, Jr., *The Arms Race and Sino-Soviet Relations* (Stanford: Hoover Institution, 1968), pp. 154-71.

122. *Scinteia*, March 13, 1968.

123. For a discussion of Romania's view of the Treaty see Jeanne K. Laux, "Small States and Inter-European Relations 1962-1968," unpublished Ph.D. Dissertation, London: London School of Economics, 1970, Appendix VII; and Clemens, *op. cit.*, pp. 159-60. For further Romanian comment on the treaty see Bucharest D.S., March 13, 1968.

124. For the Yugoslav position of the draft treaty see *Review of International Affairs*, April 11, 1968, p. 17.

125. *Bashkimi*, April 21, 1968.

126. *Pravda*, May 10, 1969; cf. *Pravda*, May 17, 1969.

127. *Trybuna Ludu*, April 27, 1967; *Rude Pravo*, April 30, 1967.

128. *Scinteia*, April 25, 1967 (as reported by Belgrade D.S.,same date).

129. *Scinteia*, May 7, 1967, p.4.

130. *Review of International Affairs*, June 20, 1967, p. 27.

131. On the Moscow conference of 1957 see Zbigniew Brzesinski, *The Soviet Bloc: Unity and Conflict* (Cambridge: Harvard University Press, 1971), pp. 298-308, 317-20. On the 1960 conference see pp. 410-12; and William E. Griffith, "The November 1960 Moscow Meeting: A Preliminary Reconstruction," in Walter Laquer and Leopold Labedz, *Polycentrism* (New York: Praeger, 1962), pp. 107-26.

132. *Zeri i Popullit*, May 5, 1967.

133. *Pravda*, November 28, 1967. For a text of the 18-party call for the consultative meeting, see *Pravda*, November 25, 1967, p. 1.

134. *Zivot Strany*, December 6, 1967.

135. Radio Zagreb, November 25, 1967.

136. Bucharest D.S., February 14, 1968.

137. *Ibid.*

138. *Ibid.*

139. *Trybuna Ludu*, January 21, 1968.

140. TASS, Moscow, December 27, 1967.

141. *Ibid.*

142. See *Kommunist* (Belgrade), January 17, 1968.

143. Sofia D.S., March 14, 1968.

144. *Scinteia*, March 1, 1968; and *Ekonomska Politika*, March 11, 1968.

145. For a description of this meeting see the *Yearbook on International Communist Affairs, 1969* (Stanford: Hoover Institution Press, 1970), pp. 997-1006.

146. *Prace*, March 4, 1968. This approach coincides with a generally more frank discussion of issues which began to appear in the Czechoslovak media during these months. See Ronald H. Linden, "Czech Foreign Policy and the Prague Spring," Paper presented at the 1976 Annual Meeting of the Northeastern Political Science Association, November 11-13, 1976, South Egremont, Massachusetts.

147. *Nepszabadsag*, March 2, 1968.

148. *Praca*, March 3, 1968 and *Rude Pravo*, March 3, 1968.

149. Tanyung, Belgrade, March 1 and 2, 1968.

150. The Romanian explanation at Budapest was carried by Tanyung, March 1, 1968; the Communique of the Central Committee was carried by Agerpress, Bucharest, March 2, 1968.

151. The Budapest meetings took place April 24, September 27-October 1, and November 18-21, 1968; the Moscow meeting occurred May 18-22, 1969.

152. Reactions at the time of the October and November Budapest meetings are combined in one event (WCM 58).

153. *Rude Pravo*, April 21, 1968.

154. *Rude Pravo*, March 24, 1969.

155. Bucharest D.S., April 28, 1968.

156. Bucharest D.S., November 22, 1968.

157. *Zeri i Popullit*, October 8, 1968.

158. *Kommunist* (Moscow), No. 4, March 3, 1969.

159. Zagreb D.S., March 25, 1969.

160. *Scinteia*, March 11, 1969.

161. Ceauşescu had spelled this out explicitly in his "Speech on the 45th Anniversary of the RCP," May 7, 1966; see text in Ceauşescu, *Romania on the Way of Completing Socialist Construction*, Vol. I (Bucharest: Meridane Publishing House, 1969), pp. 331-40.

162. For a transcript of President Johnson's news conference at which this was announced, see *New York Times*, July 29, 1965, p. 12.

163. *Borba*, July 31, 1965.

164. East Berlin Deutschlander, July 29, 1965.

165. TASS, Moscow, June 30, 1966.

166. *Borba*, December 16, 1966.

167. TASS, Moscow, July 17, 1966.

168. *Ibid.* The full text of this statement can also be found in *Pravda*, July 8, 1966, p. 1.

169. CTK, Prague, April 24, 1967.

170. *Pravda*, February 5, 1968.

171. *Magyar Nemzet*, February 1, 1968.

172. East Berlin D.S., January 31, 1968.

173. Prague D.S., January 30, 1968.

174. *Berliner Zeitung*, April 2, 1968.

175. Radio Moscow, April 2, 1968.

176. *Politika*, April 1, 1968.

177. Warsaw D.S., April 1, 1968.

178. Prague D.S., April 1, 1968.

179. *Scinteia*, November 5, 1968.

180. *Ibid.*

181. TASS, Moscow, November 2, 1968.

182. Tirana D.S., November 4, 1968.

183. Moscow D.S., January 15, 1969.

184. Warsaw D.S., January 15, 1969.

185. Zagreb D.S., January 15, 1969.

186. *Borba*, January 15, 1976.

187. *Berliner Zeitung*, April 2, 1968.

188. Moscow D.S. and Radio Peace and Progress, April 2, 1968.

189. Radio Prague, April 1, 1968.

190. Moscow D.S., January 28, 1969.

191. Bucharest D.S., January 29, 1969.

192. ATA International Service, Tirana, January 27, 1969.

193. Radio Moscow, March 3, 1969; *Izvestia*, February 25, 1969.

194. *Rabotnichesko Delo*, March 7, 1969.

195. Zagreb D.S., March 3, 1969.

196. Prague D.S., February 23, 1969.

197. *Pravda* (Bratislava), March 4, 1969.

198. Thus the variable WTOS (the WTO-without-the Soviet Union) equals BUL + CZE + GDR +HUN + POL + ROM/ 6; the variable WTOB (the WTO-without-Bulgaria) equals SOV + CZE + GDR + HUN + POL

+ ROM / 6, etc.

199. The Kendall tau is preferred here to the Spearman r due to the large number of tied ranks incurred. See the discussion in Hubert M. Blalock, *Social Statistics* (New York: McGraw-Hill, 1960), pp. 319-21. The Spearman r would tend to have greater values since it gives greater weight to extreme rank disparity.

200. Of these variables, WTOS through WTOR, only the last is plotted on the graphs.

201. These are actually computed as follows. For each event in the issue group the absolute difference between the country's score and the WTO mean is taken. These differences are then averaged to produce the variable DEV- for the issue group. Similarly, the mean difference between the country's score and the score "WTO-without-country" is the variable DIS- . The mathematical signs attached to variables DEVYUG through DISROM are explained on pp. 155-56.

202. It reflects also the Czech leadership's signature on the March Dresden communique which expressed "confidence . . . that the proletariat and all working people of Czechoslovakia under the leadership of the Communist Party of Czechoslovakia would insure further progress of socialist construction in the country." TASS, Moscow, March 23, 1968.

203. See pp. 84-90.

204. This explanation is further suggested by the fact that when the two countries' scores are eliminated from the computation of the Pact mean, the Hungarian distance grows three times more than does the Polish distance.

205. Due to its clearly aberrant position in the group, low number of cases reporting, and limited interpretive value, Albania will be excluded from this composite index, though some comment is made on its general performance (see p. 155).

206. For each issue area the distance scores are ranked, highest to lowest. These rankings are then combined into an overall ranking by computing a score for each country, giving 8 points for first highest distance, 7 points for second highest, etc. Tied ranks are given the median score. The scores for events EUR 46, the FRG note and WCM 56, the Romanian walkout of the consultative conference, are not included due to the low number of reporting cases.

207. These are combined as were the distance rankings (see note 206 above). The z-scores by issue area are given in Appendix V, Table A.4.

208. See formula, Chapter II, note 52.

209. It should be noted that in computing z-scores each country's own score is included in the computation of the mean.

210. The ranks correlate as follows: A and B, .88 (s<.01); A and C .79 (s<.01); A and D, .67 (s<.05); B and C, .95 (s<.01); B and D .76 (s<.02); C and D, .79 (s<.01).

211. Poland's Distance score on the CCP issue area is .816, third highest for the group. Hungary's mean score for the Czech events (−1.267) differs clearly from the mean scores of Bulgaria, Poland, the GDR and the Soviet Union (average of these scores = −1.988).

212. The WTO-without-Romania mean reaches +2.12; but a WTO mean without both the Soviet Union and Romania equals only +1.98.

213. See East, "Size and Foreign Policy Behavior."

214. As regards z-scores, Yugoslavia's scores are, of course, not computed into the WTO mean, since it is not a member. Rather the WTO is used as a criterion group against which to measure Yugoslavia's reactions. While its z-score is thus not strictly computed in the same way as those for Pact members, it is comparable.

215. The Distance score of .1857 for Albania on the USVAC issue group is somewhat misleading due to the low number of cases (3) for Albania included in that group. On no issue area did Albania's reaction scores fall closer than one S. D. unit from those of the other countries. Z-scores for Albania are given in Appendix VI, Table A.6.

216. The Kendall coefficients are not included in this index and, in addition, the Soviet Union is not included in the rankings in order to make the resulting evaluative index comparable to that constructed for the interaction patterns.

217. The combining of the evaluated distance and z-scores is done as it was for the unevaluated scores, (see note 206) except that the most negative receives 7 points, and the most positive 1 point. The evaluated z-scores by issue area are given in Table A.5, Appendix V; and the evaluation of the distance scores for each issue area is indicated by the signs for variables DEVYUG through DISROM in Tables III.6-III.19.

218. The three ranks correlate as follows: (all s<.01): A and B, .89; A and C, .96; and B and C, .85.

219. Kintner and Klaiber, *op. cit.*, pp. 234-36. The trade/aid dependency measure is a cumulative ranking of: 1) the degree of the state's trade dependence upon the Soviet Union as measured by Kintner and Klaiber, pp. 226-34 (see Chapter II, note 60 for items making up this measure); 2) the amount of Soviet aid (credits and grants) received per capita, 1954-65 (from Charles L. Taylor and Michael C. Hudson, *World Handbook of Political and Social Indicators* [New Haven: Yale University Press, 1972], pp. 360-65); and 3) the degree of concentration of the state's exports (as of 1965); that is, the proportion of its total exports

going to its largest buyer, which in all cases was the Soviet Union (from *World Handbook*, pp. 369-71 and *Yearbook of International Trade Statistics* [New York: United Nations, 1967]). This index of trade/aid dependency is offered as an embellishment upon the straight trade dependency measurement, in an attempt to get a more sensitive and powerful indicator. In order to be consistent, this broader index was also tested against interactional deviance. As with pure trade dependence, no significant association was found.

220. For such a statement on Yugoslavia and Albania, see H. Gordon Skilling, *Communism, National and International* (Toronto: University of Toronto Press, 1964), pp. 70-71.

221. On Romania see e.g. David Floyd, *op. cit.*, pp. 126-29; F. Stephen Larabee, "The Rumanian Challenge to Soviet Hegemony," *Orbis*, XVII, 2 (Spring, 1973),
Rumania (Cambridge: M.I.T. Press, 1967), pp. 56-8 and ff. Fischer-Galati notes, "The Rumanian leadership appeared determined to pursue policies of limited disengagement from Russia and promotion of its own schemes for Rumania's economic development. To attain this aim and resist Russian pressures, it sought to expand further and diversify the country's international relations, to secure the withdrawal of Russian troops, *and to destroy all internal opposition to Gheorghiu-Dej's plans for the socialist transformation of Rumania," (p. 68,* emphasis added). *Cf.* Gabriel Fischer, "Rumania," in Bromke, and Rakowska-Harmstone, *op.cit.*, p. 163.

On Albania see Peter Mayer, *Cohesion and Conflict in International Communism* (The Hague: Martinus Nijhoff, 1968), pp. 158-66; and Skilling, *Communism, National and International*, pp. 68-83. Similarly, in Yugoslavia, it was only after Tito's divergence from Moscow in 1948-50 that a certain degree of relaxation of political and economic control was implemented. See John C. Campbell, *Tito's Separate Road* (New York: Harper and Row Publishers, 1967), pp. 130-50. In fact, Ulam states, "What perturbed the Russians most, it is clear, was the extent of Tito's hold on his party and state." Adam Ulam, *Expansion and Coexistence* (New York: Praeger, 1968), p. 462.

222. The Czech reforms were almost totally directed at reform of the internal economic and political structure of the country. See the documents in Remington, *Winter in Prague*, especially "The Action Program" of April, 1968, and the "Draft Proposal by a Group of Economists," of May, 1968, pp. 88-136 and 173-80, respectively. See also Skilling, *Czechoslovakia's Interrupted Revolution*. As the previous analysis has shown, however, there was some Czech deviance on foreign policy as well during this period.

223. Barbara G. Salmore and Stephen A. Salmore, "Political Regimes

and Foreign Policy," in Maurice East, Stephen Salmore and Charles Hermann, *Why Nations Act* (Beverly Hills: Sage Publications, 1978) pp. 103-22.

224. Jan F. Triska and Paul M. Johnson, *Political Development and Political Change in Eastern Europe: A Comparative Study* (Denver: University of Denver, 1975), pp. 11-16. The index comprises: 1) the degree of decentralizating economic reforms, formally approved by the countries' Central Committee, as of January, 1966 (assessed by Guttman scale); 2) the degree of press censorship exercised in 1964 (assessed by panel of experts); 3) degree of autonomy permitted to empirical social research (four-point scale); 4) total internal security forces as percentage of adult population, 1962-63 (estimate); 5) percentage of total agricultural land privately tilled, 1965; and 6) the presence or absence of the custom of presenting more candidates for national office thanthe number of seats to be filled (dichotomous). The items are combined into an overall index of subsystem autonomy by computing z-scores to produce a ranking. See *Idem*, Tables 1-3.

225. The Soviet Union which ranked fifth and Albania which ranked ninth on the Triska and Johnson scale were removed and the other states filled in accordingly to make the rankings congruent.

226. Salmore and Salmore, *op. cit.*, p. 119.

227. David Easton, *A Systems Analysis of Political Life* (New York: John Wiley and Sons, Inc.. 1965); see his discussion, pp. 247ff.

228. Zvi Gitelman, "Power and Authority in Eastern Europe," in Chalmers Johnson, (ed.) *Change in Communist Systems* (Stanford: Stanford University Press, 1970), pp. 237-38. Cf. Alfred Meyer, "The Legitimacy of Power in East Central Europe" in Istvan Deak and Peter C. Ludz, *East Europe in the 1970s* (New York: Prageger, 1972), pp. 45-68.

229. Salmore and Salmore, *op. cit.*, p. 119.

230. Gitelman, *op. cit.*, pp. 253-59.

231. Rudolph Tokes characterizes the situation as a delicate balance between "indigenous legitimacy" and "derivative legitimacy", a balance which fluctuates over time. See his analysis of Meyer, *op. cit.* in Deak and Ludz, *op. cit.*, pp. 81-82.

232. Speaking of Romania, for example, Gitelman comments, "The Rumanian party seems tohave successfully established itself as an authentic link in the chain of Rumanian national history. It is around these two values–economic development and national pride–that the Rumanian regime seeks to integrate its population." (*op. cit.*, p. 255).

233. That ethnolinguistic minorities have such significance in East Europe is borne out in the substantial literature on national minorities

in the region. See for example, Robert R. King, *Minorities Under Communism* (Cambridge: Harvard University Press, 1973); George W. Simmonds *Nationalism in the USSR and Eastern Europe in the Era of Brezhnev and Kosygin* (Detroit: University of Detroit Press, 1977); Stephen E. Palmer, Jr. and Robert R. King *Yugoslav Communism and the Macedonian Question* (Hamden, Conn.: The Shoe String Press, Inc., 1971); M. George Zaninovitch, *The Development of Socialist Yugoslavia* (Baltimore: The Johns Hopkins University Press, 1968), pp. 11-17, 39-44, 168-71; Robert W. Dean, *Nationalism and Political Change in Eastern Europe: The Slovak Question and the Czechoslovak Reform Movement* (Denver: University of Denver Press, 1973); George Klein, "The Role of Ethnic Politics in the Czechoslovak Crisis of 1968 and the Yugoslav Crisis of 1971," *Studies in Comparative Communism* VIII, 4 (Winter, 1975), pp. 339-69.

234. This ranking is taken from the world ranking of the states found in the *World Handbook*, pp. 271-74.

235. For a discussion of the energy situation in East Europe, see J.G. Polach, "The Development of Energy in Eastern Europe," in Joint Economic Committee, *Economic Developments in Countries of Eastern Europe.* 91st Congress, 2nd Session (Washington: US Government Printing Office, 1970), pp. 348-434; J. Richard Lee, "Petroleum Supply Problems in Eastern Europe," in Joint Economic Committee, *Reorientation and Commercial Relations of the Economies of Eastern Europe*, 93rd Congress 2nd Session (Washington: US Government Printing Office, 1974), pp. 406-20; and Christopher Joyner, "The Energy Situation in Eastern Europe: Problems and Prospects" *East European Quarterly* X, 4 (Winter, 1976), pp. 495-516.

236. See e.g. Floyd, *op. cit.*, pp. 4, 44, 73-75; cf. the discussion below, p. 197.

237. Triska and Johnson, *op. cit.*. Table V, The states are classified as "sufficient", "deficient" or "surplus," "according to their endowments of energy resources relative to their levels of economic activity." (p. 27).

NOTES TO CHAPTER FOUR

1. The possible significance of issue areas for foreign policy was first suggested in a comparative framework in Rosenau, "Pretheories and Theories of Foreign Policy," pp. 71-92.

2. Liska, *op. cit.*, Hopmann, *op. cit.* See also the discussion below, p. 209.

3. For a complete description of negative and positive deviance for interactions, see p. 41; for attitudes, see pp. 55-58.

4. This definition of consistency, comparing simultaneous aspects of one nation's behavior, differs from those which focus upon a nation's behavioral consistency over time; for example, Richard Vengroff, "Instability and Foreign Policy Behavior: Black Africa in the U.N.," *American Journal of Political Science*, XX, 3 (August, 1976), pp. 425-38. See the discussion in Frederick F. Butler and Scott Taylor, "Toward an Explanation of Consistency and Adaptability in Foreign Policy Behavior: The Role of Political Accountability," Paper prepared for presentation at the 1975 meeting of the Midwest Political Science Association, Chicago, Illinois, May 1-3, 1975, pp. 6-7.

5. The West target group contains countries other than those in Western Europe (see Chapter II). Yet it seems logical to conclude that interactions with countries clearly allied with the Western powers can be compared with support proposals on peace, security, nuclear nonproliferation and disarmament. Furthermore, the most active states of the West group in receiving interactions from East European countries were those in West Europe.

6. In addition, Bulgaria's negative deviance position in interactions is further heightened by a large deviance toward the Nonaligned states, a target group having no comparable issue area group.

7. The total of Poland's evaluated z-scores for eleven issues (the FRG note, Romanian walkout and *Pueblo* excluded) is positive, +.801.

8. Poland was, in addition, the only Pact state to have any pre-1967 interactions with Israel.

9. See Nazli Choucri, "The Non-Alignment of Afro-Asian States: Policy Perception and Behavior," *Canadian Journal of Political Science* (March, 1969), pp. 1-17; Ole R. Holsti and John D. Sullivan, "National-International Linkages: France and China As Nonconforming Alliance Members," in James N. Rosenau (ed.), *Linkage Politics* (New York: The Free Press, 1969), pp. 147-95; and Holsti, North and Brody, *op. cit.* A particularly clear case in point occurred in October, 1974, when the Chinese People's Republic hosted a United States trade delegation while Hsinhua, the Chinese news agency blasted the United States for using foreign trade "to control and plunder other countries." *New York Times*, October 20, 1974, p. 11.

10. Robert C. Tucker, "The Deradicalization of Marxist Movements," in Tucker, *The Marxian Revolutionary Idea*, (New York: W.W. Norton, 1969), pp. 172-214.

NOTES TO CHAPTER FIVE

1. Gheorghe Gheorghiu-Dej, "Concerning the Foreign Policy of the Government of the Rumanian People's Republic," Speech delivered at the Fifth Session of the Grand National Assembly, February 22, 1955 (Bucharest: Meridane Publishing House, 1955), p. 22. Cf. Gheorghe Gheorghiu-Dej, "Raportul de Activitate al Comitetului Central al Partidului Muncitoresc Romin," section I, "Situaţia Internaţionala," in *Congresul al II-lea al Partidului Muncitoresc Romin, 23-28 decembrie, 1955* (Bucharest: Editura de stat pentru Literatura Politica, 1956), pp. 18-32; and his "Report of the CC of the RWP on the Activity of the Party in the Period Between the Second and Third Party Congresses, on the Plan of Development of the National Economy in the Years 1960-1965 and the Outline of the Long Term Economic 15-Year Plan," in Gheorghiu-Dej, *Articles and Speeches* (Bucharest: Meridane Publishing House, 1963), esp. pp. 89-90. Speaking on the fortieth anniversary of the founding of the Romanian Communist Party, Gheorghiu-Dej said, "The entire policy of the Communist Party of the Soviet Union, of its Central Committee headed by Comrade Khrushchev, represents a brilliant manifestation of creative Marxism-Leninism in action, an example of how to serve the interests of the people, the interests of the cause of peace and socialism all over the world." See "Forty Fighting Years Under the Victorious Banner of Marxism-Leninism," in Gheorghe Gheorghiu-Dej, *Articles and Speeches*, p. 193.

2. See *Rezoluţia Plenarei Comitetului Central al Partidului Muncitoresc Romin Din 28 iunie- 3 iulie, 1957* (Bucharest: Editura de stat pentru Literatura Politica, 1957).

3. See the editorial, "Sub Conducerea Partidului, Spre Desavirşirea Construcţiei socialiste in Patria Noastra," *Lupta de Clasa*, XL, 7 (June, 1960), pp. 7-14 for a favorable discussion of the 1957 Declaration: and Stefan Voicu, "Un document-program al mişcarii comuniste mondiale," *Lupta de Clasa*, XV, 12 (December, 1960), pp. 3-21 for praise of the 1960 meeting.

4. J.F. Brown, "Rumania Steps Out of Line," *Survey*, 49 (October, 1963), p. 30. See also Stephen Fischer-Galati, *The New Rumania* (Cambridge: M.I.T. Press, 1967), pp. 71-2.

5. Fischer-Galati, *op. cit.*, p. 69. Romanian praise of the 1960 meeting can, to some degree, be seen as equivocal. Voicu, *op. cit.*, for example, contains the following paragraph: "As is known, the general lines of revolution and socialist construction in each country were formulated

in the Declaration of Moscow of November 1957. The meeting of November, 1960 confirms the validity of these general laws and underlines, again, the thesis according to which the exaggeration of the role of national particularities, deviation from the universal truths of marxism-leninism prejudices the common cause of socialism. Also, it is underlined again that marxism-leninism does not permit the mechanical copying of policies and tactics by the communist parties from other countries, ignoring national particularities," (p. 7-8). On the other hand, the article levels a clear blast against revisionism, i.e. the Yugoslavs, as the principal danger, and is unequivocal in its evaluation of the position of the CPSU: "The Communist Party of the Soviet Union is a guiding beacon and a source of strength for all communist parties. The Romanian Workers Party– as the other fraternal parties– has learned and learns from the gigantic accumulated experience of the CPSU in the construction of socialism and communism, drawing precious lessons, which permit us to resolve better and more surely the complex problems regarding the construction of socialist economy and culture, the development of socialist democracy, ideological activity, internal party life." (p.19). How different this is from the future Romanian position can be seen by comparing this attitude to that expressed by Prime Minister I.G. Maurer in an article published in the *World Marxist Review* three years later, on the anniversary of the 1960 meeting. Regarding the experience of the CPSU, for example, it is only its "fundamental importance for the entire communist movement" that this article mentions. See I.G. Maurer, "The Inviolable Foundations of the Unity of the International Communist Movement," *World Marxist Review*, VI, 11 (November, 1963), pp. 12-21.

6. Fischer-Galati, *op. cit.*, p. 84. Compared with Khrushchev's attack on the Albanian party leadership at this congress, Gheorghiu-Dej's speech certainly was "restrained." He noted that "the position adopted for some time now by the leadership of the Albanian Workers Party is out of tune with the spirit of internationalism which animates the other communist and workers parties," and urged them to "renounce their wrong concepts and resume the path of unity of the other socialist countries and fraternal parties." However, the speeches of party leaders Gomulka of Poland, Kadar of Hungary, and Novotny of Czechoslovakia were also relatively mild, while those of Zhivkov of Bulgaria and Ulbricht of the GDR were strong denunciations. The text of all of the speeches can be found in "Kommunisticheskaia Partiia Sovetskogo Soiuza, 22 s"ezd, Moscow, 1961 (22nd Congress of the Soviet Communist Party) *Proceedings and Related Materials* Vol. 5, 6, 9, and 10 (Washington:

Foreign Broadcast Information Service, 1961; quote from Gheorghiu-Dej speech in Vol. 10, pp. 79-80. Certain speeches of the Albanian-Soviet dispute are in Alexander Dallin (ed.) *Diversity in International Communism* (New York: Columbia University Press, 1963). For a discussion of the subject see Peter S.H. Tang, *The Twenty-Second Congress of the Communist Party of the Soviet Union and Moscow-Tirana-Peking Relations* (Washington: Research Institute on the Sino-Soviet Bloc, 1962); William E. Griffith, *Albania and the Sino-Soviet Rift* (Cambridge: M.I.T. Press, 1963).

7. This can be seen in "Directivele Congresului al Doilea al Partidului Muncitoresc Romin Cu Privire La Cel De-Al Doilea Plan Cincinal de Dezvoltare a Economiei Naţionale pe Anii 1956-1960," in *Congresul al II-Lea*, pp. 722-59. On Romanian economic development during the 1953-57 period see John M. Montias *Economic Development in Communist Rumania* (Cambridge: M.I.T. Press, 1967), pp. 38-53.

8. This is how the period 1955-61 is labelled by F. Stephen Larabee in "The Rumanian Challenge to Soviet Hegemony," *Orbis*, XVII, 2 (Spring, 1973), pp. 227-46.

9. Quoted in Brown, *op. cit.*, p. 21.

10. On the differences between the Romanian position and that of Czechoslovakia and the GDR, see Montias, *op. cit.*, pp. 188-99, and 226-30; Brown, *op. cit.*, pp. 24-5; J. B. Thompson, "Rumania's Struggle with Comecon," *East Europe*, XII, 6 (June, 1964), pp. 4-5; and Ghiţa Ionescu, *The Reluctant Ally* (Oxford: Ampersand Ltd., 1965), *passim.*

11. For an examination of this period see John M. Montias, "Background and Origins of the Rumanian Dispute with Comecon," *Soviet Studies*, XVI, 2 (October, 1964), pp. 125-52; or pp. 187-205 of Montias, *Economic Development; Cf. Ionescu, op. cit.*, pp. 33-121.

12. Khrushchev said to the Congress, "How will the further development of the socialist countries toward communism proceed? Can one imagine one of the socialist countries attaining communism and introducing the communist principles of production and distribution, while other countries are left trailing somewhere behind in the early stages of building socialist society?

"This prospect is highly improbable if one takes into account the laws governing the economic development of the socialist system. From the theoretical standpoint it would be more correct to assume that by successfully employing the potentialities inherent in socialism, the socialist countries will enter the higher phase of communist society more or less simultaneously." See the text of Khrushchev's Report to the Congress in *Current Soviet Policies*, III, The Documentary Record of the Extraordinary 21st Congress of the Communist Party of the Soviet Union (New York: Columbia University Press, 1960), p. 68. Stepanyan's earlier article, "Oktyabrskaia revolyutsiia i stanovlenie kommunisticheskoi formatsii," appeared in *Voprosy Filosofii*, 10 (1958), pp. 19-37. For discussion see Montias, *Economic Development*, pp. 197-98; Kenneth Jowitt, *Revolutionary Breakthroughs and National Development, The Case of Romania, 1944-65*, (Berkeley, Calif.: University of California Press, 1971), p. 179.

13. The details of this plan can be found in Gheorghe Gheorhiu-Dej, "Expunere facuta la ședința plenara a CC al PMR din 26-28 noiembrie, 1958," (Bucharest: Editura Politica, 1958). For a discussion of this plenum see Montias, *Economic Development*, pp. 53-54, 199; and Floyd, *op. cit.*, pp. 56. Jowitt notes, furthermore, that Khrushchev specifically praised the Romanian development plans in this speech to the Third Congress of the Romanian Workers' Party in June, 1960. See *op. cit.*, p. 179.

14. See Montias, *Economic Development*, pp. 193-205; and Jowitt, *op. cit.*, pp. 179-84. The latter explains that ". . . at this time the Romanian elite's major effort in terms of purposive behavior was its attempt to reinforce existing Soviet ideological and policy committments precisely because they were in accord with Romanian aspirations." (p. 181).

15. I. Dudinskii, "Nekotoriye cherty razvitiia mirovogo sotsialisticheskogo rynka," *Voprosy ekonomiki*, No. 2 (1961), p. 42; quoted in Montias, *Economic Development*, p. 207.

16. The "Basic Principles" Declaration can be found in Michael Kaser, *Comecon* (London: Oxford University Press, 1967), pp. 249-55; and Khrushchev's article is, "Nasushchniie voprosy razvitie mirovoi sotsialisticheskoi sistemy," *Kommunist*, 12 (August, 1962), pp. 3-27. For discussion, see Montias, *Economic Development*, pp. 210-12; Thompson, *op. cit.*, pp. 7-9.

17. Both the plan and the projections can be found in "Directivele Congresului al III-lea al P.M.R. cu Privire la Planulde Deznoltare a Economiei Nationale pe Anii 1960-65 si la Schita Planului Economie de Prespectiva pe 15 Ani," in *Congresul al III-Lea al Partidului Muncitoresc Romin*, 20-25 iunie, 1960, (Bucharest: Editura Politica, 1961), pp. 645-697. For a detailed discussion of this plan and its fulfillment, see Montias, *Economic Development*, pp. 53-79.

18. Montias, *Economic Development*, p. 215. The Romanians also did not send anyone to the 70th birthday celebrations of East German party leader Walter Ulbricht the same month. See Brown, *op. cit.*, p. 19. Ionescu, *op. cit.*, pp. 101-2, indicated that this was more than a birthday party for the German Party leader, who had had his differences with the

Romanians as noted. It was, rather, an attempt by Khrushchev to gather all the party First Secretaries together to prepare a strong response to the Chinese "twenty five points" of disagreement with Moscow, which had just been published. Gheorghiu-Dej's failure to attend, or even send a delegation, was thus a major slap both at Ulbricht and Khrushchev.

19. See Jacques Levesque, *Le Conflit sino-sovietique et l'Europe de l'Est* (Montreal: Les Presses de l'Université de Montreal, 1970) pp. 163-65; Michael Kaser *op. cit.*, pp. 105-12.

20. For a text of this statement see William E. Griffith, *Sino-Soviet Relations 1964-65* (Cambridge: M.I.T. Press, 1967), pp. 269-96. (Hereafter, "Statement;" pages refer to this volume).

21. Virtually all of the studies of the Romanian-Soviet dispute recognize that the origins of their differences lie in the period of the mid or late fifties, and at least one traces them back to the first Romanian party program of 1945. (See Stephen Fischer-Galati, *The Socialist Republic of Rumania* [Baltimore: The Johns Hopkins University Press, 1969], pp. 51-55 and his *The New Rumania*, pp. 17-43 esp. pp. 29-31. Fischer-Galati admits, though, "Any search for clues to the origins of Rumanian policies in any way independent of the Kremlin would be frustrating if based on an analysis of events antedating Stalin's death." [*The New Rumania*, p. 441]). The authors tend to differ on how salient the need and drive for "independence" was in the calculus of the Romanian party leadership. Fischer-Galati sees the Romanians pursuing independence and sovereignty consciously ever since the Second Party Congress in 1955. He sees Gheorghiu-Dej as shrewdly and quietly manuevering Romanian goals through the pitfalls of Soviet countermoves, until the emergence of the conflict into the public view in the early sixties. See his *The New Rumania; Twentieth Century Rumania* (New York: Columbia University Press, 1970); and more briefly, *The Socialist Republic of Rumania.*

Kenneth Jowitt, on the other hand, labels the periods of the fifties and early sixties as times of "latent learning and docility" (1953-57) and "emulation" (1958-61) by the Romanian leadership. During this time, according to this analysis based on organization theory, the Romanian party leadership saw other communist parties (the Chinese, Yugoslav, Italian, Polish) assert various degrees of ideological independence and pursue certain limited national goals with some degree of success vis-a-vis the Soviet Union. In addition, within the CPSU itself several "redefinitions" of ideological positions took place. Thus, Gheorghiu-Dej—whom Jowitt sees as a "rigid, national leader"—having apparently learned that "a militant strategy might be more efficacious than a moderate one in achieving a workable equality with the Soviet Union," merely continued to pursue an economic development course and only came into conflict with the Soviets when the latter changed their orientation to take a more critical view of autarchic development. "If my analysis is correct," Jowitt states, "the goal of independence was not a major aspiration of Gheorghiu-Dej until sometime in 1962 or later." (p. 210). See Jowitt, *op.cit.* An earlier article containing Jowitt's view in embryonic form is Brown, *op.cit.*

The definitive study of the economics of the dispute is contained in Montias, *Economic Development* (which includes within it the substance of the "Background and Origins" article), and all post-1967 discussions of the issues draw heavily on it. Montias sees the conflict as having its origins in the early fifties, becoming more serious in terms of hardened positions and Romanian reorientation of trade, in 1958, and escalating into a full public dispute by 1961. See. pp. 187-230. Writing in 1964, J.B. Thompson, *op. cit.* sketched a similar argument, focusing especially on the Soviet decision to strengthen Comecon in the later fifties. As Montias' purpose is to discuss the economics of the dispute, he does not make explicit judgement as to the goal of broader, political "independence." But the implication of his analysis are those drawn by Jowitt.

F. Stephen Larabee, while eschewing picking a particular date for the start of the conflict, states, ". . . the important point is that from 1955 to 1961 the issues that would later lead to open conflict began to emerge and crystallize." (*op.cit.,* p. 228) He sees the "institutionalization" of the conflict as coming much later, with the emergence of Ceaușescu and the explicit ideological framework with which he shrouded Romanian policy. (See pp. 231ff.) Robert L. Farlow agrees to some extent but gives Gheorghiu-Dej more credit for ". . . placing all these particular policies within a coherent ideological framework." ("Romanian Foreign Policy: A Case of Pàrtial Alignment," *Problems of Communism* [November-December, 1971] , pp. 54 and ff.) This assessment is based mostly on the appearance of the party "Statement" in April 1964, a year before Gheorghiu-Dej's death. (See his longer discussion in "Alignment and Conflict: Romanian Foreign Policy, 1958-69" unpub. Ph.D. dissertation, Cleveland: Case Western Reserve University, 1971 pp. 32-33.) This suggests that an interesting subject for further scrutiny might be the politics and personalities surrounding the formulation and publication of the "Statement." In terms of chronology, Farlow see Romania as "partially aligned" since 1958, though his analysis strangely lacks detailed investigation of possible earlier conflict. (See pp. 63-152.)

Similarly R.V. Burks dates the dispute from 1958, seeing the important November plenum of that year, the beginning of the trade reorganization, and the withdrawal of Soviet troops from Romania *inter alia*, as key factors at this time. See his "The Rumanian National Deviation: An Accounting" in Kurt London (ed.) *Eastern Europe in Transition* (Baltimore: The Johns Hopkins University Press, 1966), pp. 93-113, and his review of Jowitt's book, "Romania and the Theory of Progress," *Problems of Communism* (May-June, 1972), pp. 83-5. In this review Burks specifically takes issue with Jowitt's contention that Gheorghiu-Dej did not pursue independence until after 1962. He cites, in addition to the above, Gheorghiu-Dej's slow or nonexistent de-Stalinization after 1956 (for more on this see H. Gordon Skilling, *Communism National and International* [Toronto: University of Toronto Press, 1964], pp. 52-68); his consistent pursuing of industrialization, at times conflicting with Soviet views, even before 1962; and his purge of inner party rivals (especially Miron Constantinescu and Iosif Chisinevski) whom Burks says were Khrushchev's men in the RCP.

David Floyd (*Rumania: Russia's Dissident Ally*) also lays great emphasis upon the 1958 period, saying that while the elements of the conflict were indeed "latent" before then, they emerged only during this period due to: 1) the more compromising mood of the Soviets, chagrined by the Polish and Hungarian events of 1956; 2) the weakness of the Romanian economy until this time; and very significantly 3) the presence of Russian troops in Romania until 1958. Floyd says the growing strength of the Romanian economy was signalled at the November, 1958 plenum, at which the Romanians "raised their economic sights." He tends to agree with Fischer-Galati on the diplomatic skills of Gheorghiu-Dej and all of Romania's economic and diplomatic negotiators, but with Jowitt that it was basically a Comecon dispute that become "more than economics."

Ghiţa Ionescu's *The Reluctant Ally* is subtitled "A Study of Communist Neo-Colonialism." His emphasis is on the economics of the dispute and he also sees the November plenum as a "turning point" in Romanian economic development and hence in the prehistory of the public conflict. While he focuses on the period from 1960 on, he, like Floyd, does not fail to note postwar Soviet domination and exploitation of the Romanian economy and the events of 1956 as actions of an earlier time period that would have lasting significance. Finally, he is sufficiently impressed by Gheorghiu-Dej to note his "cunning and shrewd political instincts."

Two other studies of East Europe which discuss the Romanian deviation lay even greater emphasis upon 1956. Both Paul Lendvai, *Eagles in*

Cobwebs (New York: Doubleday, 1969) and François Fejtö, *A History of the People's Democracies* (London: Praeger, 1971) say that while key events occurred in 1958 and 1959, consideration must be extended back to the events in Hungary in 1956 and their impact on Romania, if one is to understand the genesis of the conflict. Lendvai says, "It is very difficult, if not impossible, to date the beginning of the coordinated efforts to gain independence of action. To pick out particular events, such as the withdrawal of Soviet troops in 1958, and generalize from them would be as unwise as to regard subsequent Rumanian moves purely as dazzling improvisations. Many important aspects are still shrouded in secrecy and, at best, obscure. But if the past is re-examined with the benefit of present knowledge, based partly on "off-the-record" conversations with informed Rumanian communists, it seems probable that the 1956 crises in Hungary and Poland were the precipitating factors in laying the domestic foundations for the anticipated shift by the Gheorghiu-Dej group." (p. 295) [The present writer himself received the same impression in just such an "off-the-record" conversation in Bucharest in the fall, 1975]. The criticism of the "improvisation" view is apparently directed at J.F. Brown who saw the Romanian party as "going from brilliant improvisation to brilliant improvisation"; (quoted in Fejtö, *op. cit.*, p. 197). Fejtö agress with Lendvai and Burks: "One should guard against anticipating the sudden revelation in 1963 of a Gheorghiu-Dej fiercely resisting Soviet integrationism. At the end of 1956 Rumania was still occupied by Soviet troops, who were even strengthened after demonstrations of sympathy for the Hungarian rebels in Cluj, Targu Mureş, Timişoara and even in Bucharest. Nevertheless, there is some justification for believing that the events of late 1956 promoted Dej to reduce his dependence on Moscow by enlarging the popular basis of his power and improving his country's international status." (p. 87).

22. Montias, *Economic Development*, pp. 200-1.

23. *Anuarul Statistic al Republicii Socialiste Romania, 1970*. (Bucharest: Direcţia Centrala de Statistica, 1971), pp. 563-67. "Capitalist countries" or "Developed capitalist countries" refers to Western Europe, the United States, Canada and Japan.

24. Based on data in Paul Marer, *Soviet and East European Foreign Trade, 1946-69, Statistical Compendium and Guide* (Bloomington, Ind.: Indiana University Press, 1972), pp. 30, 40. Clark and Farlow, *op. cit.*, report percentages for these years as 21.93% and 37.96%, while a Romanian source reports a growth from 27.4% to 40.4% for the period 1960-1966. See Gheorghe Surpat and Nicolae Ionel, "Relaţiile economice

externe ale Romaniei in perioada construirii socialismului," *Probleme Economice* XX, 12 (December, 1967), p. 172. For a discussion of trade statistics, see Marer, *Idem.*, pp. 347-59.

25. Thad P. Alton, "Economic Structure and Growth in Eastern Europe," in Joint Economic Committe, *Economic Developments in Countries of Eastern Europe: A Compendium of Papers*, 91st Congress, 2nd Session (Washington: US Government Printing Office, 1970), pp. 41-68. On some measures Bulgarian economic growth was higher, see pp. 59-60; cf. Montias, *Economic Development*, pp. 67-68. For a discussion of the different measures of growth see John M. Montias, "Romania's Foreign Trade: An Overview" in Joint Economic Committee, *East European Economies Post-Helsinki: A Compendium of Papers*, 95th Congress, 1st Session (Washington: US Government Printing Office, 1977) pp. 867-68. It is interesting to note that the other CMEA country having a high growth rate, Bulgaria, also demonstrated deviant international interactions during 1965-70, though its trade remained closely tied to the Soviet Union. See Marer, *Soviet and East European Foreign Trade*, pp. 25, 35.

26. During the period 1958-68, 58.5% of all imports from capitalist countries were machines, equipment or complete installations. See Gheorghe Barbu, "Colaborarea economica cu ţarile capitaliste dezvoltate-parte inseparabila a participarii Romaniei la circuitul mondial de valori materiale şi spirituale," unpub. Ph.D. dissertation (Bucharest: Academia de studii economice, 1970), p. 107. An interesting further point is that for 1965 Romania and Bulgaria had the highest level of importation of western machines and equipment of any CMEA member, including the Soviet Union, when figures are "normalized" for per capita national income. See Montias, *Economic Development*, p. 238.

27. Further, it is clear that it was the smaller CMEA countries, not the Soviet Union, who were hurt by the shifting of the Romanian market. While the non-Soviet CMEA share of this trade fell from 39.2% in 1960 to 24.6% in 1965, the Soviet share actually rose, from 34% to 37.7% for the same period. (Compiled from data in Montias, "Romania's Foreign Trade," pp. 876, 885.)

28. See Montias, "Romania's Foreign Trade," p. 872.

29. *Anuarul Statistic, 1970*, p. 564. Just how significant this trade was to Romanian economic development can be seen by the fact that as of 1965 the FRG was Romania's leading supplier of: electro-technical equipment; equipment for the chemical industry, for the manufacturing of wool and paper products; railroad rolling stock; anthracite coal; metal tubing; and auxiliary textiles for the important leather and rubber

industries. It was Romania's second leading supplier of energy equipment and tractors and farm equipment; and an important source of equipment for metallurgy and mining. See *Comerţul Exterior al Republicii Socialiste Romania* (Bucharest: Direcţia Centrala de Statistica, 1973), pp. 117-63. Over 25% of the goods imported from the FRG—the largest portion—were equipment and materials for complete installations. See T. Cristureanu, *Roumaine Commerce Exterieur* (Bucharest: Editura Meridane, 1969), p. 64. Romanian-West German trade did level off after 1967—when it reached a full 12% of total Romanian trade—due largely to Romania's persistent and growing deficit position. For figures, see *Anuarul Statistic, 1970*, p. 564. For discussion, see Edwin M. Snell, "Eastern Europe's Trade and Payments with the Industrial West" in Joint Economic Committee, *Reorientation and Commercial Relations of the Economies of Eastern Europe: A Compendium of Papers*, 93rd Congress, 2nd Session (Washington: US Government Printing Office, 1974), pp. 716-17; and Robert W. Dean, *West Germany's Trade with the East: The Political Dimension* (New York: Praeger, 1974), pp. 182-84, 196-197.

30. I. Rachmuth, "Tendinţa de egalizare a economiei RPR la nivelui ţarilor socialiste mai dezvoltate," in *Dezvoltarea Economica a Rominiei 1944-1964*, Institutul de cercetari economice, (Bucharest: Editura Academiei RPR, 1964), pp. 711-22. Rachmuth also apparently expressed these views to his counterparts in other CMEA countries. See Montias, *Economic Development*, p. 218.

31. See Table V.1. For a breakdown by country, see *Comerţul Exterior*, pp. 21-25. This levelling and the reasons behind it are discussed in Snell, *op. cit.*, pp. 713-18; and Marvin R. Jackson, "Industrialization, Trade and Mobilization in Romania's Drive for Economic Independence," in *East European Economies*, pp. 886-940. Not surprisingly, the CMEA share of Romania's importation of machines and equipment recovered somewhat, from roughly 40% in 1967 to over 50% in 1969. (See Table V.2).

32. See, e.g. Ilie Verdeţ, "Expunerea la Consfatuirea privind activitatea in domeniul Comerţul Exterior," *Viaţa Economica*, V, 10 (March, 1967), pp. 12-14. Ceauşescu, in a speech at a conference on foreign trade activity, in February 1967 said, "It is also necessary for us to pay greater heed to the import of complex equipment and installations, to delivery terms and *especially to technical standard*." See Nicolae Ceauşescu, "Speech Made at the Conference on Foreign Trade Activity," in Ceauşescu, *Romania on the Way of Completing Socialist Construction*, Reports, Speeches, Articles, Vol. 2 (Bucharest: Meridane Publishing House, 1969), p. 209 (emphasis added). Cf. his speech at a session of

of the Grand National Assembly devoted to foreign policy, held in July 1967, "Speech Concerning the Foreign Policy of the Party and of the Government," in *Ibid.*, p. 413. In addition, see Costin Murgescu, "Ştiinţa contemporana şi cooperarea internaţionala," in *Revoluţia Ştiinţifica şi Technica Contemporana* (Bucharest: Editura Politica, 1967), pp. 333-55; Ion Olteanu, "Comerţul Exterior in Contextul Dezvoltarii Economiei Noastre Naţionale," *Lupta de Clasa*, XLVIII, 8 (August, 1968), pp. 40-49; Surpat and Ionel, *op. cit.*, p. 177; and Gheorghe Dolgu, "Romania in Concertul Naţionilor Lumii," *Viaţa Economica*, V, 30 (July 28, 1967), pp. 2-3. Barbu, *op. cit.*, in discussing why trade with Western Europe has constituted such a large part (97%) of Romanian trade with the West, refers to an unnamed American study which appeared in 1967 and reportedly showed that in the last twenty-five years more inventions have appeared in Europe than in the United States. (p. 124n). Romanian concern with attaining the highest efficienty in its foreign trade was reflected in the attention paid to the foreign trade sector in 1966-67, such as in the above conference, and in significant decentralization and reforms effected during 1967-69. For a discussion, see Josef C. Brada and Marvin R. Jackson, "Strategy and Structure in the Organization of Romanian Foreign Trade Activities, 1967-75," in *East European Economies*, pp. 1260-76.

33. See J.G. Polach, "The Development of Energy in East Europe" in *Economic Developments*, pp. 348-434 (esp. pp. 428-9); and J. Richard Lee, "Petroleum Supply Problems in Eastern Europe," in *Reorientation*, pp. 406-20.

34. See Montias, *Economic Development*, pp. 181-82.

35. For statistics on iron ore imports see *Comerţul Exterior*, p. 136; for cotton, see pp. 152-53; for coke, see p. 135; for ferrous metals, see pp. 139-40. Cf. M.K. Dziewanowski, "The Pattern of Rumanian Independence," *East Europe*, XVIII, 6 (June, 1969), p. 9.

36. "Speech Made at the Conference on Foreign Trade Activity," February 23, 1967, p. 208.

37. Olteanu, *op. cit.*, p. 45.

38. *Ibid.*, p. 46. For this and other reasons, long-term trade agreements wre considerd particularly desirable. See T. Pavel and I. Kun, "Importanta acordurilor de lunga durata pentru comerţul internaţional," *Probleme Economice* XVII, 8 (August, 1965), pp. 78-90.

39. Levesque, *op. cit.*, pp. 260-63. For further discussion of the internal situation of the party, see pp. 197-99 below.

40. The growth in exports to developed capitalist countries, was largely accounted for by the exportation of raw materials, semi-manufactures,

and foodstuffs, with CMEA continuing to take the dominant proportion of Romanian production of machines and equipment, less marketable in the West. See John M. Montias, "The Structure of Comecon Trade and the Prospects for East-West Exchanges," in *Reorientation*, pp. 670-676; and Montias, "Romania's Foreign Trade," p. 872. For a discussion of the growth and nature of Romanian exports of machines and equipment, see Israel Burstein, Virgil Popescu, Mariana Malinschi, "Structura şi eficienta in exportul Romanesc de maşini si utiliaje," in *Eficienta si Crestere Economica*, Institutul de Cerectarii Economice (Bucharest: Editura Acad. Rep. Soc. Rom., 1972).

41. M. Horovitz, I. Burstein, I. Lemnij, "Schimbarea locului Rominiei in economia mondiala," in *Dezvoltarea Economica a Rominia 1944-1964*, p. 673.

42. Rachmuth, *op. cit.*, p. 711. "Historical conditions" refers *inter alia* to the pre-war, i.e. capitalist, international economic situation in which Romania served as the "grocery store and gasoline station of Europe." The Romanians were not loathe to remind the Soviets of this situation and by implicaiton stress that they (the Soviets) would certainly not want to emulate *that* international division of labor! See C. Arnautu, Gh. P. Apostol, "Dezvoltarea Bazei Tehnice-Materiale: A Socialismului in Republicii Populara Rominia," *Lupta de Clasa*, XLI, 2 (February, 1961), pp. 18-32; Horovitz, Burstein and Lemnij, *op. cit.*; and Mircea Maliţa, Costin Murgescu and Gheorge Surpat, *Romania Socialista şi Cooperarea Internaţionala* (Bucharest: Editura Politica, 1969), pp. 109-11, 233, 242. I.G. Maurer wrote in 1963, "International socialist division of labor, then, helps to eradicate the aftermath of the previous, capitalist, division of labor which fettered the productive forces of our countries and tended to perpetuate their divisions into advanced industrial and backward agrarian countries." *(op.cit.*, p. 17).

43. Rachmuth, *op. cit.*, p. 719.

44. That this included not only the Soviet Union, but also the GDR and Czechoslovakia, and to a lesser extent Poland, can be seen from Montias, *Economic Development*, pp. 188-230. Fejtö, *op. cit.* writes, " According to her economists, acceptance of the Soviet plans for integration would have kept Rumania in an unfavourable position in relation to the more developed bloc countries, which were accused of selling the Rumanians industrial products and equipment, which were frequently inferior, at higher than world market prices," p. 108.

45. "Statement" p. 284. Note especially the phrase "like the other socialist states" by which the Romanians remind the Soviet Union that they are, after all, only emulating the action of the first socialist state.

46. Surpat and Ionel, *op. cit.*, p. 173.

47. "A XX-a aniversare a eliberarii patriei," *Probleme Economice*, XVII, 8 (August, 1964), p. 12.

48. Alexandru C. Aureliu, *Principii ale relaţiilor dintre state* (Bucharest: Editura Politica, 1966), p. 43; Cf. his strong emphasis on economic independence, pp. 53-63. The economic and political sovereignty link is fundamental to the 1964 "Statement." See pp. 282-83. This link was supported also with a legalistic framework; see Traian Ionascu and Eugen Barasch, "Suveranitatea in Domeniul Economiei- Element Constitutiv al Suveranitaţii Statului Socialist," *Studii şi Cercetari Juridice* X, 3 (1965), pp. 375-405. In addition, sovereignty itself became an increasingly salient conern in Romanian writings on foreign policy. See the discussion below.

49. See e.g. Gheorge Gheorghiu-Dej, "Raportul de activitate al Comitetului Central al Partidului Muncitoresc Romin," section I, "Situaţia Internaţionala," in *Congresul al II-lea*, pp. 18-32; and his "Raportul CC al PMR cu privire la activitatea Partidului in perioada dintre Congresul al II-lea şi Congresul al III-lea al Partidului cu privire la planul planului de dezvoltare a economiei naţionale pe anii 1960-1965 şi la schiţa economic de prespectiva pe 15 anii," section V, "Situaţia Internaţionala," in *Congresul al III-lea*, pp. 94-106. Earlier discussions of international relations of a "new type" put heavier emphasis on the proletarian internationalism aspect, along with mutual and fraternal aid. See Gheorghiu-Dej, "Internaţionalismul— ideologia prieteniei intre popoare" in Gheorghiu-Dej, *Atricole şi Cuvintari*, decemberie, 1955-iulie, 1959 (Bucharest: Editura Politica, 1960), pp. 159-168; and "Vom dezvolta colaborarea frateasca Romino-Cehoslovaca spre folosul popoarelor noastre şi al intregului lagar socialist," ("Speech . . . in Prague," October 24, 1958), in *Ibid.*, pp. 510-24.

50. Gheorghiu-Dej, "Concerning the Foreign Policy," p. 22.

51. See Gheorghe Gheorghiu-Dej, "Speech on the Fifteenth Anniversary of the RPR" at the Grand National Assembly, December 29, 1962, in *Articles and Speeches*. Cf. two editorials which appeared in *Lupta de Clasa* at the time of the Third Party Congress: "Sub conducerea partidului, spre desavirsirea construcţiei socialiste in patria noastra," *Lupta de Clasa*, XL, 7 (June, 1960), pp. 7-14; and "Pentru resolvarea paşnica a problemelor internaţionale vitale şi instaurarea unui climat de pace in lume," *Lupta de Clasa*, XL, 9 (September, 1960), pp. 3-9.

52. Gheorghe Gheorghiu-Dej, "Raportul," in *Congresul al III-lea*, p. 102. See also Gheorghe Gheorghiu-Dej, "Situaţia Internationala şi Politica Externa a Republicii Populare Romine," (Report presented

to an extraordinary session of the Grand National Assembly, August 30, 1960) (Bucharest: Editura Politica, 1960).

53. "Statement", p. 287.

54. The "Statement" points out: "Even when socialism has triumphed on a world scale or at least in most countries, the diversity of the peculiarities of these countries, of the distinctive national and state features which, as Lenin pointed out, will prevail for a long time even after the victory of the proletariat on a world scale, will make it an extremely complex task to find the organization forms of economic cooperation. Life, experience, will shape these forms, the concrete methods of cooperation. To establish now these forms linked to the setting up of a single world economy, a problem of a future historical stage, lacks any real basis." *Ibid.*, p. 283. Cf. pp. 285-86.

55. *Ibid.*, pp. 282-83.

56. Edwin Glaser, "Politica Externa a Republicii Populare Romine, Politica intemiate pe principiile progresiste ale dreptului internaţional contemporan," in *Studii Juridice*, Institutul de cerecetari juridice, Academia RPR (Bucharest: Editura Academia RPR, 1960), p. 512. An article by Glaser commemorating the ninetieth anniversary of Lenin's birth develops a similar theme; see "Triumful Ideilor Leniniste in Domeniul Dreptului Internaţional," *Justiţia Noua*, XVI, 2 (1960), pp. 213-28.

57. Edwin Glaser, "Sovereign Equality- Basic Principle of Contemporary International Law," *Revue Roumaine de Sciences Sociales*, Serie de Sciences Juridiques, IX, 1 (1965), p. 54.

58. See Dan Ciobanu, "Principul Egalitaţii Suverane a Statelor," *Justiţia Noua* XXI, 4 (1965), pp. 10-26; Edwin Glaser, "Contribuţia Politicii Externe a Republicii Socialiste Romania la Dezvoltarea Progresista a Drepului Internaţional Contemporan," *Studii şi Cercetari Juridice*, XX, 3 (1965), pp. 441-456.

59. Glaser, "Contribuţia," p. 442.

60. A similar illustration of the evolution of Romanian views on foreign policy can be seen in the writings of Marţian Niciu. Writing in 1956, he offers, "Only in the new conditions of the construction of socialism, when at the helm of the state stands the working class which promotes on an international plane a policy of peace and collaboration with all states, above all with the Soviet Union and the other states of peoples' democracy, [only then] does our country have a real independence and sovereignty." "Politica externa a Republice Populare Romine- politica de aparare a pacii şi suveranitaţii ţarii noastre," *Buletinul* Universitatilor "V. Babes" si "Bolyai," Cluj, Seria Sţiintelor Sociale, I, 1-2 (1956), p. 174. In an article in 1962 in *Studia*, the theme is "The Principle of

Peaceful Coexistence—a Fundamental Principle of Contemporary International Law" (Principul Coexistenţei Paşnice—Principiu Fundamental al Dreptului Internaţional Contemporan") *Studia* Universitatis Babeş-Bolyai, Series Iurisprudentia, VII (1962), pp. 31-39.
By 1973 Niciu is stressing the importance of full equality of rights for states, respect for national sovereignty and independence, noninterference in internal affairs, and quotes Ceauşescu as follows, "Respect for the right of each people to decide their destiny themselves, to choose without interference their path of development, constitutes the cornerstone of international collaboration. Any damage or harm to this sacred right, to the sovereign prerogatives of peoples, endangers security and peace." See Niciu and Vladimir Hanga, "Un Sfert de Veac de Politica Externa a Romaniei Socialiste," *Studia* Universitatis Babeş-Bolyai, Series Iurisprudentia, XVIII (1973), pp. 29-30 (Ceauşescu quote, p. 30). For world historical principles, as for people, *sic transit gloria mundi.*

For a lucid discussion of the Romanian view of the relationship between independence and socialist unity see Kenneth Jowitt, "The Romanian Communist Party and the World Socialist System: A Redefinition of Unity," *World Politics*, XXIII, 1 (October, 1970), pp. 38-60. For a discussion of the Soviet perspective on "international relations of a new type," see Nish Jamgotch, Jr., "Alliance Management in Eastern Europe (The New Type of International Relations)" *World Politics* XXVII, 3 (April, 1975), pp. 405-29.

61. "Speech Made at Kremlin Palace Meeting dedicated to the Romanian-Soviet Friendship, organized on the occasion of the visit of the Party and Government delegation of the Socialist Republic of Romania to the USSR," in Ceauşescu, *Romania*, V. 1, p. 138. Cf. "Raportul Comitetului Central al Partidului Comunist Roman cu privere la Activitatea Partidului in Perioada Dintre Congresul al VIII-lea şi Congresul al IX-lea al PCR," in *Congresul al IX al Partidului Comunist Roman*, 19-24 iulie, 1965 (Bucharest: Editura Politica, 1966), p. 84; "Partidul Comunist Roman- continuator al luptei revoluţionare şi democractice a poporului Roman, al tradiţiilor mişcarii muncitoreşti şi socialiste din Romania," (Speech delivered on the 45th anniversary of the RCP, May 7, 1966), *Scinteia*, May 8, 1966, pp. 1-6; and "Speech Concerning the Foreign Policy of the Party and of the Government," p. 405.

62. "Partidul Comunist Roman - continuator," p. 6. This speech, Ceauşescu's most forceful statement on the abolition of military blocs, received much attention in the western press. See *New York Times*, May 14, 1966, pp. 1, 5 (excerpts were also reprinted, p. 4); *Le Monde*, May 19, 1966, pp. 1, 3; and the London *Times*, May 13, 1966, p. 1. But Ceauşescu had previously articulated such a viewpoint—once in

Moscow; see his "Kremlin" speech, *op. cit.*, pp. 41-2. Cf. his speech to the party organization of Bucharest on June 12, 1965, before his election as party General Secretary had been officially endorsed by the party congress, *Revista Romana de studii internaţionale*, I, 1966, p. 267. The difference in his May 7 statement was that in previous speeches his references to the abolition of blocs had been made either as a passing reiteration of formal WTO positions, or had been encased in a general condemnation of "aggressive" western (US) alliances. Neither of these provided the context in his May 7 speech.

This assertion, given on a major national day (the 45th anniversary of the RCP), was followed by even clearer elaborations, offered in a speech to the Argeş region party *aktiv* on June 13 (Agerpress, 0900 GMT, June 13, 1966) and in an interview given by Ceauşescu to a group of Italian journalists on June 18. At the time he stated, ". . . the military blocs, the Warsaw Treaty included, will no longer have a place in Europe in which the relations will be set on the basis of the respect for the people's sovereignty and independence." See "Interview granted to a Group of Italian Jorunalists," in Ceauşescu, *Romania*, V. 1, p. 468. Finally, when such a call was included in the "Bucharest Declaration" of the Warsaw Pact Political Consultative Committee in July, 1966, Ceauşescu took special note of it in his summary of the meeting. See his "Speech concerning the Bucharest meeting of the Political Consultative Committee of the Warsaw Treaty member-states, and the meeting of the party leaders and government heads of the CMEA member-countries," in Ceauşescu, *Romania*, V. 1, pp. 527-28.

63. "Speech Concerning the Foreign Policy," p. 391. Cf. "Partidului Comunist Roman- continuator," p. 6.

64. Each of these aspects of the Romanian assertion of sovereignty in international relations expounded by Ceauşescu was backed up and elaborated on by a variety of theoretical legal treatises during the decade. On national distinctiveness, see a remarkable series of articles on the concept of the nation: El. Florea, "Cu Privire la evoluţia conceptului marxist naţiune," *Analele* Institutului de Studii Istorice şi Social-Politice de pe linga CC al PCR, XII, 6 (1967), pp. 66-79; M. Marian, "Comunitatea de naţiune," *Ibid.*, pp. 54-65; Ana Gavrila, "Naţiunea socialista-etapa superioara in viaţa naţiunilor," *Analele* Institutului de Studii Istorice şi Social-Politice de pe linga CC al PCR, XIV, 5 (1968), pp. 101-108. See also a volume published in 1970 on the one hundredth anniversary of the birth of Lenin, *Forţa Creatoarea Ideilor Leniniste*, Institutul de Studii Istorice şi Social-Politice de pe linga CC al PCR (Bucharest: Editura Politica, 1970), especially articles by Ervin Hutira,

"Aplicarea creatoarea de catre Partidul Comunist Roman a teoriei Leniniste despre faurirea economiei socialiste multilateral dezvoltare," pp. 64-107; and Constantin Vlad, "Internaţionalismul, suveranitatea şi patriotismul," pp. 155-73. See also "Intarirea Unitaţii Mişcarii Comunist şi Muncitoresti- Indatorire Suprema," *Scinteia*, February 28, 1967, pp.1, 5.

On military blocs and bases see Mircea Maliţa, "Speech Delivered at the XXII Session of the UN General Assembly," *Revue Roumaine d' etudes Internationales*, 1-2 (1967), pp. 57-64; and Al. Campeanu, *State in State* (Bucharest: Editura Politica, 1968).

On the role of small and medium-sized states see Mircea Maliţa, "Statele mici şi mijlocii in relaţiile internaţionale," *Lupta de Clasa*, XLVIII, 2 (February, 1968), pp. 44-57; and Edwin Glaser, *Statele Mici si Mijlocii in Relaţiile Internaţionale* (Bucharest: Editura Politica, 1971). In addition, the first Romanian text on international law appeared in 1965, written by Grigore Geamanu, *Dreptul Internaţional Contemporan* (Bucharest: Editura Didactica şi Pedagogica, 1965), and replaced a Romanian translation of 'the standard Russian text by Fedor Kozhevnikov, *Mezhdunarodnoe Pravo* (Moscow: Gos. uzd-bo iurid. litry, 1957). The volume and the official review of it lays heavy emphasis on, *inter alia*, noninterference in internal affairs, equality of rights, state sovereignty, as well as "the fundamental principles of contemporary international law and the relations between socialist states;" see Chapter 3 and the review by Constantin Flitan, *Lupta de Clasa*, XLV, 10 (October, 1965), pp. 108-112.

65. See "Eforturi intense ale Partidului Comunist şi popoarelor din Cehoslovacia pentru perfecţionarea vieţii socialie pe calea socialismului," *Scinteia*, March 29, 1968, p. 1; "The Communist Party—Leader of the Process of Renewal and Perfection of the Socialist Society," Agerpress, 1120 GMT, July 15, 1968; "Calea spre intarirea unitaţii ţarilor socialiste," *Scinteia*, July 31, 1968, pp. 1, 8; "All Efforts to Strengthen the Unity of the Socialist Countries," Bucharest Domestic Service, 0600 GMT, August 7, 1968; "Speech delivered by Nicolae Ceauşescu from the balcony of the Party Central Committee headquarters in Bucharest," Bucharest Domestic Service, 1113 GMT, August 21, 1968; "Speech by Nicolae Ceauşescu . . . at the Grand National Assembly's extraordinary session on 22 August," Bucharest Domestic Service, 0842 GMT, August 22, 1968.

On "limited sovereignty" Ceauşescu was unequivocal: "The thesis that some are attempting to accredit, of late, and according to which the joint defence of the socialist countries against an imperialist attack would presuppose limitation or renunciation of the sovereignty of any Treaty member-state does not accord with the principles of relationships between the socialist countries and can by no means be accepted. Not only does membership in the Warsaw Treaty not render questionable the sovereignty of the member-countries, not only does it not "limit" in any way their State independence, but on the contrary, as provided for in the Treaty itself, it serves for strengthening the independence and national sovereignty of each State." See "Expose made at the Jubilee of the Union of Transylvania with Romania," in Ceauşescu, *Romania*, V. 3, pp. 682-83. A similarly clear denunciation can be found in "Speech delivered at the All-Country Conference of the Teaching Staff," in Ceauşescu, *Romania*, V. 3, p. 826. The "limited sovereignty" doctrine was originally floated in an article by S. Kovalev, "Sovereignty and the Internationalist Obligations of Socialist Countries," *Pravda*, September 26, 1968, p. 4; and pronounced by Brezhnev at the Fifth Congress of the Polish United Workers' Party on November 12, 1968, (*Pravda*, November 13, 1968, pp. 1-2).

66. Nicolae Ceauşescu, "Raportul Comitetului Central al Partidului Comunist Roman cu Privire la Activitatea PCR in Perioada dintre Congresul al IX-lea şi Congresul al X-lea si Sarcinile de Viitor ale Partidului," section VI, "Romania—factor activ in lupta pentru triumful socialismului şi pacii in lume," in *Congresul al X-lea al Partidului Comunist Roman*, 6-12 august, 1969 (Bucharest: Editura Politica, 1969), pp. 77-8.

67. Prime Minister Ion Gheorghe Maurer's speech on the twenty-fourth anniversary of the liberation of Romania, August 23, 1968 (also two days after the Czech invasion), is quoted partly as follows, "Social and individual liberty is not conceivable where national independence and sovereignty do not exist or, [where] in one form or another, they are prejudiced," Maliţa, Murgescu, Surpat, *op. cit.*, p. 33. In addition, respect for national sovereignty is illustrated to have been a basic part of the thinking of pre-war Romanian diplomat and statesman, Nicolae Titulescu; see pp. 34-5. For other references to Titulescu, see note 79.

68. *Ibid*,, pp. 33, 35.

69. *Ibid.*, p. 68.

70. *Ibid.*, p. 75.

71. *Ibid.*, p. 112.

72. *Ibid.*, pp. 187-188. In addition, the authors voice other criticisms of CMEA: "The perfecting and stimulating of the development of the process of specialization and cooperation in production implies, with regard to future years, the necessity of studying more and more deeply some specific aspects connected with general economic problems, hard currency-financing, and foreign trade; and elaboration of some suitable

proposals, directed toward increasing the contribution of specialization and cooperation to the enlarging of mutual exchange and assimilation of some types of machines and installations of a high technical level which are not produced or are not produced in sufficient quantities yet in the member countries of CMEA." (p. 189). "It is necessary also, in perfecting the forms and methods of collaboration between the member countries of CMEA, to pay attention to the fact that the organization, planning and leadership of the national economy constitutes an exclusive attribute of each communist party and socialist state. The determination of the directions, proportions, levels and rhythms of economic growth, the establishment of shares of accumulation and consumption in the national income, of the volume and the structure of investments, the application of a pricing policy corresponding to the tasks of national economic growth and to the standard of living of the population, the development of foreign trade and of relations of collaboration and cooperation with other countries, etc. are factors which have profound influence on the course of the economic and social-political development of the respective nations, and for which the responsibility of decision rests exclusively on the party and the government of each socialist country, invested with power by the people of the respective country and responsible to them." (pp. 197-98).

73. *Ibid.*, p. 231. While illustrating the growth of Romanian trade with the capitalist countries, the author (Gheorghe Murgu) explains: "Springing from the necessity of assuring, on the one hand, the covering of some requirements of the process of socialist industrialization in conditions of promoting continually the achievement of the highest of contemporary science and technology, and on the other hand, from the ever more active engagement of Romania in the world economic circiut, relations between our country and the developed capitalist countries have registered an ascendant evolution." (p. 235).

74. See the discussion in Floyd, *op. cit.*, pp. 1-36, and Ionescu, *op. cit.*, pp. 17-26. Floyd reports one estimate that between 1944-1948 the Soviets "absorbed", in war booty and reparations, no less than 86% of the total Romanian national income. (p. 32) Marer estimates Romania's reparations and subvention transfers to the USSR through 1960 to have been $1.7 billion, second only to those of the GDR ($19.5 billion). See Paul Marer, "Has Eastern Europe Become a Liability to the Soviet Union? (III) The Economic Aspect," in Charles Gati (ed.) *The International Politics of Eastern Europe* (New York: Praeger, 1976), pp. 63-64.

75. Lendvai, *op. cit.*, p. 265. One of the more curious aspects of this de-Russification process was the changing of the spelling of the

name of the country from the slavic-looking "Romînia" to the Latinate "România". The Romanian letter "â" appearing in this word and its derivative adjective is not used elsewhere in the language, the letter "î" being retained to represent this distinctly Slavic sound. The significance of this de-slavification of the Romanian language is discussed in George Schopflin, "Rumanian Nationalism," *Survey* XX, 2/3 (Spring/Summer, 1974), pp. 84-7.

76. See e.g., Niciu, "Politica externa;" Glaser, "Politica externa."

77. Examples of this are Dan Berindei, *Din Inceputurile Diplomaţiei Romaneşti Moderne* (Bucharest: Editura Politica, 1965); Virgil Candea, *Pagini din Trecutul Diplomaţiei Roman* (Bucharest: Editura Politica, 1966); and Aureliu, *op. cit.*. This thesis found its fullest expression in Mircea Maliţa, *Romanian Diplomacy: A Historical Survey* (Bucharest: Meridane Publishing House, 1970), in which the author is at some pains to explain how Romania's tributary status vis-a-vis the Turkish Sultan through the end of the nineteenth century qualifies as an independent, self-determined foreign policy. In a latter volume, the same author, a one-time Deputy Foreign Minister and later (presently) personal counsellor to Ceauşescu, even uncovers the existence of a distinctive "Romanian school" of diplomacy. See his *Diplomatia* (Bucharest: Editura Didactica şi Pedagogica, 1975), esp. pp. 77-83 and 395-475.

78. M. Ghelmegeanu, "Un diplomat patriot, promovator al ideii de pace şi securitate colective: Nicolae Titulescu," *Contemporanul*, 37 (September 14, 1962), pp. 1-2.

79. The following are examples. Tudor Teodorsecu Branişte, "L-am cunoscut pe Titulescu," *Lumea*, November 1, 1963, p. 20 (This was the inaugural issue of a Romanian weekly dealing with foreign relations which replaced a Romanian translation of the Soviet foreign affairs weekly *New Times*.); Dan Berindei, "Balcescu Diplomat," *Lumea*, November 28, 1963, p. 22 (Nicolae Balcescu was one of the leaders of the 1848 republican revolution in Wallachia, a then separate part of Romania); Virgil Candea, "Ienachiţa Vacarescu- Diplomat" *Lumea*, February 4, 1965, p. 21 (Vacarescu was a Wallachian diplomat of the late eighteenth century); Virgil Candea, "Politica externa a lui Stefaniţa Voda" *Lumea*, April 1, 1965, pp. 21-22 (A sixteenth century Moldavian *voevod* [baron] who fought against the Turks, Hungarians, and Poles); Mircea Augustin "Conferenţa de la Montreaux," *Lumea*, April 15, 1965, pp. 2-22 (The article praises Titulescu's work at this conference in 1936); Grigore Geamanu and Gheorghe Moca, "Nicolae Titulescu- susţinator al dreptului international ca drept al pacii şi colaborarii internaţionale," *Justitia Noua*, XXII, 3 (1966), pp. 31-33.

An entire issue of *Revue Roumaine d'Histoire* (VII, 7 [1966]) was devoted to Titulescu, commemorating the twenty-fifth anniversary of his death. In one of the articles, Gheorghe Macovescu, the head of the Institute for the Study of International Economic Relations and later Foreign Minister, wrote: "Titulescu's thinking and activity had undoubtedly certain limitations, determined by the historical circumstances to which every personality is, to a smaller or greater extent, tributary . . . . But prevalent in his life and work are neither his limitations, nor his errors. What is prevalent and goes beyond these limitations is that he perceived the outstanding and permanent interests of the Romanian people." A remarkable statement about a bourgeois politician. "Nicolae Titulescu—A Progressive Romanian Diplomat" *loc. cit.*, p. 393. At the same time, a book-length study of Titulescu appeared: Ion Oprea, *Nicolae Titulescu* (Bucharest: Editura Ştiinţifica, 1966); and the next year two volumes of Titulescu's documents and speeches were published Robert Deutsch, *Nicolae Titulescu: Discursuri* (Bucharest: Editura Ştiinţifica, 1967) and G. Macovescu, *Nicolae Titulescu: Documente Diplomatice* (Bucharest: Editura Politica, 1967). In 1969 another book appeared on Titulescu by Vasile Netea, with a forward by then Deputy Foreign Minister Mircea Maliţa, who wrote: "Titulescu was an unrelenting and passionate champion of the role of small and medium-sized nations in the international community . . . . The message of Titulescu's life and work is to this day still as compelling and stimulating as it was in his time. The lofty spirituality and pathos in which this message is clad render it still more engaging and direct. How could it be otherwise, when Titulescu explored the same themes and wielded the same instruments with which, today, we are striving with ever greater energy to defend the cause of peace and further international understanding and cooperation, respecting the specific contribution and individuality of every nation." *Nicolae Titulescu* (Bucharest: Meridane Publishing House, 1969), p. 11. Maliţa wrote a similar foreward to a work by Netea on Titulescu's predecessor and mentor, *Take Ionescu* (Bucharest: Editura Meridane, 1971).

80. See Cristian Popisteanu, *Romania şi Antanta Balcanica* (Bucharest: Editura, 1968), with a foreward by Gheorghe Macovescu; Eliza Campus, *Mica Inţelegere* (Bucharest: Editura Ştiinţifica, 1968); for a favorable review of this volume stressing the collective security aspects of this alliance—with obvious inferences to be drawn regarding Czechoslovakia during the crucial time of the Prague Spring—see *Revista romana de studii internaţionale*, 3-4 (1968), pp. 171-75; Eliza Campus, *Inţelegere Balcanica* (Bucharest: Editura Academiei Republici Socialiste Romania, 1972);

and Gheorghe Colţ, *Romania şi Colaborarea Balcanica* (Bucharest: Editura Politica, 1973).

81. In Macovescu's foreward to the Popisteanu volume he notes, "The Balkan theme has a past and a present in the development of Romanian diplomacy." *Loc. cit.*, p. 14. On Soviet opposition to formalized Balkan cooperation see Brzezinski, *op. cit.*, pp. 55-57 and John C. Campbell, *Tito's Separate Road* (New York: Harper and Row, 1967), pp. 112-13. A Balkan pact was signed in 1954 by Greece, Turkey and Yugoslavia; see Robert L. Wolff, *The Balkans in Our Time* (New York: W.W. Norton Company, 1967), pp. 416-17.

In 1957 Romania sent a note to the governments of the other Balkan states and introduced a proposal at the United Nations for a permanently operating conference of the Balkan heads of state to discuss issues, improve relations and stimulate exchanges. Though this never came to fruition, the General Assembly in 1965 unanimously approved the Romanian proposal. For a text of the note, see *Scinteia*, September 17, 1957. In the same year Bucharest proposed the establishment of a Balkan nuclear free zone. See the discussion in Colţ, *op. cit.*

82. This version is told, in varying levels of detail in: Gheorghe Ştefan, *Momente din Istoria Poporul Roman* (Bucharest: Editura Politica, 1966); Vasile Anescu, Eugen Bantea, Ion Cupsa, *Participarea Armatei Romane la Razboiul Antihitlerist* (Bucharest: Editura Militara, 1966). This volume is complete with dated maps, showing that virtually all of the geopolitically significant territory of Rumania–including the capital, the Danube delta, Black Sea coast, and Ploieşti oil fields–was being liberated by Romanians, and that during most of this key activity the Soviet Army was engaged only in the northeast part of the country (in Moldavia). Cf. *Romania in Razboiul AntiHitlerist*, Institutul de Studii Istorice şi Social-Politice de pe linga CC al PRC, (Bucharest: Editura Militara, 1966). For a bibliography of recent (1958-1971) Romanian works on this subject see Petre Ilie, Gheorghe Stoean, *Romania in Razboiul AntiHitlerist*, Contribuţii bibliografice (Bucharest: Editura Militara, 1971). For a discussion of Romanian historiography, see Schopflin, *op. cit.* pp. 87-92 also.

83. Lendvai, *op. cit.*, p. 280. The pre-war history of the Romanian Communist Party is found in Ionescu, *Communism in Rumania* (London: Oxford University Press, 1964), pp. 1-68.

84. As of the 1966 census there were 1.6 million Hungarians (9% of the total population) in Romania, the largest minority in East Europe. See George Schopflin, *The Soviet Union and Eastern Europe, A Handbook* (New York: Praeger Publishers, 1970), pp. 33, 99, 247. For a discussion

of their political and cultural status see pp. 247-49. For their reaction to the 1956 revolution in Hungary see Ionescu, *Communism*, pp. 267-73; Fischer-Galati, *Twentieth Century Rumania*, p. 148; and note 79.

85. During the Hungarian uprising of 1956 there had been some limited agitation and demonstrating in Romania by Hungarians in Transylvania and in border cities. These were directed both against regime economic failures and the dominance of Russian culture in the schools and universities. Writing of the reactions to the Hungarian events, Ionescu comments, "They were certainly short-lived, superficial, and sporadic. But still Rumania was the country in which solidarity with Hungary was more openly and more strongly expressed than in any other satellite, with the exception, of course, of Poland. Everything consistently pointed towards a similarity of the Rumanian and Hungarians attitudes to the regimes imposed upon them by Soviet Russia." (*Communism*, p. 271.) Cf. the discussion, pp. 267-75. On Ceauşescu's courting and exploitation of Hungarian minority feelings after the Czech invasion see Cioranescu, *op. cit.*, pp. 4-6.

86. Fejtő, *op. cit.*, p. 110.

87. See Fischer-Galati, *The New Rumania*, esp. pp. 44-77; and *Twentieth Century Rumania*, pp. 167-74 and 190-92. Fischer-Galati writes of Gheorghiu-Dej, "Gheorghiu-Dej propounded the alternate formulas of socialist patriotism and party paternalism, both designed to convey the image of himself as architect of the country's "national" and "social" liberation in 1944 and Rumania's subsequent and future progress. By implication he also sought to alter the popular image of him as a Moscovite agent by depicting himself as a Communist leader acting in terms of specific Rumanian conditions." The author lists "wider recognition of the legitimacy of his regime" as one of the goals of Gheorghiu-Dej's policies. (*The New Rumania*, p. 57). See also Farlow, "Alignment and Conflict," pp. 62, 92-5. This author writes that Ceauşescu "has reintegrated a once alienated population with the RCP in an anti-Soviet coalition backing autonomous development." See "Models of Monism and Pluralism in the East European Developmental Context," paper prepared for delivery at the 1974 Meeting of the American Political Science Association, August 29- September 28, 1974, Chicago, p. 17. Cf. Ghiţa Ionescu, *The Breakup of the Soviet Empire in Eastern Europe* (Baltimore: Penguin Books, 1965), pp. 130-131; Fejtő, p. 87.

88. One limitation of this policy, in relation to the Hungarian minority, is pointed out by Schopflin, who notes that a degree of "spillover" of resurgent Romanian nationalism produced certain

anti-minority policies within Romania, such as the absorption of the Hungarian University in Cluj, and the reduction of the autonomy of the Hungarian region. *Handbook*, p. 248. These would not be likely to improve the regime's standing with its minority populations.

89. Ionescu, *The Reluctant Ally*, p. 77. On this see also Floyd, *op. cit.*, pp. 62-99, 128. and Graeme J. Gill, "Rumania: Background to Autonomy," *Survey* XXI, 3 (Summer, 1975), pp. 106-8.

90. See pp. 72-84 above.

91. George Cioranescu, *op. cit.* relates a joke circulating in 1969 in which Ceauşescu takes out a classified ad reading: "Wish to exchange excellent foreign policy for a better geographical position." (p. 7).

92. Peter Bender, *East Europe in Search of Security* (Baltimore: The Johns Hopkins University Press, 1972), p. 112. Similar statements can be found in Larabee, *op. cit.*, p. 244, and Vernon V. Aspaturian, "The Soviet Union and Eastern Europe: The Aftermath of the Czechoslovak Invasion," in I. William Zartman, *Czechoslovakia: Intervention and Impact* (New York: New York University Press, 1970), p. 36, and Fejtö, p. 331.

93. Before the discoveries in the North Sea, Romania's reserves of both of these resources were reportedly the largest in Europe. See Floyd, *op. cit.*, p. 4. Cf. Polach, *op. cit.*, esp. pp. 428-29; Lee, *op. cit.*; Joyner, *op. cit.*

94. Montias, *Economic Development*, pp. 235, 243.

95. Floyd, *op. cit.*, p. 4, 44.

96. See pp. 160-61 above.

97. Larabee, *op. cit.*, p. 244. Robert Farlow writes, "The RCP has not delegated much decision-making power to lower level groups. If anything there has been a growing concentration of power in the hands of the upper-most elite, but especially Ceauşescu himself." See "Models of Monism and Pluralism," p. 17. See also Gabriel Fischer, "Rumania," in Bromke and Rakowska-Harmstone, *op. cit.*, pp. 158-79; Gill, *op. cit.*, p. 108.

98. In addition to *Forţa Creatoarea, op. cit.*, see *Marx, Engels, Lenin despre rolul industriei in dezvoltarea societaţii* (Bucharest: Editura Politica, 1965).

99. The rise of Gheorghiu-Dej and his intra-party maneuvers are detailed in Fischer-Galati, *The New Rumania*, pp. 17-78. Levesque, *op. cit.*, p. 183 relates an unconfirmed report that Gheorghiu-Dej successfully resisted an attempt by Khrushchev to remove him after the publication of the 1964 "Statement".

100. At the first party congress after the death of Gheorghiu-Dej held in July, 1965, (labelled the Ninth Congress of the RCP corresponding to the return to the old party appellation and numbering of congresses) Gheorghiu-Dej's politburo, though renamed the Permanent Presidium of the Party, was reelected with only two exceptions. Peter Borila, described by Farlow as "one of the last remaining Muscovites in the party leadership," ("Alignment and Conflict," p. 240), and Alexandru Moghioros were not reelected. Ceauşescu himself was elected "General Secretary" (changed from "First Secretary"). However, an unprecedented Executive Committee of the party was created, to "deal with important tasks in all fields of party activity." It was made up of the seven Presidium members plus eight others, of whom six were Ceauşescu associates. In addition, all ten candidate members of the Executive Committee and more than half of the enlarged Secretariat were Ceauşescu associates.

Between the Ninth and Tenth Congresses Ceauşescu succeeded in discrediting and/or removing three of Dej's closest associates: Alexandru Draghici, in the Politburo since 1955; discredited and expelled in 1968 for alleged complicity in the persecution of loyal party members during the fifties, especially Lucretiu Patrascanu (the Patrascanu affair is described in Ionescu, *Communism*, pp. 151-56); Gheorghe Apostol, in the Politburo since 1952 and a Deputy Premier in the government until 1968 when he was demoted to the Chairmanship of the Central Council of Trade Unions, leaving Ilie Verdeţ, a Ceauşescu protege, the only First Deputy Premier; Apostol was himself attacked, discredited and removed from this position in August, 1969; Chivu Stoica, in the Politburo since 1945, one of Gheorghiu-Dej's oldest and closest associates; replaced as well by Ceauşescu as President of the Council of State in December 1967.

In addition, Ilie Verdeţ, and Paul Niculescu-Mizil were added to the party Presidium in June, 1966 and Verdeţ and Maxim Berghianu were added to the Executive Committee the same year. Janos Fazekas and Florian Danalache joined the Committee in 1967 and Virgil Trofin replaced the expelled Draghici on the Presidium and the Executive Committee in August, 1968. In December of that year Borila and Moghioros, as well as Alexandru Birladeanu, left the Executive Committee (with Birladeanu leaving the Presidium as well) and were replaced.

Thus the Executive Committee and Presidium that were elected at the Tenth Congress of the Party in August, 1969 retained only one of Gheorghiu-Dej's close associates, Emil Bodnaraş, a politburo member since 1952. In fact, he along with Ion Gheorghe Maurer (the Prime

Minister) were the only members of either body—exclusive of Ceauşescu himself- who were in the party politburo at Gheorghiu-Dej's last congress (1960). By this time also, all top government posts were in the hands of Ceauşescu's "new guard." In less than five years Ceauşescu gained a totality of party, personnel and government control unequalled by any party leader, save perhaps Enver Hoxha of Albania.

Information on Romanian personnel changes is drawn from: *Congresul al III-lea, Congresul al IX-lea, Congresul al X-lea; Keesings Contemporary Archives*, reports of August 14-21, 1965, pp. 20905-06; February 3-10, 1968, pp. 22509-12; and July 18-25, 1970, pp. 24085-87; Farlow, "Alignment and Conflict," pp. 180, 238-46; and Cioranescu, *op. cit.*, pp. 5-6.

101. This has been accompanied, especially more recently, by the increasing predominance of the personality of Ceauşescu himself, as leader, director, purveyor of all programs, and personal embodiment of Romanian internal and external policies.

102. Jowitt, "The Romanian Communist Party," p. 51. Jowitt considers the "political maturity" of the Romanian party to have been a "critical factor" in the pursuit of and success in the RCP's independent line. See *Ibid.*, pp. 49-53. Others who note party unity as a significant factor are Farlow, "Alignment and Conflict," pp. 178-86, 211-18. 238-46; Fejtö, *op. cit.*, p. 197; Thompson, *op. cit.*, pp. 3, 9; Lendvai, *op. cit.*, p. 295; and Floyd, *op. cit.*, pp. 48-55, 68-69; and Levesque, *op. cit.*, pp. 183, 258-60.

103. On the Soviet use of factions in Hungary in 1956, see Ferenc A. Vali, *Rift and Revolt in Hungary* (Cambridge: Harvard University Press, 1961), pp. 358-80; on Poland, see Brzezinski, *op. cit.*, pp. 239-68. In addition, Skilling notes the significance in the Albanian and Yugoslav cases, of the inability of the Soviets to develop or manipulate such factions. (See *Communism National and International,* p. 721.) On factions in Czechoslovakia during 1968, see Skilling, *Czechoslovakia's Interrupted Revolution,* passim.

104. See Carl A. Linden, *Khrushchev and the Soviet Leadership* (Baltimore: The Johns Hopkins University Press, 1966).

105. Jowitt, *Revolutionary Breakthroughs*, pp. 200-01. For similar opinions see Fischer-Galati, *Twentieth Century Rumania*, pp. 171-72; Fejto, *op. cit.*, p. 109. Floyd writes of the Romanian leaders; "It would perhaps not be too much to suggest that their contacts with the Soviet leaders had led them to believe that Khrushchev was unlikely to last long in power and that there would be an adjustment of Soviet policy which would meet Rumanian demands." (*Op. cit.*, p. 118). In support

of this suggestion, he offers certain evidences of the Romanian's ignoring or downplaying Khrushchev's actions during 1964; see pp. 108, 119-20. Levesque, *op. cit.*, pp. 188-90; Floyd (p. 118) and Ionescu (*The Reluctant Ally*, p. 108) see Khrushchev's failure to bring the Romanians to heel as one of the key factors precipitating his downfall. In addition, Floyd takes notice of a personal coolness, if not outright hostility between Gheorghiu-Dej and Khrushchev; see pp. 107-08, 125.

106. Gill, *op. cit.*, describes the immediate post-Khrushchev leadership as "characterized by hesitancy, *immobilisme* and a sense of drift." (p. 105). Fischer-Galati notes, "In the period of adjustment Brezhnev and Kosygin were judged more vulnerable than the deposed leader and hence more likely to condone if not to ratify, the consolidation of Rumania's independent course." (*The New Rumania*, p. 109).

107. On this point, see Levesque, *op. cit.*, pp. 273-74.

108. Robert R. King, "Rumania and the Sino-Soviet Conflict," *Studies in Comparative Communism*, IV, 4 (Winter, 1972), p. 374.

109. Fejtö, *op. cit.*, p. 109; Lendvai, *op. cit.*, p. 302; Levesque, *op. cit.*, pp. 161-63. Fejtö goes on to say of the April 1964 "Statement": "The Rumanian document echoed the Peking 'propositions' of a year earlier." (p. 110). On this see also, p. 101. For a text of the Chinese twenty-five points, see William E. Griffith (ed.) *The Sino-Soviet Rift* (Cambridge: M.I.T. Press, 1964) pp. 249-88.

110. Fejtö notes, "In the speeches and articles of the most prominent Rumanian leaders of the time (1956) there is a striking number of references to China, which Gheorghiu-Dej had visited in September 1956 on the occasion of the Chinese Party Congress." (p. 87). In 1964 I. Rachmuth made the argument that advancement and equalization of Romania's development required higher rates of growth than those of the developed socialist countries, and cited as an example, China's accelerated rhythm of growth (*op. cit.*, p. 712). Praise for China's "Successes . . . in raising industrial and farm output, in developing the national economy, and in the continuous improvement in the working peoples living standard," ("Statement" p. 281, in which the Soviet Union was the only ohter country specifically singled out for praise) was notable in that it continued after such encomia had generally been dropped by the other East European parties. See e.g. Ceaușescu, "Speech Concerning the Foreign Policy of the Party and of the Government," pp. 402-03.

111. Montias, *Economic Development*, pp. 215-16, Ionescu, *Breakup*, p. 132; Brown, *op. cit.*, p. 31; Levesque, *op. cit.*, pp. 161-63; and Fejtö, *op. cit.*, pp. 109-10, note explicit Chinese support of the Romanian position in the "twenty-five points." Also Levesque discusses CPR support

for Rumania on the issue of Bessarabia (pp. 205-16); at the time of the Third Congress of the Romanian Party, 1960 (p. 127); in 1962 during the CMEA dispute (pp. 150-51); after the 1964 "Statement" (p. 186); and after the Soviet invasion of Czechoslovakia, 1968 (p. 253).

112. Gill, *op. cit.*, p. 102.

113. Fischer-Galati, *The New Rumania*, p. 97. J.B. Thompson, *op. cit.* agrees: "Rather than using the 'ideological leading role of the CPSU' as an instrument of control over his European satellites, Khrushchev suddenly found himself seeking the support of those very satellites in his struggle with the Chinese communists." (p. 6).

114. See Montias, *op. cit.*, p. 225; Floyd, *op. cit.*, pp. 61-63; Burks, *op. cit.*, pp. 96-97; Brown, *op. cit.*, pp. 30-31; Lendvai, *op. cit.*, pp. 299-328; Farlow, "Romanian Foreign Policy," p. 63 and "Alignment and Conflict," pp. 1, 200-01, 226-27; Levesque, *op. cit.*, pp. 129-36, 163. Fischer-Galati suggests that an explicit bargain was made by the Romanians in 1960 and again in 1962, trading their support for the Soviet international position in return for Soviet support for Romanian industrialization plans and goals (see *The New Rumania*, pp. 79-80, 90).

115. Lendvai, *op. cit.*, p. 299. Levesque, *op. cit.*, pp. 139-43, 150-63, 218. Khrushchev's desire for firmer control was stimulated, evidently, by fear of losing allies not only to the East. Floyd, *op. cit.*, writes: "Khrushchev's plan to bind the east European countries irrevocably to the Soviet Union was a gesture of despair and an admission that the Russians could see no prospect of retaining the economic allegiance of their allies by other means. It was in fact an admission that they did not really believe that they could win the economic competition with the Western world, as they boasted they could, and that their east European partners would inevitably drift away through economic attration if they were not restrained by force." (p. 81; cf. pp. 124-25). Ionescu (in *Breakup*, p. 125 and *The Reluctant Ally*, pp. 27, 31, 58, 65-66) agrees with Floyd's assessment of the significance of the strong Western counter example, as does Adam Ulam, *Expansion and Coexistence* (New York: Praeger Publishers, 1968), p. 711. In addition, both Ionescu (*Breakup*, p. 125, *The Reluctant Ally*, p. 65) and Kaser (*op. cit.*, p. 197), note the importance of the Soviet desire to placate their more developed allies who were growing increasingly displeased at the lack of efficiency of the CMEA structure. The original initiative, for example, for strengthening the CMEA structure came from the Polish Party.

116. A comprehensive review of Sino-Rumanian relations, including a discussion of a substantial cooling off in these relations between 1966

and 1968, can be found in Levesque, *op. cit.*, pp. 97-281; and King, *op. cit.*

117. Fischer-Galati says, "For Bucharest COMECON was an instrument of Soviet political imperialism, an economic subterfuge for maintaining and extending control over Romania." *Twentieth Century Rumania*, p. 160.

118. Indeed, Farlow ("Alignment and Conflict") sees this "idiosyncratic" variable as oneof the primary explanatory factors underlying the Romanian deviation. He notes (p. 195) "The nativists' [Gheorghiu-Dej, Ceauşescu and others] sense of national identity and their negative experiences with the USSR produced a predisposition on their part to resist, from 1958 on, all Soviet policies which were interpreted as a continuation of the exploitive pattern that had dominated Russo-Romanian relations for several decades and that had contributed to undermining the efficacy of the Romanian communist movement. When conditions were opportune, this basic idiosyncratic element of the nativists became a significant factor in the shaping of foreign policy." See his discussion pp. 186-95 and 219-24. This point is more briefly touched on in his "Romanian Foreign Policy," pp. 62-63. For more on the character of the Romanian leadership, see Floyd, *op. cit.*, pp. 45-69.

119. Lendvai, *op. cit.*, p. 3.

120. As Gill, *op. cit.* puts it: ". . . the Rumanian population was made aware of the regime's differences with Moscow, and because the root of these differences lay in an assertion of Rumanian national interest and Rumanian nationalism, the population was easily able to identify with the regime. Consequently, the Rumanian situation was a mutually re-inforcing process: the regime's enactment of nationalist and anti-Russian measures led to a greater identification of the people with the regime, which thereby gave the Party leadership a strengthened base from which to pursue further national objectives." (p. 108).

NOTES TO CHAPTER SIX

1. George Modelski, *The Communist International System* (Princeton: Center of International Studies, Princeton University, 1960); George Modelski, "Communism and the Globalization of Politics," *International Studies Quarterly*, XII, 4 (December, 1968), pp. 380-94; Edward L. Miles and John S. Gillooly, "Processes of Interaction Among the Fourteen Communist Party States: An Exploratory Essay," in Jan F. Triska,

(ed.) *Communist Party States* (Indianapolis: Bobbs-Merrill, Co., 1969), pp. 106-34; Harvey J. Tucker, "Measuring Cohesion in the International Communist Movement," *Political Methodology II*, 1 (1975), pp. 83-112; cf. Bruce M. Russett, *International Regions and the International System* (Chicago: Rand McNally, 1967).

2. Barry Hughes and Thomas Volgy, "Distance in Foreign Policy Behavior: A Comparative Study of Eastern Europe," *American Journal of Political Science*, XVI, 3 (August, 1970), pp. 459-92; P. Terrence Hopmann, "International Conflict and Cohesion in the Communist System," *International Studies Quarterly*, XI, 3 (September, 1967), pp. 212-36; Vilho Harle, "Actional Distance Between the Socialist Countries in the 1960s," *Cooperation and Conflict*, 14, 3 & 4 (1971), pp. 201-22; and William R. Kintner and Wolfgang Klaiber, *Eastern Europe and European Security* (New York: Dunellen, 1971).

3. Modelski, *The Communist International System*, and "Communism and the Globalization of Politics".

4. Miles and Gillooly, *op. cit.*

5. The present IP data is not drawn from a period contemporaneous with that used by Modelski or Miles and Gillooly; therefore this procedure is necessary to produce a roughly simultaneous comparison. Modelski compares percentages of the states' trade with East Europe and the Soviet Union for 1948, 1950, 1953, and 1956-58 in *The Communist International System*; and for 1965 in "Communism and the Globalization of Politics." Miles and Gillooly, *op. cit.* use 1952 (3)-1962 (3). The present test does the same for the median year of the study, 1967.

6. The mean trade percentage is 63.2%; the mean IP, 40.9%.

7. See pp. 37 above.

8. P. Terrence Hopmann and Barry B. Hughes, "The Use of Events Data for the Measurement of Cohesion in International Political Coalitions: A Validity Study," in Edwar E. Azar and Joseph D. Ben-Dak, *Theory and Practice of Events Research* (New York: Gordon and Breach, 1975), pp. 81-94.

9. Cal Clark and Robert L. Farlow, *Comparative Patterns of Foreign Policy and Trade: The Communist Balkans in International Politics* (Bloomington, Ind.: International Development Research Center, Indiana University, 1976), pp. 101-02.

10. Hopmann and Hughes, *op. cit.*, p. 94, make a similar suggestion.

11. The World Communist System (WCS) IP includes, in addition to the Soviet Union and East Europe: Albania, China, Yugoslavia and the Other Communist target groups.

12. Hughes and Volgy, *op. cit.*

13. Harle, *op. cit.*

14. Kintner and Klaiber, *op. cit.*

15. Hopmann, *op. cit.*; Harle, *op. cit.*; George Liska, *Nations in Alliance* (Baltimore: The Johns Hopkins University Press, 1962).

16. Cal Clark, "The Study of East European Integration: A 'Political' Perspective," *East Central Europe*, II, 2 (1975), p. 148.

17. David W. Moore, "National Attributes and Nation Typologies: A Look at the Rosenau Genotypes," in James Rosenau (ed.) *Comparing Foreign Policies* (Beverly Hills, Calif.: Sage Publications, 1974), pp. 251-67.

18. James G. Kean and Patrick J. McGowan, "National Attributes and Foreign Policy Participation: A Path Analysis," in McGowan (ed.) *Sage International Yearbook of Foreign Policy Studies*, Vol. I (Beverly Hills, Calif.: Sage Publications, 1973), pp. 219-52.

19. *Ibid.*, p. 242.

20. *Ibid.*, pp. 243, 246.

21. Patrick J. McGowan and Klaus-Peter Gottwald, "Small State Foreign Policies," *International Studies Quarterly*, XIX, 4 (December, 1975), pp. 469-500.

22. *Ibid.*, p. 475.

23. *Ibid.*, p. 474.

24. Edward L. Morse, "The Transformation of Foreign Policies: Modernization, Interdependence, and Externalization," *World Politics*, XXII, 3 (April, 1970), pp. 371-92.

25. Maurcie A. East, "National Attributes and Foreign Policy,"in East, Stephen A. Salmore, and Charles F. Hermann (eds.) *Why Nations Act* (Beverly Hills, Calif.: Sage Publications, 1978), p. 136.

26. *Ibid.*, p. 139.

27. *Ibid.*, pp. 140-41.

28. *Ibid.*, p. 141.

29. *Ibid.*, pp. 139-40.

30. Harle, *op. cit.*, p. 288, and Hughes and Volgy, *op. cit.*, p. 491, report similar findings.

31. Maurice Sendak, *Where the Wild Things Are* (New York: Harper and Row, 1963), n.p.

APPENDIX I. NATIONAL INTERACTION SCORES

TABLE A.1: NATIONAL INTERACTION SCORES

A. BULGARIA (IS=883)[1], N=310[2])

	Alb	CPR	Other Comm	USSR	EE	Yugo	Non	AR/IS	West	US	FRG
1965			11	38	26	13	23	14/	37	1	3
1966	5	2	8	9	32	12	2	10/	62	3	7
1967		1	15	32	40	22	67	30/	97	5	2
1968				27	44		60	1/	45	3	1
1969							65		8		
Total (883)	5	3	34	106	142	47	217	55/	249	12	13

B. CZECHOSLOVAKIA (IS=694, N=198)

	Alb	CPR	Other Comm	USSR	EE	Yugo	Non	AR/IS	West	US	FRG
1965		3	12	14	4	10	21		33	6	
1966		7	11	22	23	9	26	5/	35	8	
1967				20	43	11	19	29/	63		5
1968				86	81	38					
1969				24	15	2			9		
Total (694)		10	23	166	166	70	66	34/	140	14	5

C. GERMAN DEMOCRATIC REPUBLIC (IS=284, N=168)

	Alb	CPR	Other Comm	USSR	EE	Yugo	Non	AR/IS	West	US	FRG
1965			6	27		8	1	4/		1	
1966	3		3	12	21	3		12/			
1967			9	10	61	5	15	15/			
1968				3	23						
1969				14	10		10	5/			3
Total (284)	3		18	66	115	16	26	36/		1	3

1. IS = total interaction score all partners, for the period.
2. N = total number of interactions, uncoded, for the period.

D. HUNGARY (IS=554, N=181)

	Alb	CPR	Other Comm	USSR	EE	Yugo	Non	AR/IS	West	US	FRG
1965			9	21	9	9	12		3	5	
1966			11	11	48	11	27	8/	32	4	
1967		2		34	33	21	8		44		1
1968				33	57	13	6		17		
1969			3	14	28	6			11	2	1
Total (554)		2	23	113	175	60	53	8/	107	11	2

E. POLAND (IS=552, N=163)

	Alb	CPR	Other Comm	USSR	EE	Yugo	Non	AR/IS	West	US	FRG
1965			8	25		5	25	10/	51	8	2
1966		5	15	8	35	9		/7	44	3	5
1967	2	2	3	8	46	2	26	9/	114	2	3
1968			7		28			5/	6		
1969				10	11					3	
Total (552)	2	7	33	51	120	16	51	24/7	215	16	10

F. ROMANIA (IS=970, N=288)

	Alb	CPR	Other Comm	USSR	EE	Yugo	Non	AR/IS	West	US	FRG
1965			3	8	18	4	16		3	4	5
1966		10	16	16	64	27	10	5/	82	12	18
1967		2	23	16	29	7	29	4/26	82	11	23
1968		2	2	16	34	40	33	12/11	169	9	3
1969				10	11	8	5		31	1	
Total (970)		14	44	66	156	86	93	21/37	367	37	49

G. YUGOSLAVIA (IS=816, N=219)

	Alb	CPR	Other Comm	USSR	EE	Yugo	Non	AR/IS	West	US	FRG
1965				6	49		20	10/	29	9	
1966	3			12	71		55	14/	52	11	
1967	2			25	68	XX	52	23/	29	18	5
1968			6	14	91		48	12/	31	13	5
1969		3	6		16				7		5
Total (816)	5	3	12	57	295		175	59/	148	51	11

H. ALBANIA (IS=86, N=25)

	Alb	CPR	Other Comm	USSR	EE	Yugo	Non	AR/IS	West	US	FRG
1965		14					4				
1966		16	3		8	3		2/			
1967	XX	12			2	2			5		
1968		12							3		
1969											
Total (86)		54	3		10	5	4	2/	8		

APPENDIX II. ATTITUDE SOURCES

The following are the radio and printed sources of the attitudes expressed by the various East European governments on the issues examined in Chapter III, searched and evaluated as described on pp. 58-59. Official government statements, presented as such, were most often carried by the domestic or international radio service of the country. The printed sources are roughly in descending order of value and frequency.

Albania
Tirana Domestic Service
ATA International Service
*Zeri i Popullit**
Bashkimi

Bulgaria
Sofia Domestic Service
BTA International Service
*Rabotnichesko Delo**
Pogled
Otechestven Front
Trud
Zemedelsko Zname
Narodna Armiya

Czechoslovakia
Prague Domestic Service
CTK International Service
Prague Diplomatic Information Service
Radio Czechoslovakia
Bratislava Domestic Television Service
*Rude Pravo**
Pravda (Bratislava) **
Prace; Praca
Zivot Strany
Mlada Fronta
Zemedelske Noviny
Smena
Svobodne Slovo
Rolnicke Noviny
Lud
Lidova Democracie

* Communist Party daily
** Slovak Communist Party daily

GDR
East Berlin ADN Domestic Service

East Berlin Deutschlandsender
*Neues Deutschland**
Horizont
Berliner Zeitung
Tribune
Junge Welt
Neue Zeit
Bauern-Echo
National Zeitung

Hungary

Budapest MTI Domestic Service
MTI International Service
*Nepszabadsag**
Magyar Nemzet
Magyar Hirlap
Nepszava

Poland

Warsaw Domestic Service
PAP International Service
*Trybuna Ludu**
Glos Pracy
Zycie Warszawy
Zolnierz Wolnosci
Dziennik Ludowy
Sztandar Moldych
Trybuna Mazowiecka

Soviet Union

Moscow Domestic Service
Tass International Service
Moscow Radio Peace and Progress
*Pravda**
Izvestiya
Kommunist
Soviet Russia
Red Star
Trud

Yugoslavia

Belgrade Domestic Service
Tanyung International Service
Zagreb Domestic Service
Radio Zagreb
*Borba**
Politika
Review of International Affairs
Kommunist
Vjesnik (Zagreb)
Nedeljne Informativne Novine
Ekonomska Politika

Romania

Bucharest Domestic Service
Agerpress International Service
*Scinteia**
Lumea

*Communist Party daily

APPENDIX III. SAMPLE CODING SCHEMES

The following are examples of coding schemes used to rank the reactions of the East European governments to several types of events.

AIS 01
Item: Outbreak of war in Middle East
Date: June 6, 1967

Rank as follows:

+3	Strongly support Israel; condemn Arabs
+2	Support Israel; criticize Arabs
+1	Tend to support Israel; tend to criticize Arabs
0	Neutral; report only; ambiguous
−1	Tend to support Arabs; tend to criticize Israel
−2	Support Arabs; criticize Israel
−3	Strongly support Arabs; condemn Israel

CPR 11
Item: 11th Plenum of the Central Committee of the Chinese Communist Party, ending in a communique re: the Soviet Union
Date & Place: Peking; plenum ended 8/13/66

Rank reactions:

+3	Strongly positive; strongly support CPR
+2	Positive; support CPR
+1	Tend to support CPR; mildly positive
0	Neutral; ambiguous
−1	Mildly negative; tend to criticize CPR
−2	Negative; criticize CPR; support Sov.
−3	Strongly negative; strongly support Sov.; condemn CPR

CZE 29
CZE 30
Item: Meeting of five party leaders (and delegations) with Czech party leader (and delegation)
Date & Place: August 3, 1968; Bratislava, Czechoslovakia
Rank the views that appeared at the time of this meeting on two items, if expressed, as follows:

a) Their view of the Czech events:

+3 Strongly positive; strongly support activities
+2 Positive; encourage activities
+1 Mildly positive; tend to encourage activities
0 Neutral; ambiguous; report only
−1 Mildly negative; tend to criticize activities
−2 Negative; criticize and discourage activities
−3 Strongly negative; strongly oppose activities

b) Their view of the Bratislava meeting itself:

+3 Strongly positive
+2 Positive
+1 Mildly positive
0 Neutral; report only
−1 Mildly negative
−2 Negative
−3 Strongly negative

WCM 52

Item: Eighteen parties issue a call for the holding of a consultative meeting on the holding of an international conference of communist and workers' parties.

Date & Place: November 24, 1967; Moscow

Rank reactions as follows:

+3 Strongly positive; strongly favor conference
+2 Positive; support conference
+1 Mildly positive; tend to favor conference
0 Neutral; ambiguous
−1 Mildly negative; tend to oppose conference
−2 Negative; oppose conference
−3 Strongly negative; strongly oppose conference

APPENDIX IV. INTER-CODER RELIABILITY BY COUNTRY

Below are listed the average correlations (Pearson r and Kendall tau) between the three coders' scores, computed by country, over a maximum of fifty-seven cases. All are significant at the level of .001, except those for Albania and Yugoslavia, which are significant at below .003.

TABLE A.2: AVERAGE INTER-CODER CORRELATIONS

	Pearson r	Kendall t
SOV	.9414	.7718
BUL	.9389	.7695
CZE	.8434	.7294
GDR	.9206	.7601
HUN	.8841	.7549
POL	.8952	.7542
ROM	.8212	.7306
YUG	.7318	.5868
ALB	.9013	.5791
Average Overall Correlation	.8753	.7152

In addition the coders matched exactly in the scores they assigned over all cases as follows:

TABLE A.3: PERCENTAGE OF EXACT MATCHES OF CODERS' SCORES

Coder A with Coder B	40.9%
Coder B with Coder C	39.2%
Coder A with Coder C	42.8%
Average Percentage of Exact Matches	41.0%

APPENDIX V. Z-SCORES OF REACTIONS BY ISSUE AREA

TABLE A.4: ABSOLUTE Z-SCORES BY ISSUE AREA (RANKED HIGHEST TO LOWEST)

AIS		*CPRCO*		*CPRBR*		*CZEEV*	
ROM	2.206	ROM	2.058	ROM	1.910	ROM	1.510
SOV	.636	(YUG)	1.491	(YUG)	1.910	(YUG)	1.441
POL	.467	BUL	.820	SOV	1.351	CZE	1.382
HUN	.467	GDR	.820	CZE	.497	SOV	.811
BUL	.467	SOV	.509	BUL	.264	BUL	.643
GDR	.261	POL	.355	POL	.264	GDR	.577
(YUG)	.242	CZE	.131	HUN	.264	POL	.567
CZE	.093	HUN	.131	GDR	.264	HUN	.294

CZEAC		*EUR*		*WCMCF*		*USVAC*	
(YUG)	1.700	(YUG)	5.102	(YUG)	2.903	(YUG)	3.938
ROM	1.490	ROM	2.210	ROM	2.134	CZE	1.471
CZE	1.394	GDR	.577	SOV	.816	POL	1.120
BUL	.771	SOV	.565	BUL	.626	GDR	1.069
GDR	.771	HUN	.479	POL	.474	BUL	.933
SOV	.604	BUL	.466	GDR	.251	SOV	.706
HUN	.446	POL	.130	CZE	.241	ROM	.234
POL	.290	CZE	.008	HUN	.208	HUN	.117

USVO		*USLBJ*		*USPUEB*		*USNIX*	
BUL	1.884	GDR	1.481	HUN	1.906	GDR	1.733
SOV	1.973	CZE	1.102	GDR	1.314	(YUG)	1.467
POL	.778	(YUG)	1.072	SOV	.394	CZE	1.270
CZE	.702	ROM	1.058	BUL	.394	BUL	.802
ROM	.304	BUL	.791	ROM	.394	ROM	.618
GDR	.304	SOV	.703	POL	.296	HUN	.422
(YUG)	.245	HUN	.511	CZE	.296	POL	.390
HUN	.127	POL	.304	(YUG)	.296	SOV	.166

NOTE: Yugoslavia's reaction scores are not included in the calculation of the WTO mean from which the z-score is computed; see further discussion, Chapter III, note 214.

TABLE A.5. EVALUATED Z-SCORES BY ISSUE AREA [1] (RANKED MOST NEGATIVE TO MOST POSITIVE)

AIS		*CPRCO*		*CPRBR*		*CZEEV*	
ROM	−2.206	ROM	−2.058	(YUG)	−1.910	ROM	−1.510
CZE	− .093	(YUG)	−1.491	ROM	−1.910	(YUG)	−1.441
(YUG)	+ .242	POL	− .355	CZE	− .497	CZE	−1.382
GDR	+ .261	HUN	+ .131	BUL	+ .264	HUN	+ .294
BUL	+ .467	CZE	+ .131	HUN	+ .264	POL	+ .567
HUN	+ .467	BUL	+ .820	POL	+ .264	GDR	+ .577
POL	+ .467	GDR	+ .820	GDR	+ .264	BUL	+ .643

CZEAC		*EUR*		*WCMCF*		*USVAC*	
(YUG)	−1.700	(YUG)	−5.102	(YUG)	−2.903	(YUG)	−3.938
ROM	−1.490	ROM	−2.210	ROM	−2.134	CZE	−1.471
CZE	−1.394	CZE	− .008	CZE	− .241	POL	−1.120
POL	+ .291	POL	+ .130	HUN	+ .208	ROM	− .234
HUN	+ .446	BUL	+ .466	GDR	+ .251	HUN	+ .117
GDR	+ .771	HUN	+ .479	POL	+ .474	BUL	+ .933
BUL	+ .771	GDR	+ .577	BUL	+ .626	GDR	+1.069

USVO		*USLBJ*		*USPUEB*		*USNIX*	
BUL	−1.884	CZE	−1.102	HUN	−1.906	(YUG)	−1.467
CZE	− .702	(YUG)	−1.072	CZE	− .296	CZE	−1.270
HUN	+ .127	ROM	−1.058	POL	− .296	ROM	− .618
(YUG)	+ .245	HUN	− .511	(YUG)	− .296	HUN	− .422
ROM	+ .304	POL	− .304	ROM	+ .394	POL	− .390
GDR	+ .304	BUL	+ .791	BUL	+ .394	BUL	+ .802
POL	+ .778	GDR	+1.481	GDR	+1.314	GDR	+1.733

1. For the manner in which the evaluations (signs) were determined, see pp. 155-56. The scores of the Soviet Union are not included in these rankings in order to make the final evaluative index comparable to that constituted for interaction patterns.

APPENDIX VI. Z-SCORES FOR ALBANIA

TABLE A.6: Z-SCORES FOR ALBANIA (ABSOLUTE) BY ISSUE AREA

Issue Area	*Score*
AIS	XXX[1]
CPRCO	XXX
CPRBR	5.171
CZEEV	XXX
CZEAC	3.142
EUR	8.785
FRG (note)	XXX
WCMCF	3.883
WCMR	XXX
USVAC	1.412
USVO	1.886
USLBJ	2.271
USPUEB	1.313
USNIX	2.124

NOTE: As with Yugoslavia, Albania's score is not part of the mean (WTO) from which the z-score is calculated. See discussion, Chapter III, note 214.

---

1. No cases reporting.

APPENDIX VII NUMBER OF REPORTING CASES PER COUNTRY

FIGURE A.7 REPORTING AND NON-REPORTING CASES BY COUNTRY
(MAXIMUM = 57)

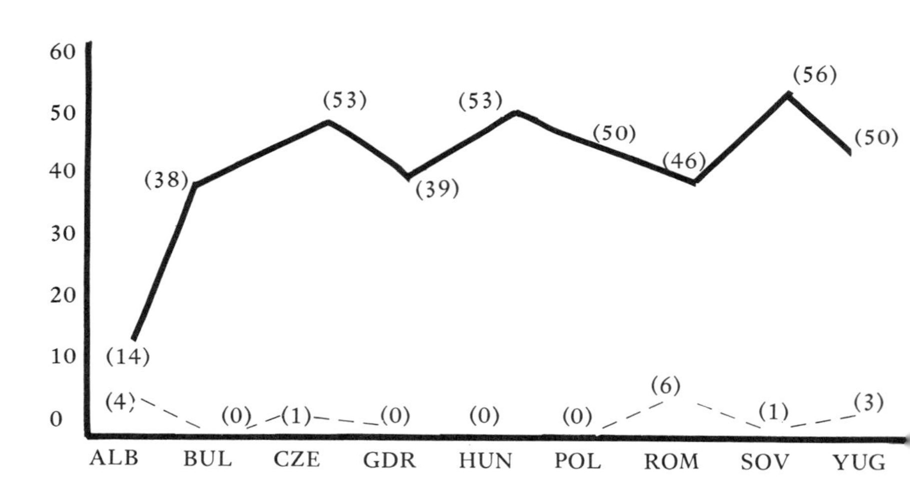

———————— = Reactions issued (Coded +3 to −3)

– – – – – – = Reactions not issued (Coded +7.0)

Balance equals number of reactions not found.

BIBLIOGRAPHY

A. International Relations and Foreign Policy

I. ARTICLES

Andrews, Bruce. "Social Rules and the State as a Social Actor." *World Politics*, XXVII, 4 (July, 1975). 521-540.

Angell, Robert C. "Content Analysis of Elite Media." *Journal of Conflict Resolution*, VIII, 4 (December, 1964). 330-386.

Axline, W. Andrew. "Common Markets, Free Trade Areas, and the Comparative Study of Foreign Policy." *Journal of Common Market Studies*, X, 2 (December, 1971). 163-176.

Azar, Edward E. "Analysis of International Events." *Peace Research Reviews*, IV, 1 (November, 1970). Entire issue.

Baldwin, David A. "The Power of Positive Sanctions." *World Politics*, XXIV, 1 (October, 1971). 19-39.

Bernstein, Robert A. "International Integration: Multidimensional or Unidimensional?" *Journal of Conflict Resolution*, XVI, 3 (September, 1972). 403-408.

Brams, Steven J. "The Structure of Influence Relationships in the International System." In James N. Rosenau (ed.), *International Politics and Foreign Policy*, New York: The Free Press, 1969. 583-599.

Brodin, Katherine. "Belief Systems, Doctrine and Foreign Policy." *Cooperation and Conflict*, VII, 2 (1972). 97-112.

Brody, Richard A. "Some Systemic Effects of the Spread of Nuclear Weapons Technology." *Journal of Conflict Resolution*, VII, 4 (December, 1963). Entire issue.

Butler, Frederick F. and Taylor, Scott. "Toward an Explanation of Consistency and Adaptability in Foreign Policy Behavior: The Role of Political Accountability." Paper prepared for presentation at the 1975 Meeting of the Midwest Political Science Association, Chicago, Illinios, May 1-3, 1975.

Calhoun, Herbert L. "Exploratory Applications to Scaled Event Data." Dallas: paper prepared for delivery at the 1972 International Studies Association Convention, March 15-17, 1972.

Choucri, Nazli. "The Non-Alignment of Afro-Asian States: Policy Perception and Behavior." *Canadian Journal of Political Science*, (March, 1969). 1-17.

Deutsch, Karl W., et al. "Political Community and the North Atlantic Area." In *International Political Communities*, Garden City, New York: Doubleday and Co., 1966. 1-93.

Deutsch, Karl W. "A Comparison of French and German elites in the European Political Environment." In Deutsch, et al., *France, Germany, and the Western Alliance*, New York: Scribner's, 1967. 213-314.

Galtung, Johann. "East-West Interaction Patterns." *Journal of Peace Research*, III, 2 (1966). 146-177.

Haas, Ernst B. and Schmitter, Philippe C. "Economics and Differential Patterns of Political Integration." In *International Political Communities*. Garden City, New York: Doubleday & Co., 1966. 259-301.

Hansen, Roger D. "European Integration: Forward March, Parade Rest, or Dismissed?" *International Organization*, XXVII, 3 (Spring, 1973). 225-254.

Henkin, Louis. "Force, Intervention and Neutrality in Contemporary International Law." In Falk, Richard a. and Mendlovitz, Saul H. (eds.), *The Strategy of World Order*, II, *International Law*. 335-352.

Hill, Gary. "Dyad Coverage in *New York Times*." World Event Interaction Survey (WEIS), Project Memorandum, September 10, 1973.

______ and Fenn, Peter H. "Comparing Event Flows–*The New York Times* and *The Times* of London: Conceptual Issues and Case Studies." *International Interactions*, 1 (1974). 163-186.

Hoffmann, Stanley. "International Sytems and International Law." In Klauss Knorr and Sidney Verba (eds.), *The International System*, Princeton: Princeton University Press, 1961. 205-239.

Holsti, Ole R. "The Belief System and National Images: A Case Study." In James N. Rosenau, *International Politics and Foreign Policy*, revised edition, New York: The Free Press, 1969. 543-551.

______, Brody, Richard A., and North, Robert C. "Measuring Affect and Action in International Reaction Models: Empirical Materials From the 1962 Cuban Crisis." In James Rosenau (ed.), *International Politics and Foreign Policy*, revised edition, New York: The Free Press, 1969. 679-696.

______, North, Robert C., and Brody, Richard A. "Perception and Action in the 1914 Crisis." In J. David Singer, *Quantitative Inter-National Politics*, New York: The Free Press, 1968. 123-159.

______, and Sullivan, John D. "National-International Linkages: France and China as Nonconforming Alliance Members." In James N. Rosenau (ed.), *Linkage Politics*, New York: The Free Press, 1969. 147-195.

Hopmann, P. Terrence and Hughes, Barry. "The Use of Events Data for the Measurement of Cohesion in International Political Coalitions: A Validity Study." In Edward E. Azar and Joseph Ben-Dak (eds.), *Theory and Practice of Events Research*, London and New York: Gordon and Breach, 1975. 81-95.

Howell, Llewellyn J., Jr. "Attitudinal Distance in Southeast Asia." *Southeast Asia*, III, 1 (Winter, 1974). 577-605.

Jervis, Robert. "Hypotheses on Misperception." In James N. Rosenau (ed.). *International Politics and Foreign Policy*, revised edition, New York: The Free Press, 1969. 239-254.

———. "The Costs of the Quantitative Study of International Relations." In Klauss Knorr and James N. Rosenau (eds.) *Contending Approaches to International Politics*, Princton: Princeton University Press, 1969. 177-217.

Kaplan, Morton and Katzenbach, Nicholas B. "Law in the International Community." In Falk, Richard A. and Mendlovitz, Saul H. (eds.), *The Strategy of World Order*, II, *International Law*. New York: World Law Fund, 1966. 19-45.

Kean, James G. and McGowan, Patrick J. "National Attributes and Foreign Policy Participation: A Path Analysis." In Patrick J. McGowan (ed.), *SAGE International Yearbook of Foreign Policy Studies*, I (Beverly Hills, California: SAGE Publications, 1973). 219-251.

Kegley, Charles W., Salmore, Stephen A., and Rosen' D. "Convergences in the Measurement of Interstate Behavior." in Patrick J. McGowan (ed.), *SAGE International Yearbook of Foreign Policy Studies*, II, Beverly Hills, California: SAGE Publications, 1974. 309-339.

Kegley, Charles W., Salmore, Stephen A., Rosen, David, and Howell, Llewellyn Jr. "The Dimensionality of Regional Integration: Construct Validation in the Southeast Asian Context." *International Organization*, XXIX, 4 (Autumn, 1975). 997-1020.

Lindberg, Leon. "Political Integration as a Multidimensional Phenomenon Requiring Multivariate Measurement." In Leon Lindberg and Stuart A. Scheingold, *Regional Integration Theory and Research*, Cambridge, Massachusetts: Harvard University Press, 1971. 45-127.

McClellan, Charles A. and Hoggard, Gary D. "Conflict Patterns in the Interaction Among Nations." In James N. Rosenau (ed.), *International Politics and Foreign Policy*, New York: The Free Press, 1969. 711-724.

McDougal, Myres S. "Some Basic Theoretical Concepts About International Law; A Policy Oriented Framework of Inquiry." In Falk, Richard A. and Mendlovitz, Saul H. (eds.), *The Strategy of World Order*, II, *International Law*, New York: World Law Fund, 1966. 116-134.

McGowan, Patrick J. and Gottwald, Klaus-Peter. "Small State Foreign Policies." *International Studies Quarterly* XIX, 4 (December, 1975). 469-500.

McWhinney, Edward. "Soviet and Western International Law and the Cold War in the Era in Bipolarity." In Falk, Richard A. and Mendlovitz, Saul H. (eds.), *The Strategy of World Order*, II, *International Law*, 189-231.

Merritt, Richard L. "Distance and Interaction Among Political Communities." *General Systems Yearbook*, IX (1964). 255-263.

Moses, Lincoln, Brody, Richard, Holsti, Ole, Kadane, Joseph, and Milstein, Jeffrey. "Scaling Data on Inter-Nation Action." *Science*, May 26, 1967. 1054-1059.

Morse, Edward L. "The Transformation of Foreign Policies: Modernization, Interdependence, and Externalization." *World Politics*, XXII, 3 (April, 1970). 371-392.

Morse, Edward L. "Defining Foreign Policy for Comparative Analysis: A Research Note." Unpublished mimeo, Princeton: Princeton University, June, 1971.

Mueller, John E. "The Use of Content Analysis in International Relations." In George Gerbner, Ole R. Holsti, Laus Krippendorff, William J. Paisley and Philip J. Stone, *The Analysis of Communication Content*, New York: John Wiley and Sons, 1969. 187-197.

Olson, Mancur, and Zeckhauser, Richard. "An Economic Theory of Alliances." *Review of Economics and Statistics*, XLVII, 3 (August, 1966). 266-279.

Paige, Glen D. "The Korean Decision." In James N. Rosenau (ed.), *International Politics and Foreign Policy*, New York: The Free Press, 1969. 461-473.

Puchala, Donald J. "Integration and Disintegration in Franco-German Relations, 1954-65." *International Organization*, XXIV, 2 (Spring, 1970). 183-209.

Puchala, Donald J. "International Transactions and Regional Integration." In Leon N. Lindberg and Stuart A. Scheingold, *Regional Integration Theory and Research*, Cambridge, Massachusetts: Harvard University Press, 1971. 128-159.

Reisman, W. Michael. "Sanctions and Enforcement." In Cyril E. Black and Richard A. Falk (eds.), *The Future of the International Legal Order*, Vol. III: *Conflict Management*, Princeton: Princeton University Press, 1971. 273-335.

Rosenau, James N. "Pre-theories and Theories of Foreign Policy." In R. Barry Farel (ed.), *Approaches to Comparative and International Politics*, Evanston, Ill.: Northwestern University Press, 1966. 27-92.

———— . "Comparative Foreign Policy: Fad, Fantasy, or Field?" *International Studies Quarterly*, XII (September, 1968). 296-329.

Rosenau, James N. "Moral Fervor, Systematic Analysis and Scientific Consciousness." In James N. Rosenau, *The Scientific Study of Foreign Policy*, New York: The Free Press, 1971. 23-65.

———. "Comparing Foreign Policies: Why, What, How." In James N. Rosenau (ed.), *Comparing Foreign Policies*, Beverly Hills, California: SAGE Publications (John Wiley Distrib.), 1974. 3-22.

———. "Comparative Foreign Policy: One-Time Fad, Realized Fantasy, and Normal Field." In Charles W. Kegley, Jr., Gregory A. Raymond, Robert M. Rood, and Richard A. Skinner, *International Events and the Comparative Analysis of Foreign Policy*, Columbia, South Carolina: University of South Carolina Press, 1975. 3-38.

———. "Comparison as a State of Mind." *Studies in Comparative Communism*, VIII, 1&2 (Spring/Summer, 1975). 57-61.

Rummel, Rudolph. "Testing Some Possible Predictors of Conflict Within and Between Nations." Peace Research Society, *Papers*, 1 (1964). 79-111.

———. "Some Dimensions in the Foreign Behavior of Nations." *Journal of Peace Research*, III (1966). 201-224.

———. "Dimensions of Conflict Behavior Within and Between Nations." In Jonathan Wilkenfeld (ed.), *Conflict Behavior and Linkage Politics*, New York: David McKay, 1973. 59-106.

Sigler, John H. "Cooperation and Conflict in United States-Soviet-Chinese Relations, 1966-71." Peace Research Society, *Papers*, XIX, (1971). 107-128.

Singer, J. David. "The Level-of Analysis Problem in International Relations." In Klaus Knorr and Sidney Verba (eds.), *The International System*, Princeton: Princeton University Press, 1961. 77-93.

———. "Content Analysis of Elite Articulations." *Journal of Conflict Resolution*, VIII, 4 (December, 1964). 424-485.

Teune, Henry and Synnestvedt, Sig. "Measuring International Alignment." In Julian R. Friedman, Christopher Bladen and Steven Rosen, *Alliance in International Politics*, Boston: Allyn and Bacon, 1970. 316-332.

Vengroff, Richard. "Instability and Foreign Policy Behavior: Black Africa At the U.N." *American Journal of Political Science*, XX, 3 (August, 1976). 425-438.

Wallensteen, Peter. "Characteristics of Economic Sanctions." in William D. Coplin and Charles W. Kegley, Jr. (eds.), *A Multimethod Introduction to International Politics*, Chicago: Marham Publications, 1971. 128-154.

Wilkenfeld, Jonathan. "Domestic and Foreign Conflict Behavior of Nations." In William Coplin and Charles Kegley, Jr. (eds.), *A Multi-Method*

Introduction to International Politics, Chicago: Markham Publishers, 1971. 189-204.

Zimmerman, William. "Hierarchical Regional Systems and the Politics System Boundaries." *International Organization*, XXVI, 1 (Winter, 1972). 18-36.

Zinnes, Dian A. "The Expression and Perception of Hostility in Prewar Crisis: 1914." In J. David Singer, *Quantitative International Politics*, New York: The Free Press, 1968. 85-123.

II. BOOKS

Allison, Graham T. *Essence of Decision*. Boston: Brown & Co., 1971.

Azar, Edward E. *Probe for Peace*. Minneapolis: Burgess Publishing Co., 1973.

Azar, Edwar and Ben-Dak, Joseph, (eds.). *Theory and Practice of Events Research*, London and New York: Gordon and Breach, 1975.

Burgess, Philip M. and Lawton, Raymond W. *Indicators of International Behavior: An Assessment of Events Data Research*. Beverly Hills, California: SAGE Publications, 1972.

D'Amato, Anthony A. *The Concept of Custom in International Law*. Ithaca: Cornell University Press, 1971.

East, Maurice A., Salmore, Stephen A., and Hermann, Charles F. *Why Nations Act*, Beverly Hills, California: Sage Publications, 1978.

Falk, Richard A. and Mendlovitz, Saul H. (eds.). *The Strategy of World Order*, Volume II *International Law*, New York: World Law Fund, 1966.

Goldman, Kjell. *International Norms and War Between States*. Stockholm: Läromedelsförlagen, 1971.

Gould, Wesley L. and Barkun, Michael. *International Law and the Social Sciences*. Princeton: Princeton University Press, 1970.

Henkin, Louis. *How Nations Behave*. New York: Praeger Publishers, 1968.

Holsti, Ole R., Hopmann, P. Terrence and Sullivan, John D. *Unity and Disintegration in International Alliances: Comparative Studies*. New York: John Wiley, 1973.

International Political Communities. Garden City, New York: Doubleday, 1966.

Kegley, Charles W., Jr., Raymond, Gregory A., Rood, Robert M., and Skinner, Richard. *International Events and the Comparative Analysis of Foreign Policy*. Columbia, South Carolina: University of South Carolina Press, 1975.

Kelsen, Hans. *Principles of International Law*, 2nd ed., revised and edited by Robert W. Tucker, New York: Holt, Rinehart & Winston, 1966.

Laux, Jeanne K. "Small States and Inter-European Relations 1962-1968." Unpublished Ph.D. dissertation. London: London School of Economics, 1970.

Lindberg, Leon N. and Scheingold, Stuart A. *Europe's Would-Be Polity*, Englewood Cliffs, New Jersey: Prentice-Hall, 1970.

Lindberg, Leon N. and Scheingold, Stuart A. *Regional Integration Theory and Research*. Cambridge, Massachusetts: Harvard University Press, 1971.

Liska, George. *Nations in Alliance*. Baltimore: The John Hopkins Press, 19662.

McDougal, Myres S. and Associates. *Studies in World Public Order*. New Haven: Yale University Press, 1960.

McGowan, Patrick J. (ed.). *SAGE International Yearbook of Foreign Policy Studies*, Volume II, Beverly Hills, California: SAGE Publications, 1974.

McGowan, Patrick J. and Shapiro, Howard B. *The Comparative Study of Foreign Policy*. Beverly Hills, California: SAGE Publications, 1973.

Morse, Edward L. *A Comparative Approach to the Study of Foreign Policy: Notes on Theorizing*. Princeton: Center of International Studies, 1971.

Nye, Joseph S. *Peace in Parts*. Boston: Little, Brown, 1971.

Rosenau, James N. *The Scientific Study of Foreign Policy*. New York: The Free Press, 1971.

Rosenau, James N. (ed.). *Comparing Foreign Policies*. Beverly Hills, California: Sage Publications, 1974.

Rothstein, Robert. *Alliances and Small Powers*. New York: Columbia University Press, 1968.

Rubin, Theodore J. and Hill, Gary A. *Experiments in the Scaling and Weighting of International Event Data*. Brentwood, California: Consolidated Analysis Center Inc., 1973.

Russett, Bruce M. *International Regions and the International System*. Chicago: Rand McNally, 1967.

Scott, Andrew M. *The Functioning of the International Political System*. New York: Macmillan, 1967.

Synder, R.C., Bruck, H.W. and Sapin, B. *Foreign Policy Decision Making: An Approach to the Study of International Politics*. New York: The Free Press, 1962.

Vital, David. *The Inequality of States*. Oxford: Oxford University Press, 1967.

von Glahn, Gerhard. *Law Among Nations*. 2nd. ed., New York: Macmillan, 1970.

B. Communist States

I. ARTICLES

Aspaturian, Vernon V. "The Soviet Union and Eastern Europe: The Aftermath of the Czechoslovak Invasion." In I. William Zartman (ed.), *Czechoslovakia: Intervention and Impact*, New York: New York University Press, 1970. 15-46.

Bromke, Adam. "Polycentrism in Eastern Europe." In Adam Bromke and Teresa Rakowska-Harmstone, (eds.), *The Communists States in Disarray, 1965-1971*. Minneapolis: University of Minnesota Press, 1972. 3-21.

Caldwell, Lawrence, T. "The Warsaw Pact: Directions of Change." *Problems of Communism*, XXIV (September/October, 1975). 1-19.

Cary, Charles D. "Patterns of Soviet Treaty-Making Behavior with Other Communist Party States." In Triska, Jan F. (ed.), *Communist Party States*, New York: Bobbs-Merrill, 1969. 135-160.

Clark, Cal. "Foreign Trade as an Indicator of Political Integration in the Soviet Bloc." *International Studies Quarterly*, XV, 3 (September, 1971). 259-295.

Clark, Cal. "The Study of East European Integration: A Political Perspective." *East Central Europe*, II, 2 (1975). 133-151.

Costello, Michael. "Bulgaria." In Adam Bromke and Teresa Rakowska-Harmstone, *The Communist States in Disarray, 1965-1971*, Minneapolis: University of Minnesota Press, 1972. 135-158.

Dorrill, William F. "Power, Policy and Ideology in the Making of the Chinese Cultural Revolution." In Thomas W. Robinson (ed.), *The Cultural Revolution in China*, Berkeley, California: University of California Press, 1971. 21-112.

Eckhardt, William and While, Ralph K. "A Test of the Mirror-Image Hypothesis: Kennedy and Khrushchev." In Erik P. Hoffman and Frederic J. Fleron, Jr., *The Conduct of Soviet Foreign Policy*, Chicago: Aldine, 1971. 308-318.

Farlow, Robert F. "Models of Monism and Pluralism in the East European Developmental Context." Paper prepared for delivery at the 1974 Meeting of the American Political Science Association, August 29-September 28, 1974, Chicago.

Finley, David D. "Integration Among the Communist Party-States: Comparative Case Studies." In Jan F. Triska, *Communist Party-States*, New York: Bobbs-Merrill, 1969. 57-81.

Finley, David D. "What Should We Compare, Why and How?" *Studies in Comparative Communism*, VIII, 1&2 (Spring/Summer, 1975). 12-19.

Gati, Charles. "Area Studies and International Relations: Introductory Remarks." *Studies in Comparative Communism*, VIII, 1&2 (Spring/Summer, 1975). 5-11.

Gehlen, Michael P. "The Integrative Process in East Europe: A Theoretical Framework." *Journal of Politics*, XXX, 1 (February, 1968). 90-114.

Gitelman, Zvi. "Toward a Comparative Foreign Policy of Eastern Europe." In Peter Potichnyj and Jane Shapiro (eds.). *From the Cold War to Detente*. New York: Praeger, 1976. 144-165.

Gitelman, Zvi. "Power and Authority in Eastern Europe." In Chalmers Johnson (ed.). *Change in Communist Systems*. Stanford: Stanford University Press, 1970. 235-263.

George, Alexander L. "The Operational Code: A Neglected Approach to the Study of Political Leaders and Decision-Making." In Erik P. Hoffman and Frederic J. Fleron, Jr., *The Conduct of Soviet Foreign Policy*, Chicago: Aldine, 1971. 165-191.

Griffith, William E. "The November 1960 Moscow Meeting: A Preliminary Reconstruction." In Walther Laquer and Leopold Labedz, *Polycentrism*, New York: Praeger, 1962. 107-126.

Harle, Vilho. "Actional Distance Between the Socialist Countries in the 1960s." *Cooperation and Conflict*, 14, 3-4 (1971). 201-222.

Hempel, Kenneth S. "Comparative Research on Eastern Europe: A Critique of Hughes and Volgy's 'Distance in Foreign Policy Behavior' " *American Journal of Political Science*, XVII, 2 (May, 1973). 367-393.

Hopmann, P. Terrence. "International Conflict and Cohesion in the Communist System." *International Studies Quarterly*, XI, 3 (September, 1967). 212-236.

Hopmann, P. Terrence. "The Effects of International Conflict and Detente on Cohesion in the Communist System." In Kanet, Roger E. (ed.), *The Behavioral Revolution and Communist Studies*, New York: The Free Press, 1971. 301-338.

Hughes, Barry and Volgy, Thomas. "Distance in Foreign Policy Behavior: A Comparative Study of Eastern Europe." *Midwest Journal of Political Science*. XVI, 3 (August, 1970). 459-492.

Jamgotch, Nish Jr. "Alliance Management in Eastern Europe (The New Type of International Relations)." *World Politics*, XXVII, 3 (April, 1975). 405-429.

Joyner, Christopher. "The Energy Situation in Eastern Europe: Problems and Prospects." *East European Quarterly*, X, 4 (Winter, 1976). 495-516.

Kanet, Roger E. "Integration Theory and the Study of Eastern Europe." *International Studies Quarterly*, XVIII, 18 (September, 1974). 368-392.

Kanet, Roger E. "Is Comparison Useful or Possible?" *Studies in Comparative Communism*, VIII, 1&2 (Spring/Summer, 1973). 20-27.

Kolkowicz, Roman. "The Military." In Gordon Skilling and Franklyn Griffiths, *Interest Groups in Soviet Politics*. Princeton: Princeton University Press, 1971. 131-171.

Korbonski, Andrzej. "COMECON: The Evolution of COMECON." In *International Political Communities*, Garden City, New York: Doubleday, Inc., 1966. 351-405.

Korbonski, Andrzej. "Theory and Practice of Regional Integration: The Case of Comecon." In Leon N. Lindberg and Stuart A. Scheingold, *Regional Integration Theory and Research*, Cambridge, Massachusetts: Harvard Universtiy Press, 1971. 338-374.

Klein, George. "The Role of Ethnic Politics in the Czechoslovak Crisis of 1968 and the Yugoslav Crisis of 1971." *Studies in Comparative Communism*, VIII, 4 (Winter, 1975). 339-369.

Klein, George. "Some Undone Jobs." *Studies in Comparative Communism*, VIII, 1&2 (Spring/Summer, 1975). 28-35.

Larrabee, F. Stephen. "Bulgaria's Politics of Conformity." *Problems of Communism*, XXI, (July/August, 1972). 42-52.

Marer, Paul. "Prospects for Integration in the Council for Mutual Economic Assistance (CMEA)." *International Organization*, XXX, 4 (Autumn, 1976). 631-648.

Meyer, Alfred. "The Legitimacy of Power in East Central Europe." In Istvan Deak and Peter C. Ludz (eds.). *East Europe in the 1970s*, New York: Praeger, 1972. 45-68.

Miles, Edward, with Gillooly, John S. "Processes of Interaction Among the Fourteen Communist Party-States: An Exploratory Essay." In Jan F. Triska (ed.), *Communist Party-States*. New York: The Bobbs-Merrill Co., Inc., 1969. 106-134.

Modelski, George. "Communism and the Globalization of Politics." *International Studies Quarterly*, XII, 4 (December, 1968). 380-394.

Rubinstein, Alvin Z. "Comparison or Confusion?" *Studies in Comparative Communism*, VIII, 1&2 (Spring/Summer, 1975). 42-46.

Shoup, Paul. "Comparing Communist Nations: Prospects for an Empirical Approach." In Roger E. Kanet (ed.), *The Behavioral Revolution and Communist Studies*. New York: The Free Press, 1971. 15-47.

Spechler, Dina. "The Council for Mutual Economic Assistance: An Analysis of Soviet Motives in the Making of Foreign Policy." *Slavic and Soviet Series*, II, 3 (Fall, 1977). 3-39.

Stankovic, Slobodan. "Yugoslavia: Ideological Conformity and Political-Military Nonalignment." In Robert R. King and Robert W. Dean (eds.).

East European Perspectives on European Security and Cooperation, New York: Praeger Publishers, 1974. 191-210.

Szulc, Tad. *Czechoslovakia Since World War II*. New York: The Viking Press, 1971.

Tucker, Harvey J. "Measuring Cohesion in the International Communist Movement." *Political Methodology*, II, 1 (1975). 83-112.

Tucker, Robert C. "The Deradicalization of Marxist Movements." In Robert C. Tucker, *The Marxian Revolutionary Idea*. New York: W.W. Norton, 1969. 172-214.

Uren, Philip E. "Patterns of Economic Relations." In Adam Bromke and Teresa Rakowska-Harmstone, (eds.). *The Communist States in Dissary 1965-1971*, Minneapolis: University of Minnesota Press, 1972. 307-22.

Welsh, William A. "Towards An Empirical Typology of Socialist Systems." In Carl Beck and Carmelo Mesa-Lago (eds.). *Comparative Socialist Systems*, Pittsburgh: University of Pittsburgh, 1975. 52-91.

Yankovitch, Paul. "La Conference des pay communistes sur l'aide aux etats arabes s'est abstenue de condamner 'l'agression isrealienne'." *Le Monde*, September 8, 1967. 4.

II. Books

Beck, Carl and Mesa-Lago, Carmelo. *Comparative Socialist Systems*, Pittsburgh: University Center for International Studies, University of Pittsburgh, 1975.

Bender, Peter. *East Europe in Search of Security*, Baltimore: The Johns Hopkins University Press, 1972.

Bromke, Adam and Rakowska-Harmstone, Teresa, (eds.). *The Communist States in Disarray, 1965-71*, Minneapolis: The University of Minnesota, 1972.

Brzezinski, Zbigniew. *The Soviet Bloc: Unity and Conflict*. revised edition, Cambridge: Harvard University Press, 1971.

Campbell, John C. *Tito's Separate Road*, New York: Harper and Row, 9167.

Clark, Cal and Farlow, Robert. *Comparative Patterns of Foreign Policy and Trade: The Communist Balkans in International Politics*. Bloomington, Indiana: International Development Research Center, Indiana University, 1976.

Clemens, Walter C., Jr. *The Arms Race and Sino-Soviet Relations*, Stanford: Hoover Institution, 1968.

Dean, Robert W. *Nationalism and Political Change in Eastern Europe: The Slovak Question and the Czechoslovak Reform Movement*, Denver: University of Denver, 1973.

Dean, Robert W. *West German Trade with the East: The Political Dimension*. New York: Praeger, 1974.

Dutt, G. and Dutt, V.P. *China's Cultural Revolution*, New York: Asia Publishing House, 1970.

Fetjö, Francois. *A History of the People's Democracies*. New York: Praeger, 1971.

Fleron, Frederic J., Jr. (ed.). *Communist Studies and the Social Sciences*, Chicago: Rand McNally, 1969.

Freedman, Robert O. *Economic Warfare in the Communist Bloc*, New York: Praeger, 1970.

Gati, Charles. (ed.). *The Politics of Modernization in Eastern Europe*, New York: Praeger, 1974.

Gati, Charles. *The International Politics of Eastern Europe*. New York: Praeger, 1976.

Griffith, William E. *Albania and the Sino-Soviet Rift*, Cambridge: M.I.T. Press, 1963.

Johnson, Chalmers. *Change in Communist Systems*, Stanford, California: Stanford University Press, 1971.

Joint Economic Committee. *Economic Development in Countries of Eastern Europe*, 91st Congress, Second Session, Washington: U.S. Government Printing Office, 1970.

Joint Economic Committee. *Reorientation and Commercial Relations of the Economies of Eastern Europe*, 93rd Congress, Second Session, Washington: U.S. Government Printing Office, 1974.

Joint Economic Committee. *East European Economies Post-Helsinki: A Compendium of Papers*, 95th Congress, First Session, Washington: U.S. Government Printing Office, 1977.

Kanet, Roger E. (ed.). *The Behavioral Revolution and Communist Studies*, New York: The Free Press, 1971.

Kaser, Michael. *Comecon: Integration Problems of the Planned Economies*. London: Oxford University Press, 1967.

King, Robert R. *Minorities Under Communism*, Cambridge: Harvard University Press, 1973.

King, Robert R., and Dean, Robert. *East European Perspectives on European Security and Cooperation*. New York: Praeger, 1974.

Kinter, William R. and Klaiber, Wolfgang. *Eastern Europe and European Security*, New York: Dunellen, 1971.

Lendvai, Paul. *Eagles in Cobwebs*. New York: Doubleday, 1969.

Levesque, Jacques. *Le Conflict sino-sovietique and l'Europe de l'Est*, Montreal: Les Presses de l'universite de Montreal, 1970.

Linden, Carl. *Khrushchev and the Soviet Leadership*, Baltimore: The Johns Hopkins University Press, 1966.

Mayer, Peter. *Cohesion and Conflict in International Communism*, The Hague: Martinus Nijhoff, 1968.

Modelski, George. *The Communist International System*, Princeton: Center of International Studies, 1960.

Oren, Nissan. *Revolution Administered: Agrarianism and Communism in Bulgaria*, Baltimore: The Johns Hopkins University Press, 1973.

Palmer, Stephen E., Jr. and King, Robert R. *Yugoslav Communism and the Macedonian Question*, Hamden, Conneticut: The Shoe String Press, Inc., 1971.

Ploss, Sidney I. *The Soviet Political Process*, Waltham: Ginn and Co., 1971.

Potichnyj, Peter and Shapiro, Jane (eds.). *From the Cold War to Detente*, New York: Praeger, 1976.

Ransom, Charles. *The European Community and Eastern Europe*, Totowa, New Jersey: Rowman and Littlefield, 1973.

Remington, Robin. *The Warsaw Pact*, Cambridge: The M.I.T. Press, 1971.

Schaefer, Henry. *Comecon and the Politics of Integration*, New York: Praeger, 1972.

Schopflin, George. *The Soviet Union and Eastern Europe, A Handbook*, New York: Praeger Publishers, 1970.

Simmonds, George W. (ed.). *Nationalism in the USSR and Eastern Europe in the Era of Brezhnev and Kosygin*, Detroit: University of Detroit Press, 1977.

Skilling, H. Gordon. *Communism-National and International*, Toronto: University of Toronto Press, 1964.

Skilling, H. Gordon. *Czechoslovakia's Interrupted Revolution*, Princeton: Princeton University, 1976.

Tang, Peter S.H. *The Twenty-Second Congress of the Communist Party of the Soviet Union and Moscow-Tirana-Peking Relations*, Washington: Research Institute on the Sino-Soviet Bloc, 1962.

Triska, Jan F. (ed.). *Communist Party States*, New York: Bobbs-Merrill, 1969.

Triska, Jan F., and Johnson, Paul M. *Political Development and Political Change in Eastern Europe*. Denver: University of Denver, 1975.

Triska, Jan F., and Cocks, Paul. (ed.). *Political Development in East Europe*, New York: Praeger, 1977.

Tigrid, Pavel. *Why Dubcek Fell*, London: Mac Donald and Co., 1971.

Ulam, Adam. *Expansion and Coexistence*, New York: Praeger, 1968.

Vali, Ferenc A. *Rift and Revolt in Hungary*, Cambridge: Harvard University Press, 1961.

Wiles, Peter. *Communist International Economics*, Oxford: Basil Blackwell, 1968.

Wolff, Robert L. *The Balkans in Our Time*, New York: W.W. Norton Company, 1967.

Zaninovitch, M. George. *The Development of Socialist Yugoslavia*, Baltimore: Johns Hopkins Universtiy Press, 1968.

Zimmerman, William. *Soviet Perspectives on International Relations*, Princton: Princeton University Press, 1969.

C. Romania—Non-Romanian Works

I. ARTICLES

Brown, J.F. "Rumania Steps Out of Line." *Survey*, 49 (October, 1963). 19-34.

Burks, R.V. "The Rumanian National Deviation: An Accounting." In Kurt London (ed.). *Eastern Europe in Transition*, Baltimore: The Johns Hopkins University Press, 1966. 93-113.

Burks, R.V. "Romania and the Theory of Progress." *Problems of Communism*, May-June, 1972. 83-85.

Cioranescu, George. "Rumania After Czechoslovakia: Ceausescu Walks the Tightrope." *East Europe*, XVIII, 6 (June, 1969). 2-7.

Dziewanowski, M.K. "The Pattern of Rumanian Independence." *East Europe*, XVIII, 6 (June, 1969). 8-12.

Farlow, Robert L. "Romanian Foreign Policy: A Case of Partial Alignment." *Problems of Communism*, (November-December, 1971). 54-63.

Fischer, Gabriel. "Rumania." In Adam Bromke and Teresa Rakowska-Harmstone, *The Communist States in Disarray, 1965-1971*. Minneapolis: University of Minnesota Press, 1972. 158-179.

Gill, Graeme, "Rumania: Background to Autonomy." *Survey*, XXI, 3 (Summer, 1975). 94-113.

Jowitt, Kenneth. "The Romanian Communist Party and the World Socialist System: A Redefinition of Unity." *World Politics*, XXIII, 1 (October: 1970). 38-60.

King, Robert R. "Rumania and the Sino-Soviet Conflict." *Studies in Comparative Communism*, IV, 4 (Winter, 1972). 373-412.

King, Robert R. "Rumania: The Difficulty of Maintaining An Autonomous Foreign Policy." In Robert R. King and Robert W. Dean (eds.)., *East European Perspectives on European Security and Cooperation*, New York: Praeger Publishers, 1974. 168-191.

Larabee, F. Stephen. "The Rumanian Challenge to Soviet Hegemony." *Orbis*, XVII, 2 (Spring, 1973). 227-246.

Laux, Jeanne K. "Intra-Alliance Politics and European Detente: The Case of Poland and Romania." *Studies in Comparative Communism*, VIII, 1&2 (Spring/Summer, 1975). 98-122.
Montias, John M. "Background and Origins of the Rumanian Dispute with Comecon." *Soviet Studies*, XVI, 2 (October, 1974). 125-152.
Schopflin, George. "Rumanian Nationalism." *Survey*, XX, 2/3 (Spring/Summer, 1974). 77-104.
Thompson, J.E. "Rumania's Struggle with Comecon." *East Europe*, XII, 6 (June, 1964). 2-9.

II. Books

Farlow, Robert L. "Alignment and Conflict: Romaian Foreign Policy 1958-69." Unpublished Ph.D. dissertation, Cleveland: Case Western Reserve University, 1971.
Fischer-Galati, Stephen. *The New Rumania*, Cambridge: M.I.T. Press, 1967.
Fischer-Galati, Stephen. *The Socialist Republic of Rumania*, Baltimore: The Johns Hopkins University Press, 1969.
Fischer-Galati, Stephen. *Twentieth Century Rumania*, New York: Columbia University Press, 1970.
Floyd, David. *Rumania: Russia's Dissident Ally*, New York: Praeger, 1965.
Ionescu, Ghiţa. *Communism in Rumania*, London: Oxford University Press, 1964.
Ionescu, Ghiţa. *The Breakup of the Soviet Empire in Eastern Europe*, Baltimore: Penguin Books, 1965.
Ionescu, Ghiţa. *The Reluctant Ally*, Oxford: Ampersand Ltd., 1965.
Jowitt, Kenneth. *Revolutionary Breakthroughs and National Development, The Case of Romania, 1944-1965*, Berkeley, California: University of California Press, 1971.
Montias, John M. *Economic Development in Communist Rumania*, Cambridge: M.I.T. Press, 1967.

D. Romania—Romanian Works

I. ARTICLES

"A XX-a Aniversare a eliberarii patriei." *Probleme Economice*, XVII, 8 (August, 1964). 3-16.
Arnautu, C., Apostol, Gh. P. "Dezvoltarea Bazei Technice-Materiale A Socialismului in Republicii Populara Rominia." *Lupta de Clasa*, XLI, 2 (February, 1961). 18-32.

Aigistom, Micrea. "Conferenta de la Montreaux." *Lumea*, April 15, 1965. 2-22.

Berindei, Dan. "Balcescu Diplomat." *Lumea*, November, 28, 1963. 22.

Branişte, Tudor Teodorsecu. "L-am cunoscut pe Titulescu." *Lumea*, November 1, 1963. 20.

Burstein, Israel, Popescu, Virgil and Malinschi, Mariana. "Structura şi eficienţa in exportul Romanesc de maşini şi utilaje." in *Eficienţa şi Crestere Economica*, Institutul de Cercetarii Economice, Bucharest: Editura Acad. Rep. Soc. Rom., 1972.

Candea, Virgil. "Ienachiţa Vacarescu—Diplomat." *Lumea*, February 4, 1965. 21.

Candea, Virgil. "Politca externa a lui Stefaniţa Voda." *Lumea*, April 1, 1965. 21-22.

Ciobanu, Dan. "Principul Egalitaţii Suverane a Statelor." *Justiţia Noua*, XXI, 4 (1965). 10-26.

Dolgu, Gheorghe. "Romania in Concertul Naţiunilor Lumii." *Viaţa Economica*, V, 30 (July 28, 1967). 2-3.

Flitan, Constantin. Review of Geamanu, Grigore, *Dreptul Internaţional Contemporan* [Bucharest: Editura Didactica şi Pedagogica, 1965]. *Lupta de Clasa*, XLV, 10 (October, 1965). 108-112.

Florea, El. "Cu Privire la evoluţia conceptului marxist de naţiune." *Analele*, Institutului de studii Istorice şi Social-Politice de pe linga CC al PCR, XII, 6 (1967). 66-79.

Gavrila, Ana. "Naţiunea socialista-etapa superioara in viata naţiunilor." *Analele*, Institutului de Studii Istorice şi Social-Politice de pe linga CC al PCR, XIV, 5 (1968). 101-108.

Geamanu, Grigore, and Moca, Gheorghe. "Nicolae Titulescu—susţinator al dreptului internaţional ca drept al pacii şi colaborarii internaţionale." *Justiţia Noua*, XXII, 3 (1966). 3-13.

Ghelmegeanu, M. "Un diplomat patriot, promovator al ideii de pace şi securitate colective: Nicolae Titulescu." *Contemporanul*. 37 (September 14, 1962). 1-2.

Glaser, Edwin. "Politica Externa a Republicii Populare Romine, Politica intemiate pe principiile progresiste ale dreptului internaţional contemporan." In *Studii Juridice*, Institutul de Cercetarii Juridice, Academia RPR, Bucharest: Editura Academia RPR, 1960. 509-535.

Glaser, Edwin. "Rriumful Ideilor Leniniste in Domeniul Dreptului Internaţional." *Jusiţia Noua*, XVI, 2 (1960). 213-228.

Glaser, Edwin. "Conţibuţia Roliticii Externe a Republicii Socialiste Romania." *Studii şi Cercetarii Juridice*, XX, 3 (1965). 441-456.

Glaser, Edwin. "Sovereign Equality—Basic Principle of Contemporary International Law." *Revue Roumaine de Sciences Sociales*, Serie de Sciences Juridiques, IX, 1 (1965). 51-65.

Horovitz, M., Burstein, I. and Lemnij, I. "Schimbarea locului Rominiei in ecomomia mondiala." In *Dezvoltarea Economica a Romania 1944-1964*, Bucharest: Editura Academiei Republicii Populare Romine, 1964. 671-692.

Hutira, Ervin. "Aplicarea creatoarea de catre Partidul Comunist Roman a teoriei Leniniste despre faurirea economiei socialiste multilateral dezvoltare." In *Forţa Creatoarea Ideilor Leniniste*, Institutul de Studii Istorice şi Social-Politice de pe linga CC al PCR, Bucharest: Editura Politica, 1970. 64-107.

Ionascu, Traian and Barasch, Eugen. "Suveranitatea in Domeniul Economiei—Element Constitutive al Suveranitaţii Statului Socialist." *Studii şi Cercetari Juridice*, X, 3 (1965). 375-405.

Macovescu, Gheorghe. "Nicolae Titulescu—A Progressive Romanian Diplomat." *Revue Roumaine d'Histoire*, VII, 7 (1966). 391-440.

Maliţa, Mircea. "Speech Delivered at the XXII Session of the UN General Assembly." *Revue Roumaine d'etudes Internationales*, 1-2, 1967. 57-64.

Maliţa, Mircea. "Statele mici şi mijlocii in trlaţiile internaţionale." *Lupta de Clasa*, XLVIII, 2 (February, 1968). 44-57.

Marian, M. "Comunitatea de Naţiune." *Analele*, Institutului de studii Istorice şi Social-Politice de pe linga CC al PCR, XII, 6 (1967). 54-65.

Maurer, I.G. "The Inviolable Foundations of the Unity of the International Communist Movement." *World Marxist Review*, VI, 11 (November, 1963). 12-21.

Murgescu, Costin. "Ştiinţa contemporana şi cooperarea internaţionala." In *Revoluţia Ştiinţifica şi Techica Contemporana*, Bucharest: Editura Politica, 1967. 333-355.

Niciu, Marţian. "Politica externa a Republice Populare Romine—politica de aparare a pacii şi suveranitaţii ţarii noastre." *Buletinul*, Universitatilor "V. Babeş" şi "Bolyai," Cluj, Seria Ştiinţelor Sociale. 1, 1-2 (1956). 153-183.

Niciu, Marţian. "Principiul Coexistenţei Paşnice—Principiu Fundamental al Dreptului Internaţional Contemporan." *Studia*, Universitatis Babeş-Boylai, Series Iurisprudentia, VII, 1962. 31-39.

Niciu, Marţian, and Hanga Vladmir. "Un sfert de Veac de Politica Externa a Romaniei Socialiste." *Studia*, Universitatis Babeş-Bolyai, Series Iurisprudentia, XVIII, 1973. 29-39.

Olteanu, Ion. "Comerţul Exterior in Contextul Dezvoltarii Economiei Noastre Naţionale." *Lupta de Clasa*, XLVIII, 8 (August, 1968). 40-49.

Pavel, T. and Kun, I. "Importanţa acordurilor de lunga durata pentru comerţul internaţional." *Probleme Economice*, XVIII, 8 (August, 1965). 78-90.

"Pentru resolvarea pasnice a problemelor internaţionale vitale şi instaurarea unui climat de pace in lume." *Lupta de Clasa*, XL, 9 (September, 1960). 3-9.

Rachmuth, I. "Tendinţa de egalizare a economiei RPR la nivelul ţarilor socialiste mai dezvoltate." In *Dezvoltarea Economica a Romaniei 1944-1964*, Institutul de Cercetari Economice, Bucharest: Editura Academia RPR, 1964. 711-722.

"Sub conducerea partidului, spre desavirşirea construcţiei socialiste in partia noastra." *Lupta de Clasa*, XL, 7 (June, 1960). 7-14.

Surpat, Gheorghe and Ionel, Nicolae. "Relaţiile economice externe ale Romaniei in perioada construirii sosialismului." *Probleme Economice*, XX, 12 (December, 1967). 165-178.

Verdeţ, Ilie. "Expunerea la Consfatuirea privind activitatea in domeniul Comerţul Exterior." *Viaţa Economica*, V, 10 (March 10, 1967). 12-14.

Vlad, Constantin. "Internaţionalismul, suveranitatea şi patriotismul." In *Forţa Creatoarea Ideilor Leniniste*, Institutul de Studii Istorice şi Social-Politice de pe linga CC al PCR, Bucharest: Editura Politica, 1970. 155-173.

Voicu, Stefan. "Un document-program al mişcarii comuniste mondiale." *Lupta de Clasa*, XV, 12 (December, 1960). 3-21.

II. Books

Anescu, Vasile, Bantea, Eurgen and Cupsa, Ion. *Participarea Armatei Romane la Razboiul AntiHitlerist*. Bucharest: Editura Militara, 1966.

Aureliu, Alexandru. *Principii ale relaţiilor dintre state*, Bucharest: Editura Politica, 1966.

Barbu, Gheorghe, "Colaborare economica cu tarile capitaliste dezvoltate—parte inseparabila a participarii Romaniei la circuitul mondial de valori materiale şi spirituale." Unpublished Ph.D. dissertation, Bucharest: Academia de Studii Economice, 1970.

Berindei, Dan. *Din Inceputurile Diplomaţiei Romaneşti Moderne*, Bucharest: Editura Politica, 1965.

Campeanu, Al. *State in State*, Bucharest: Editura Politica, 1968.

Campus, Eliza. *Mica Inţelegere*, Bucharest: Editura Ştiinţifica, 1968.

Campus, Eliza. *Inţelegere Balcanica*, Bucharest: Editura Academiei Republice Socialiste Romania, 1972.

Candea, Virgil. *Pagini din Trecutul Diplomaţiei Roman*, Bucharest: Editura Politica, 1966.

Colţ, Gheorghe. *Romania şi Colaborarea Balcanica*, Bucharest: Editura Politica, 1973.

Cristureanu, T. *Roumaine Commerce Exterieur*, Bucharest: Editura Meridane, 1969.

Deutsch, Robert. *Nicolae Titulescu: Discursi*, Bucharest: Editura Ştiinţifica, 1967.

Dezvoltarea Economica a Rominiei 1944-1964, Institutul de Cercetari Economice, Bucharest: Editura Academiei RPR, 1964.

Forţa Creatoarea Ideilor Leniniste, Institutul de Studii Istorice şi Social-Politice de pe linga CC al PCR, Bucharest: Editura Politica, 1970.

Geamanu, Grigore. *Dreptul Internaţional Contemporan*, Bucharest: Editura Didactica şi Pedagogica, 1965.

Glaser, Edwin. *Statele Mici şi Mijlocii in Relaţiile Internaţionale*, Bucharest: Editura Politica, 1971.

Ilie, Petre and Stoean, Gheorghe. *Romania in Razboiul AntiHitlerist. Contribuţii Bibliografice*, Bucharest: Editura Militara, 1971.

Macovescu, Gheorghe. *Nicolae Titulescu: Documente Diplomatice*, Bucharest: Editura Politica, 1967.

Maliţa, Mircea. *Romanian Diplomacy: A Historical Survey*, Bucharest: Meridane Publishing House, 1970.

Maliţa, Mircea. *Diplomaţia*, Bucharest: Editura Didactica si Pedagogica, 1975.

Maliţa, Mircea, Murgescu, Coscin and Surpat, Gheorghe. *Romania Socialista şi Cooperarea Internaţionala*, Bucharest: Editura Politica, 1969.

Marx, Engels, Lenin despre rolul industriei in dezvoltarea societaţii, Bucharest: Editura Politica, 1965.

Netea, Vasile. *Nicolae Titulescu*, Bucharest: Meridane Publishing House, 1969.

Netea, Vasile. *Take Ionescu*, Bucharest: Editura Meridane, 1971.

Oprea, Ion. *Nicolae Titulescu*, Bucharest: Editura Ştiinţifica, 1966.

Popisteanu, Cristian. *Romania si Antanta Balcanica*, Bucharest: Editura Politica, 1968.

Romania in Razboiul AntiHitlerist, Institutul de Studii Istorice şi Social-Politice de pe linga CC al PRC, Bucharest: Editura Militara, 1966.

Ştefan, Gheorghe. *Momente din Istoria Poporul Roman*, Bucharest: Editura Politica, 1966.

III. SPEECHES, COLLECTED WORKS AND DATA VOLUMES

Anuarul Statistic al Republicii Socialiste Romania, 1970. Bucharest: Directia Centrala de Statistica, 1971.

Ceauşescu, Nicolae. *Romania on the Way of Completing Socialist Construction*, Vols.: I-III, Bucharest: Meridane Publishing House, 1969.

Comerţul Exterior al Republicii Socialiste Romania, Bucharest: Direcţia Centrala de Statistica, 1973.

Congresul al II-lea al Partidului Muncitoresc Romin, 23-28 decembrie, 1955, Bucharest: Editura de Stat pentru Literatura Politica, 1956.

Congresul al III-lea al Partidului Muncitoresc Romin, 20-25 iunie, 1960, Bucharest: Editura Politica, 1961.

Congresul al IX-lea al Partidului Comunist Roman, 19-24 iulie, 1965, Bucharest: Editura Politica, 1966.

Congresul al X-lea al Partidului Comunist Roman, 6-12 august, 1969, Bucharest: Editura Politica, 1969.

Gheorghiu-Dej, Gheorghe. "Concerning the Foreign Policy of the Government of the Rumanian People's Republic." Speech delivered at the fifth session of the Grand National Assembly, February 22, 1955., Bucharest: Meridane Publishing House, 1955.

Gheorghiu-Dej, Gheorghe. "Expunere facuta la şedinţa plenara a CC al PMR din 26-28 noiembrie, 1958." Bucharest: Editura Politica, 1958.

Gheorghiu-Dej, Gheorghe. *Atricole şi Cuvintari*, december 1955-iulie, 1959, Bucharest: Editura Politica, 1960.

Gheorghiu-Dej, Gheorghe. "Situaţia Internationala şi Politica Externa a Republicii Populare Romine." Report presented to an extraordinary session of the Grand National Assembly, August 30, 1960, Bucharest: Editura Politica, 1960.

Gheorghiu-Dej, Gheorghe. *Articles and Speeches*, Bucharest: Meridane Publishing House, 1963.

Rezoluţia Plenarei Comitelului Central al Partidului Muncitoresc Romin Din 28 iunie—3 iulie, 1957, Bucharest: Editura de stat pentru Literatura Politica, 1957.

E. Concepts and Methodology

I. ARTICLES

Festinger, Leon. "Informal Social Communication." In L. Festinger, K. Back, S. Schacter, H. Kelly and J. Thibaut, *Theory and Experiment in Social Communication*, Ann Arbor: Institute for Social Research, University of Michigan, 1950.

Heider, F. "Attitudes and Cognitive Organization." *Journal of Psychology*, XXI, (January, 1946). 107-112.

Holsti, Ole R. "An Adaption of the 'General Inquirer' for the Systematic Analysis of Political Documents." *Behavioral Science*, IX, 4 (October, 1964). 382-388.

Kitsuse, John I. "Societal Reaction to Deviant Behavior." In Earl Rubington and Martin Weinberg, *Deviance: The Inter-Actionist Perspective*, New York: Macmillan, 1968. 19-29.

Landecker, Warner S. "Types of Integration and Their Measurement." In Paul F. Lazarsfeld and Morris Rosenberg (eds.), *The Language of Social Research*, New York: The Free Press, 1955. 19-27.

McKinney, John C. *Constructive Typology and Social Theory*, New York: Appleton-Century Crofts. 1966.

Mitchell, Robert E. "Content Analysis for Exploratory Studies." *Public Opinion Quarterly*, XXXI, 2 (Summer, 1967). 230-241.

Newcomb, Theodore M. "An Approach to the Study of Communicative Acts." *Psychological Review*, LX, 6 (November, 1953). 393-404.

Newcomb, Theodore M. "Individual Systems of Orientations." In Sigmund Koch (ed.)., *Psychology: A Study of a Science*, III, *Formulations of the Person and the Social Context*, New York: McGraw Hill, 1959. 384-422.

II. BOOKS

Becker, Howard S. *Outsiders*, New York: The Free Press, 1963.

Blalock, Hubert M. *Social Statistics*, New York: McGraw-Hill, 1960.

Box, Steven. *Deviance, Reality and Society*, London: Holt, Rinehart, and Winston, 1971.

Clinard, Marshall. *Sociology of Deviant Behavior*, New York: Rinehart & Company, 1957.

Cohen, Albert K. *Deviance and Control*, Englewood Cliffs, New Jersey: Prentice Hall, 1966.

Durkheim, Emile. *The Rules of Sociological Method*, New York: The Free Press, 1964 (orig. pub. 1938).

Easton, David. *A Systems Analysis of Political Life*, New York: John Wiley and Sons, Inc., 1965.

Gould, Julius and Kolb, W.L. (eds.). *A Dictionary of the Social Sciences*, New York: The Free Press of Glencoe, 1964.

Holsti, Ole R. *Content Analysis for the Social Sciences and Humanities*, Reading, Massachusetts: Addison-Wesley, 1969.

Olson, Mancur. *The Logic of Collective Action*, New York: Schocken Books, 1968.

Osgood, Charles F., Suci, George J. and Tannenbaum, Percy H. *The Measurement of Meaning*, Urbana: University of Illionis Press, 1957.
Rubington, Earl and Weinberg, Martin. *Deviance: The Interactionist Perspective*, New York: Macmillan and Co., 1968.
Singer, J. David and Small, Melvin. *The Wages of War 1816-1955: A Statistical Handbook*, NewYork: John Wiley, 1972.
Teune, Henry and Przeworski, Adam. *The Logic of Comparative Social Inquiry*, New York: John Wiley, 1970.
Wright, George D. "Inter-Group Communication and Attraction in Inter-Nation Simulation." Unpublished Ph.D. dissertation, St. Louis: Washington University, 1963.

F. Document Collections

Current Soviet Policies, III, The Documentary Record of the Extraordinary 21st Congress of the Communist Party of the Soviet Union. New York: Columbia University Press, 1960.
Dallin, Alexander (ed.). *Diversity in International Communism*, New York: Columbia University Press, 1963.
Griffith, William E. (ed.). *The Sino-Soviet Rift*, Cambridge: M.I.T. Press, 1964.
Griffith, William E. *Sino-Soviet Relations 1964-65*. Cambridge: M.I.T. Press, 1967.
Kommunisticheskaia Partiia Sovetskogo Soiuza, 22 S'ezd, Moscow 1961. (22nd Congress of the Soviet Communist Party). *Proceedings and Related Materials*, Vol. 5, 6, 9 and 10. Washington: Foreign Broadcast Information Service, 1961.
Remington, Robin. *Winter in Prague*, Cambridge: M.I.T. Press, 1969.

G. Data Sources (see also Appendix II)

Current Digest of the Soviet Press
Deadline Data on World Affairs
East Europe: "Current Developments"
Foreign Broadcast Information Service: "Daily Report"
International Bank for Reconstruction and Development. *World Bank Atlas*, Washington: IBRD, 1970.
Kessing's Contemporary Archives

Marer, Paul. *Soviet and East European Foreign Trade, 1946-1969, Statistical Compendium and Guide*, Bloomington, Indiana: Indiana University Press, 1972.

Notes and Etudes Documentaires

Radio Free Europe: "Press Survey"

Starr, Richard F. *The Yearbook on International Communist Affairs 1969*, Stanford: Hoover Institution Press, 1970.

Taylor, Charles L. and Hudson, Michael C. *World Handbook of Political and Social Indicators*, New Haven: Yale University Press, 1972.

United Nations. *Yearbook of International Trade Statistics*, New York: United Nations, 1967.

Yugoslav Survey

H. Miscellaneous

Aleksandrov, I. "Attack on the Socialist Foundations of Czechoslovakia." *Pravda*, July 11, 1968.

Kovalev, S. "Sovereignty and the Internationalist Obligations of Socialist Countries." *Pravda*, September 26, 1968. 4.

Lonyai, S. "A Few Remarks on 'Two Thousand Words'." *Kisalforld*, July 7, 1968. (RFE Hungarian Press Survey, 16, July 1968).

Sendak, Maurice. *Where the Wild Things Are*, New York: Harper and Row, 1963.

INDEX

EAST EUROPEAN MONOGRAPHS

1. *Political Ideas and the Enlightenment in the Romanian Principalities, 1750–1831*. By Vlad Georgescu. 1971.
2. *America, Italy and the Birth of Yugoslavia, 1917–1919*. By Dragan R. Zivojinovic. 1972.
3. *Jewish Nobles and Geniuses in Modern Hungary*. By William O. McCagg, Jr. 1972.
4. *Mixail Soloxov in Yugoslavia: Reception and Literary Impact*. By Robert F. Price. 1973.
5. *The Historical and National Thought of Nicolae Iorga*. By William O. Oldson. 1973.
6. *Guide to Polish Libraries and Archives*. By Richard C. Lewanski. 1974.
7. *Vienna Broadcasts to Slovakia, 1938–1939: A Case Study in Subversion*. By Henry Delfiner. 1974.
8. *The 1917 Revolution in Latvia*. By Andrew Ezergailis. 1974.
9. *The Ukraine in the United Nations Organization: A Study in Soviet Foreign Policy. 1944–1950*. By Konstantin Sawczuk. 1975.
10. *The Bosnian Church: A New Interpretation*. By John V. A. Fine, Jr., 1975.
11. *Intellectual and Social Developments in the Habsburg Empire from Maria Theresa to World War I*. Edited by Stanley B. Winters and Joseph Held. 1975.
12. *Ljudevit Gaj and the Illyrian Movement*. By Elinor Murray Despalatovic. 1975.
13. *Tolerance and Movements of Religious Dissent in Eastern Europe*. Edited by Bela K. Kiraly. 1975.
14. *The Parish Republic: Hlinka's Slovak People's Party, 1939–1945*. By Yeshayahu Jelinek. 1976.
15. *The Russian Annexation of Bessarabia, 1774–1828*. By George F. Jewsbury. 1976.
16. *Modern Hungarian Historiography*. By Steven Bela Vardy. 1976.
17. *Values and Community in Multi-National Yugoslavia*. By Gary K. Bertsch. 1976.
18. *The Greek Socialist Movement and the First World War: the Road to Unity*. By George B. Leon. 1976.
19. *The Radical Left in the Hungarian Revolution of 1848*. By Laszlo Deme. 1976.
20. *Hungary between Wilson and Lenin: The Hungarian Revolution of 1918–1919 and the Big Three*. By Peter Pastor. 1976.
21. *The Crises of France's East-Central European Diplomacy, 1933–1938*. By Anthony J. Komjathy. 1976.
22. *Polish Politics and National Reform, 1775–1788*. By Daniel Stone. 1976.
23. *The Habsburg Empire in World War I*. Robert A. Kann, Bela K. Kiraly, and Paula S. Fichtner, eds. 1977.
24. *The Slovenes and Yugoslavism, 1890–1914*. By Carole Rogel. 1977.

25. *German-Hungarian Relations and the Swabian Problem.* By Thomas Spira. 1977.
26. *The Metamorphosis of a Social Class in Hungary During the Reign of Young Franz Joseph.* By Peter I. Hidas. 1977.
27. *Tax Reform in Eighteenth Century Lombardy.* By Daniel M. Klang. 1977.
28. *Tradition versus Revolution: Russia and the Balkans in 1917.* By Robert H. Johnston. 1977.
29. *Winter into Spring: The Czechoslovak Press and the Reform Movement 1963–1968.* By Frank L. Kaplan. 1977.
30. *The Catholic Church and the Soviet Government, 1939–1949.* By Dennis J. Dunn. 1977.
31. *The Hungarian Labor Service System, 1939–1945.* By Randolph L. Braham. 1977.
32. *Consciousness and History: Nationalist Critics of Greek Society 1897–1914.* By Gerasimos Augustinos. 1977.
33. *Emigration in Polish Social and Political Thought, 1870–1914.* By Benjamin P. Murdzek. 1977.
34. *Serbian Poetry and Milutin Bojic.* By Mihailo Dordevic. 1977.
35. *The Baranya Dispute: Diplomacy in the Vortex of Ideologies, 1918–1921.* By Leslie C. Tihany. 1978.
36. *The United States in Prague, 1945–1948.* By Walter Ullmann. 1978.
37. *Rush to the Alps: The Evolution of Vacationing in Switzerland.* By Paul P. Bernard. 1978.
38. *Transportation in Eastern Europe: Empirical Findings.* By Bogdan Mieczkowski. 1978.
39. *The Polish Underground State: A Guide to the Underground, 1939–1945.* By Stefan Korbonski. 1978.
40. *The Hungarian Revolution of 1956 in Retrospect.* Edited by Bela K. Kiraly and Paul Jonas. 1978.
41. *Boleslaw Limanowski (1835–1935): A Study in Socialism and Nationalism.* By Kazimiera Janina Cottam. 1978.
42. *The Lingering Shadow of Nazism: The Austrian Independent Party Movement Since 1945.* By Max E. Riedlsperger. 1978.
43. *The Catholic Church, Dissent and Nationality in Soviet Lithuania.* By V. Stanley Vardys. 1978.
44. *The Development of Parliamentary Government in Serbia.* By Alex N. Dragnich. 1978.
45. *Divide and Conquer: German Efforts to Conclude a Separate Peace, 1914-1918.* By L. L. Farrar, Jr. 1978.
46. *The Prague Slavic Congress of 1848.* By Lawrence D. Orton. 1978.
47. *The Nobility and the Making of the Hussite Revolution.* By John M. Klassen. 1978.
48. *The Cultural Limits of Revolutionary Politics: Change and Continuity in Socialist Czechoslovakia.* By David W. Paul. 1979.
49. *On the Borderof War and Peace: Polish Intelligence and Diplomacy in 1937-1939 and the Origins of the Ultra Secret.* By Richard A. Woytak. 1979.
50. *Bear and Foxes: The International Relations of the East European States 1965-1969.* By Ronald Haly Linden. 1979.